Wood Material and Processing Data

Peter Niemz · Alfred Teischinger · Dick Sandberg
Editors

Wood Material and Processing Data

The Most Relevant Data, Tables and Figures

Springer

Editors
Peter Niemz
Department Wood Physics (today, Wood Materials Science), Institute of Building Materials
ETH Zürich
Zurich, Switzerland

Dick Sandberg
Department of Manufacturing and Civil Engineering, Norwegian University of Science and Technology (NTNU)
Gjøvik, Norway

Alfred Teischinger
Institute of Wood Science Technology and Renewable Materials, University of Natural Resources and Life Sciences (BOKU)
Vienna, Tulln, Austria

ISBN 978-3-031-38328-1   ISBN 978-3-031-38329-8 (eBook)
https://doi.org/10.1007/978-3-031-38329-8

This Springer imprint is published by the registered company Springer Nature Switzerland AG
The registered company address is: Gewerbestrasse 11, 6330 Cham, Switzerland

Preface

In research, teaching and industry, specific material parameters for wood and wood-based materials and also data on technical processes such as material consumption and energy consumption are often required. Such information exists to only a very limited extent for wood, and then often only briefly as appendices in reference books such as Franz Kollmann's *Technologie des Holzes und der Holzwerkstoffe* in two volumes.

In other subject areas, such information is available and the tables are a very good supplement to the textbooks with the necessary general basic information. For wood and wood-based materials, there is a considerable need to catch up.

In the *Springer Handbook of Wood Science and Technology,* therefore, such data have been compiled in Chap. 38 *Important Data*, but for technical reasons, only a greatly abridged version with a link to the essential files was included in the printed version of the book. Most of the large amount of information can only be viewed online in the Springer Handbook of Wood Science and Technology, but the tables can also be easily expanded. Many efforts are currently underway to create a database for calculations, although the flood of data and often the difficulty in comparing results obtained by measurement methods often make the overview difficult. Much is also only published nationally and locally and in different languages, and this makes access difficult.

The editors and publisher have therefore decided to publish a separate book entitled *Wood Material and Processing Data—The Most Relevant Data, Tables and Figures,* in which the data from the Handbook is complete, printed and updated. Certain extensions have been made where appropriate. In a total of eight chapters, important parameters for solid wood, modified wood, wood-based materials, pulp and paper as well as material and energy consumption, further reference books, standards and other parameters are compiled.

Sixteen authors have been integrated into the elaboration, so that worldwide information on wood species, standardisation, etc. is available. This has made it possible to summarise expert knowledge from different parts of the world. The data should also be available to the reader in a printed form as a supplement to the Springer Handbook of Wood Science and Technology.

An attempt has been made to reduce the large amount of data to a manageable size and to structure the content. The editors are grateful to the readers for any suggestions for expansion or streamlining.

Zürich, Switzerland
Vienna, Austria
Gjøvik, Norway

Peter Niemz
Alfred Teischinger
Dick Sandberg

Acknowledgments

The editors thank all authors who have contributed to this book, for their profound expertise and also for their patience and support in the reviewing and final editing process. Many other persons, institutions, and companies have contributed to the book by providing related data, photographs, figures, specific knowledge, and supportive work, etc. The editors gratefully acknowledge the support by Springer, especially for the assistance in editorial matters of Dr. Dieter Merkle and Raf Harms.

Contents

Editors and Contributors

About the Editors

Peter Niemz Before his retirement, Peter Niemz was Professor of Wood Physics at the ETH Zurich. Throughout his professional career, he lectured on wood technology at TU Dresden, Universidad Austral de Chile, and as visting Professor at several Universities (Austria, China, Hungary, Poland, and Sweden). Since 2002, he is a fellow of the International Academy of Wood Science. He is on the editorial board of numerous wood-related journals, including Wood Research and Technology (Holzforschung).

Alfred Teischinger is a retired Professor of Wood Technology at the University of Natural Resources and Life Sciences (BOKU University) in Tulln, Vienna. He also acted as scientific director of the competence center for wood composites and wood chemistry (Wood Kplus). His research focus is on wood technology and processing. He was board member of various organisations linking wood science and technology such as InnovaWood, and he was associated to the committee of the prestigious Marcus Wallenberg Prize. Since 2013, he is a fellow of the International Academy of Wood Science and has been awarded by recognition of outstanding contribution to wood science.

Dick Sandberg is Chaired Professor of Design and Fabrication of Timber Construction at the Norwegian University of Science and Technology (NTNU). He received his Doctorate in Mechanical Engineering from the Royal Institute of Technology (KTH), Stockholm, Sweden. He has also been a wood science, wood processing, and timber engineering professor at KTH, Linnaeus University, Luleå University of Technology in Sweden, and the Czech University of Life Sciences Prague. His research covers material properties, wood scanning technology, wood machining, timber engineering and wood production systems, and studies the connections between different links in the forestry-wood chain. He has a background as a specialist and manager of several companies in the wood industry sphere and the former Swedish Institute of Wood Technology Research, and he has been a national delegate of several European COST (Cooperation in Science and Technology) Actions. He is Editor-in-Chief of the journal Wood Material Science & Engineering. He is an Elected Fellow of the International Academy of Wood Science.

Contributors

Andrade Egas Eduardo Mondlane University (UEM), Faculty of Agronomy and Forestry Engineering, Wood Science and Technology, Maputo, Moçambique

Rico Emmler Entwicklungs- und Prüflabor Holztechnologie GmbH, Dresden, Germany

Raquel Gonçalves University of Campinas (UNICAMP), School of Agricultural Engineering (FEAGRI), Nondestructive Testing Laboratory (LabEND), Sao Paulo, Brazil

Galina Gorbacheva Bauman Moscow State Technical University, Moscow, Russian Federation

Dirk Grunwald Follmann GmbH & Co. KG, Minden, Germany

Sabine Heinemann Technische Universität Dresden (TU Dresden), Wood Science and Fiber Material Science, Paper Technology Group, Dresden, Germany

Jiali Jiang Chinese Academy of Forestry, State Key Laboratory of Tree Genetics and Breeding, Research Institute of Wood Industry, Beijing, P. R. China

Andreja Kutnar University of Primorska, Koper, Slovenia and InnoRenew CoE, Izola, Slovenia

Róbert Németh University of Sopron Simonyi Károly Faculty, Institute of Wood Science, Sopron, Hungary

Peter Niemz ETH Zürich, Wood Physics (today Wood Material Science), Institute for Building Materials, Zurich, Switzerland

Robert J. Ross USDA Forest Service, Forest Products Laboratory, Madison, WI, USA

Dick Sandberg Norwegian University of Science and Technology (NTNU), Manufacturing and Civil Engineering Department, Gjøvik, Norway

Christopher Adam Senalik USDA Forest Service, Forest Products Laboratory, Madison, WI, USA

Alfred Teischinger University of Natural Resources and Life Sciences (BOKU), Institute of Wood Technology and Renewable Materials, Vienna, Tulln, Austria

Steffen Tobisch Institut für Holztechnologie Dresden (IHD), Dresden, Germany

Johannes Welling Sachverständigenbüro Holz-Expertise, Reinbek, Germany

General Data Related to Wood

1

Peter Niemz and Dick Sandberg

1.1 Introduction

This book provides information on the chemical, physical and mechanical properties of wood of the most common species from different regions of the world, and similar data for some other common bio-based materials such as bamboo, flax and hemp. Characteristic values for the most extensively used wood-based materials are given special attention, as such information is necessary in construction and is widely used by engineers, architects, researchers and other specialists. Useful data on subjects such as adhesives, coating, processing and wood modification are also presented, and an index is provided to facilitate the search for specific contents. The information is summarised in tables and graphics, which faciliate a quick search for important information on the properties of wood and wood-based materials, adhesives, material and energy consumption for important products in solid wood and wood-based materials. Standardisation and important technical books written in the first place English or German and which touch on various aspects related to the wood material, its further processing and use are presented in two separate chapters.

Common names are mainly used for the different wood species. A key to identify the common names with scientific names for the species is included in the Appendix.

The book is to be seen as a supplement to the *Springer Handbook of Wood Science and Technology* (Niemz et al. 2023), which in detail describes the basic principles of wood science (chemistry, anatomy, physics of wood), important basic processes of the wood industry (bonding, cutting, drying, wood modification, process and quality control) as well as technological processes and products (sawmilling and veneer production, wood-based materials, engineered wood products (EWPs), pulp and paper, fibre composites), as well as bio-energy use and biorefinery processes. For reasons of space, only a highly abridged version of the information here presented could be printed in the Springer Handbook, and the table information is only accessible on-line via the e-book version of the handbook. The information presented in the present book is updated and extended.

Within wood science, the moisture content in wood is most commonly calculated as the ratio:

$$\omega = \frac{m_w}{m_{dry}} \quad (1.1)$$

where ω (kg·kg−[1] or often in %) is the moisture content (MC), m_w is the mass of water in the wood sample, and m_{dry} is the oven-dry mass of the same wood. Since water can be present in wood both in the cell material and in the macro-void structure, the maximum moisture content is the sum of the amounts of water present in cell walls and in macro-voids, and may exceed 100%. When other measures ares used for water in a material, the definition of these measures is here given in connection with the data.

There is several on-line data-bases available describing various information about wood, wood-based materials and its properties. The *Holzatlas* and *macroHolzdata* databases that can be recommended as a complement to this handbook:

Wood atlas database including a wood reference book library: https://www.hanser-esolutions.de/holz

macroHolzdata: Android: https://play.google.com/store/apps/developer?id=Sven+Koch

iOS: https://apps.apple.com/de/app/macroholzdata/id1120922391?l=en

P. Niemz (✉)
Department Wood Physics (today, Wood Materials Science), Institute of Building Materials, ETH Zürich, Zurich, Switzerland
e-mail: niemzp@retired.ethz.ch

D. Sandberg
Department of Manufacturing and Civil Engineering, Norwegian University of Science and Technology (NTNU), Gjøvik, Norway
e-mail: dick.sandberg@ntnu.no

P. Niemz et al. (eds.), *Wood Material and Processing Data*, https://doi.org/10.1007/978-3-031-38329-8_1

1.2 Tree-Related Data

This section gives some data related to tree properties and on how wood reacts in the surrounding climate (temperature and relative humidity of the air) that may be useful in the use of wood and wood-based materials, and also for the planning of various wood-refining processes.

1.2.1 Bio-Mass Content in Trees

See Tables 1.1, 1.2 and 1.3, Figs. 1.1 and 1.2.

1.2.2 Structural Properties of Wood

In most species, the differences in properties between sapwood and heartwood are substantial, particularly in terms of water and moisture transport (cf. Fig. 1.10), and these differences must be taken into consideration when using wood. The normal heartwood formation in a stem does not follow the same growth-ring pattern in either the cross-section or in the longitudinal section, and therefore, the heartwood does not include the same number of growth rings

Table 1.1 Bark content by volume of logs (Mette and Korell 1989)

Species	Bark content (%)
European white birch	7–10
European ash	12–17
European beech	6–8
European hornbeam	6–8
European oak	12–14
European white elm	13
Norway maple	10–12
Norway spruce	10
Scots pine	10–16
European larch	16–22
Silver fir	10–17

Table 1.2 Proportion of dendromass (woody biomass) in a 100-year-old Scots pine stand (after Blossfeld in Mette and Korell 1989)

Tree share without bark	Dendromass		
	Total (Tonnes/hectare)	Above ground (%)	Under ground (%)
Stemwood	130.5	81.6	87.0
Branches	8.5	5.3	4.4
Needles	14.0	8.7	7.1
Stemwood and root wood	35.0		17.3
Dendromass above ground	160.0	100.0	
Dendromass total	195.0		100.0

Table 1.3 Content of essential oils, dry matter and ash in Norway spruce and Scots pine needles (Mette and Korell 1989)

Criterion	Spruce needles (%)	Pine needles (%)
Dry matter content	50.10	49.50
Essential oil content	0.32	0.81
Ash content after combustion of the needels	3.70	2.00

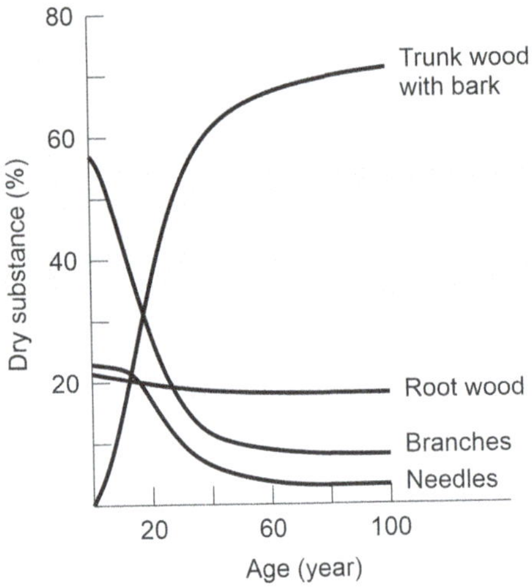

Fig. 1.1 Distribution of dendromass in Scots pine wood (after Blossfeld cited in Mette and Korell 1989)

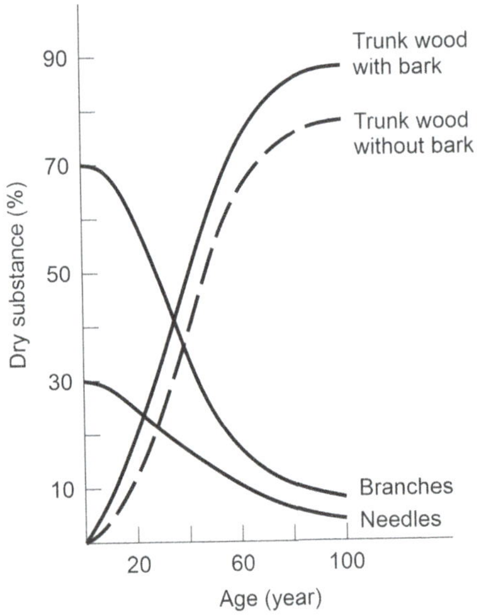

Fig. 1.2 Proportions of different tree components depending on tree age (after Blossfeld cited in Mette and Korell 1989)

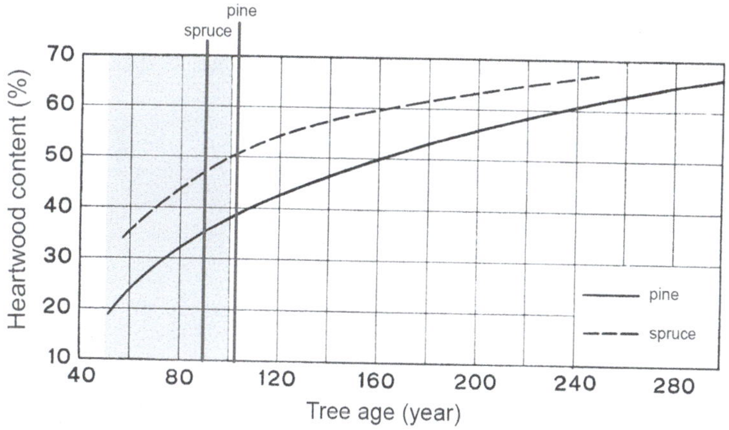

Fig. 1.3 Average heartwood content as a percentage of the cross-sectional area at 5–8 m above ground at different ages in Norway spruce and Scots pine in Central Sweden (Ekman et al. 1922). The shaded area indicates the minimum permitted age for regeneration felling according to the Swedish Forest Protection Act, which applies to the entire country and all growth indices (annual growth in m^3forest/hectare). The two vertical lines indicate the average final felling age for pine and spruce respectively in the Svealand region taken from three forestry companies operating in the region (Nilsson and Thörnqvist 2008 in Sandberg 2011).

Fig. 1.4 Juvenile wood occurs around the pith and forms a roughly cylindrical structure up the tree. In contrast to heartwood, the proportion of juvenile wood increases towards the top of the tree (Sandberg et al. 2021)

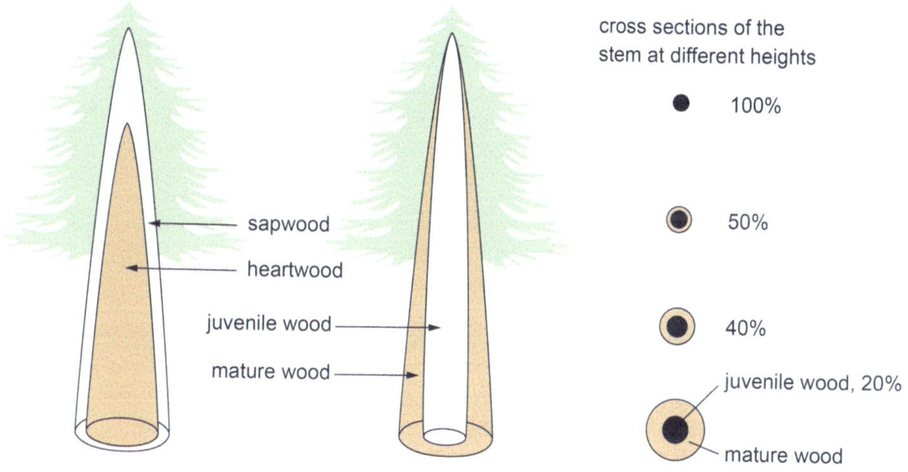

in different parts of the stem. Typically, however, the limitation of the heartwood in the cross-section is more or less circular. Some species, e.g. European aspen, European birch spp., European beech, European hornbeam, and Norway maple general do not develop heartwood; however, these tree species may exhibit discoloration around the pith due to microorganisms or frost, know as a red core or false heartwood. There are also differences in heartwood formation between species, especially between coniferous and broad-leaved trees. Fig. 1.3 shows the heartwood formation by age in Norway spruce and Scots pine from Central Sweden.

Although it is not visible in the trunk cross-section, a significant pith-to-bark gradient in density and other properties exists, which is unique to each species (Figs. 1.4, 1.5, 1.6). The fact that a relatively pronounced change in density often occurs in conifers during the first 15 to 30 years of growth has given rise to the term "juvenile wood" (Zobel and Sprauge 1998). This term can lead to confusion because this wood is not found just in young (juvenile) trees but near the pith in every tree, regardless of age.

Conifers and broad-leaved trees have adopted different strategies for forming reaction wood. The reaction wood in conifers is called *compression wood* because it forms on the compression-stress side of leaning stems. In broad-leaved trees, the reaction wood is known as *tension wood* because the increased growth occurs on the upper or tension-stress side of the leaning tree. Detailed information about compression wood can be found in Timell (1986). Table 1.4 gives some shrinkage data for compression wood in Norway spruce.

Table 1.5 presents density and fibre dimensions for some coniferous wood (softwood) and broad-leaved wood (hardwood). The wood of a coniferous tree consists of two types of cells: longitudinal tracheids (fibres) and ray parenchyma,

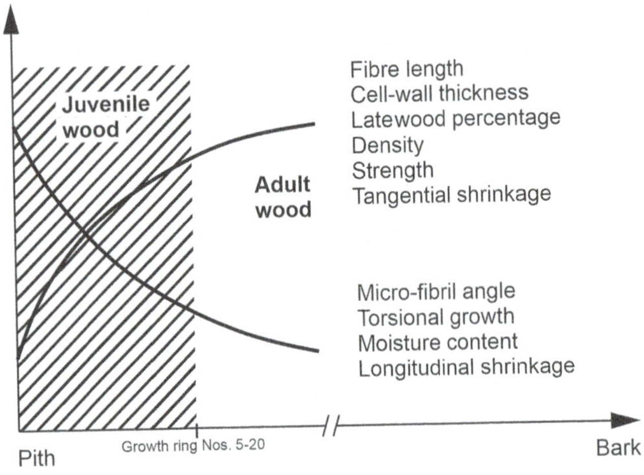

Fig. 1.5 Change in characteristics during the transition from juvenile to adult wood (after Bues 1987 cited in Bues and Fischer 2012; Bues and Bäuker 2014)

Table 1.4 Longitudinal, radial, and tangential shrinkage of normal and compression wood in Norway spruce (Rosenthal and Bäuker 2012) from green to kiln-dry conditions as a percentage of green volume

	Normal wood (%)	Compression wood (%)
Longitudinal	0.18–0.40	0.58–7.00
Radial	3.40–7.12	2.21–3.26
Tangential	6.70–8.05	3.14–4.60

oriented both axially and horizontally. Most of the tracheids are longitudinal, while the parenchyma cells have a radial orientation. In addition to these two types of cells, other elements such as epithelial cells create longitudinal and horizontal resin canals. Transverse tracheids are not present in all species. The structure of broad-leaved wood is more complex than that of conifer wood and, during its evolution, broad-leaved wood has developed special types of cells from the tracheids: vessel elements for conduction and fibres for support.

The different cell-wall layers consist of fibrils and microfibrils oriented at various angles to the fibre axis (Fig. 1.7). The microfibril angle of the S_2 layer ranges between 5–10° (latewood) and 20–30° (earlywood), that of the S_1 layer between 50° and 70° and that of the S_3 layer between 50° and 90°. As the S_2 layer is the thickest in the cell wall, it also dominates the overall microfibril orientation (Table 1.6). The microfibril orientation has an apperant

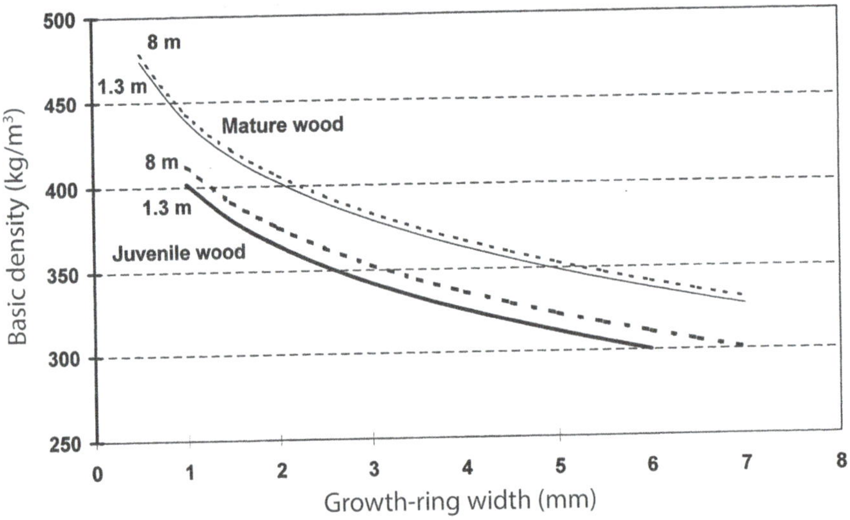

Fig. 1.6 Relationships between growth-ring width and basic density (oven-dry mass/green volume) in Norway spruce, both in juvenile wood and in mature wood and at two heights in the tree stem (Barnett and Jeronimidis 2003)

Table 1.5 Oven-dry density dimensions and volume percentages of various cells in different tree species in the temperate zones (Fengel and Wegener 1984)

	Unit	Coniferous wood			Broad-leaved wood		
		Silver fir	Norway spruce	Scots pine	European beech	European oak	Poplar spp.
Oven-dry density	(kg/m³)						
Minimum		320	300	300	490	390	
Average		410	430	490	680	650	400
Maximum		710	640	860	880	930	
Fibre[1)] length	(mm)						
Minimum		3.4	1.7	1.4	0.6	0.6	0.7
Average		4.3	2.9	3.1			
Maximum		4.6	3.7	4.4	1.3	1.6	1.6
Fibre[1)] diameter	(μm)						
Minimum		25	20	10	15	10	20
Average		50	30	30			
Maximum		65	40	50	20	30	40
Vessels, length	(mm)				0.3–0.7	0.1–0.4	0.5
Vessels, diameter	(μm)				5–100	10–400	20–150
Cell proportion[2)]	(%)						
Tracheids		90	95	93	38	44–58	62
Vessels					31	40	27
Parenchyma		scarce	1.4–5.8	1.4–5.8	4.6	4.9	scarce
Ray cells		9.6	4.7	5.5	27	16.2–29.3	11.3

1) Fibre – tracheids or libriform fibres. The following terminology misusage was pointed out by Zobel and Buijtenen (1989). The term "fibre" is commonly used for both the true fibres of broad-leaved woods and the tracheids of conifer woods. Although this is botanically incorrect, the general use of the term fibre must be recognised, since numerous publications refer to the fibre characteristics of coniferous woods as well as to those of the real fibres of broad-leaved woods.
2) Average values on volume.

Table 1.6 Microfibril orientation in cell walls of selected species (Barnett and Bonham 2004). MFA – microfibril angle, EW – earlywood, LW – latewood

Species	Method	MFA (degree)	Reference
Anginosperms			
Alpine ash	XRD	8.5–20	Evans and Illic (2001)
Blue gum (eucalyptus)	Confocal microscopy	0.3±0.4–6.6±1.0	French et al. (2000)
European oak	SAXS	23.3 (second growth ring)	Lichtenegger et al. (1999)
Palasan cane	Iodine and XRD	23–34	Abasolo et al. (1999)
Shining gum (eucalyptus)	XRD	14–25	Evans (1998)
	XRD	20–30 (pith)–10 (bark)	Evans et al. (2000)
	Confocal microscopy	1.6±0.7–13±1.6	French et al. (2000)
	XRD	10–20	Stuart and Evans (1994)
Silver birch	XRD	9.4–18.3	Bonham and Barnett (2001)
Small-leaved Lilly Pilly	XRD	5.5 (mature wood)	Watson and Dadswell (1964)
Tulipwood	XRD	21.8–26.0	Yamamoto et al. (1993)
Gymnosperms			
Douglas fir	XRD	15.6–27.3 (EW), 10–18.4 (LW)	Lotfy et al. (1973)
Silver fir	Iodine	2–8 (adult wood)	Gorisek and Torelli (1999)
Japanese cedar	XRD	3.4–11.8	Yamamoto et al. (1993)
	Pit aperture	22–44 (juvenile wood)	Yamashita et al. (2000)
Loblolly pine	PLM	32.2 (core)–29.9 (outer)	McMillin (1973)
	Pit aperture	15±4.4	Senft and Bendtsen (1985)
	Iodine	16±4.2	Senft and Bendtsen (1985)

(Continued)

Table 1.6 (Continued)

Species	Method	MFA (degree)	Reference
Norway spruce	XRD	9.0 (EW), 8.1(LW)	Sahlberg et al. (1997)
	XRD	23 (compression wood)	Sahlberg et al. (1997)
	Iodine	19 (compression wood)	Sahlberg et al. (1997)
	Iodine	2–8 (adult)	Sahlberg et al. (1997)
	Pit aperature	29 (fast grown), 21 (slow grown)	Herman et al. (1999)
	SAXS	<5 (EW)–20 (LW)	Reiterer et al. (1998)
	Iodine	2–8 (adult wood)	Gorisek and Torelli (1999)
	SAXS	0 (EW) 19.9 (LW)	Lichtenegger et al. (1999)
	WAXS and SAXS	39 (S2 layer) 89 (S1 layer)	Andersson et al. (2000)
Radiata pine	XRD	22–46	Wardrop (1951)
	XRD	37 (inner wood)	Watson and Dadswell (1964)
	PLM	9–55	Donaldson (1992)
	PLM	9.4–47.1	Donaldson and Burdon (1995)
Scots pine	SAXS	29 (juvenile wood), 15 (mature wood)	Lichtenegger et al. (1999)
	Mercury impregnation	18.6–28.2 (EW), 13.9–15.8 (LW)	Lotfy et al. (1973)
Western hemlock	XRD	8.2–9.0	Lotfy et al. (1973)
	Mercury impregnation	7.5–9.1 (EW)	Lotfy et al. (1973)

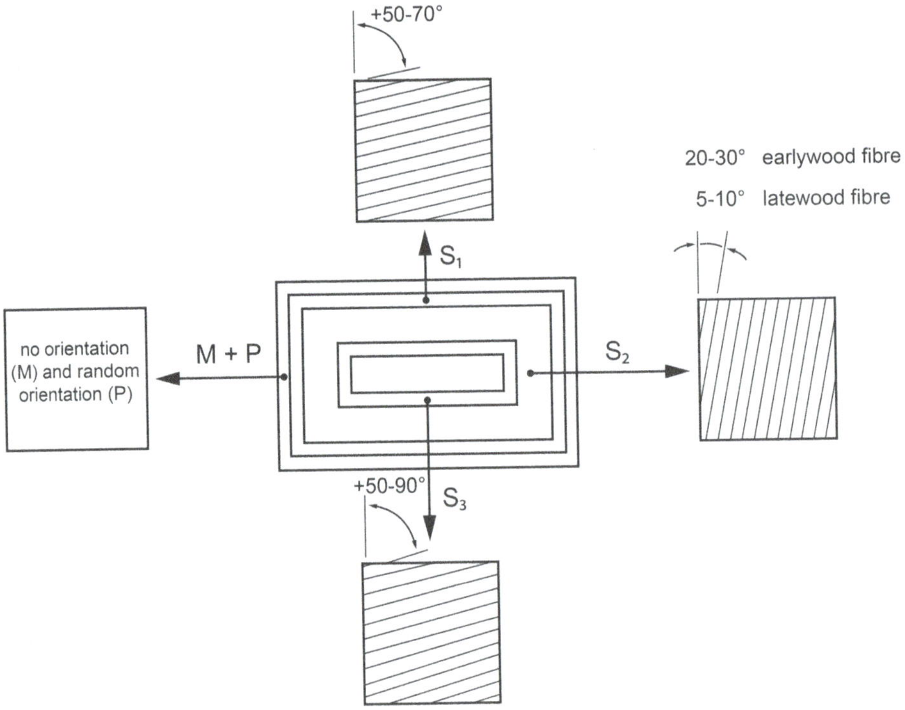

Fig. 1.7 Schematic diagram of the microfibril-angle arrangements within the primary wall, S_1, S_2 and S_3 layers. The middle lamella consist of an essential quantity of lignin (approx. 100%) and no microfibrils. (from Sandberg et al. 2021)

effect on the mechanical properties of the wood material (Figs. 1.8 and 1.9).

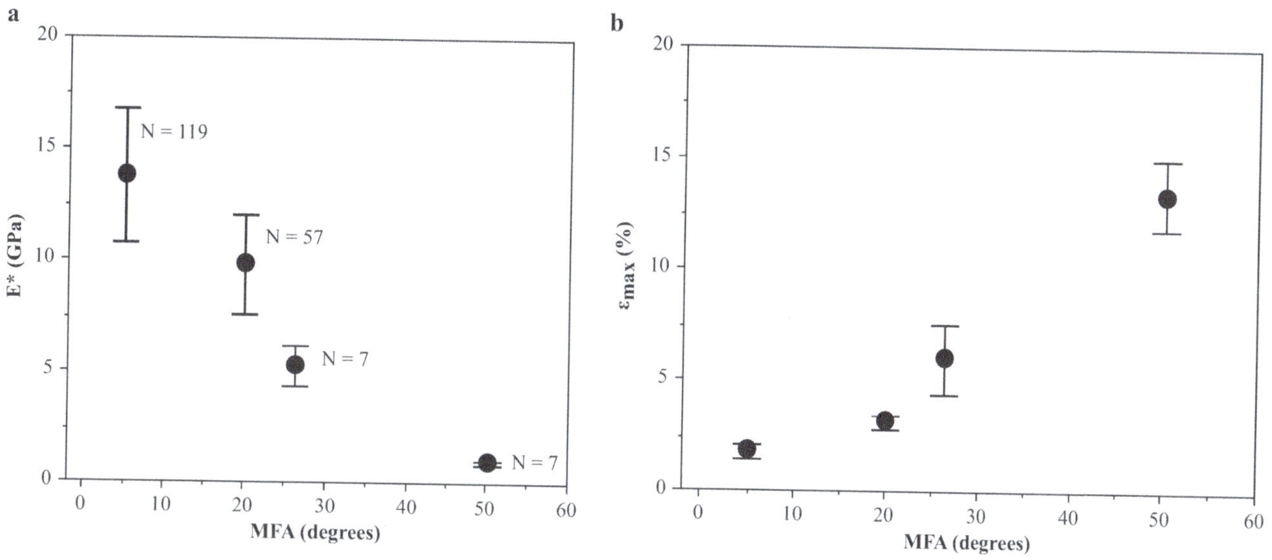

Fig. 1.8 Effect of microfibril angle (MFA) on mechanical properties of Norway spruce: (a) normalised modulus of elasticity E*, and (b) strain at maximum stress (ε_{max}) (Reiterer et al. 1999). N – number of specimens tested for each data point, bars – standard deviation

Fig. 1.9 Normalised stress–strain curves for earlywood and latewood, and branch wood (Reiterer et al. 1999)

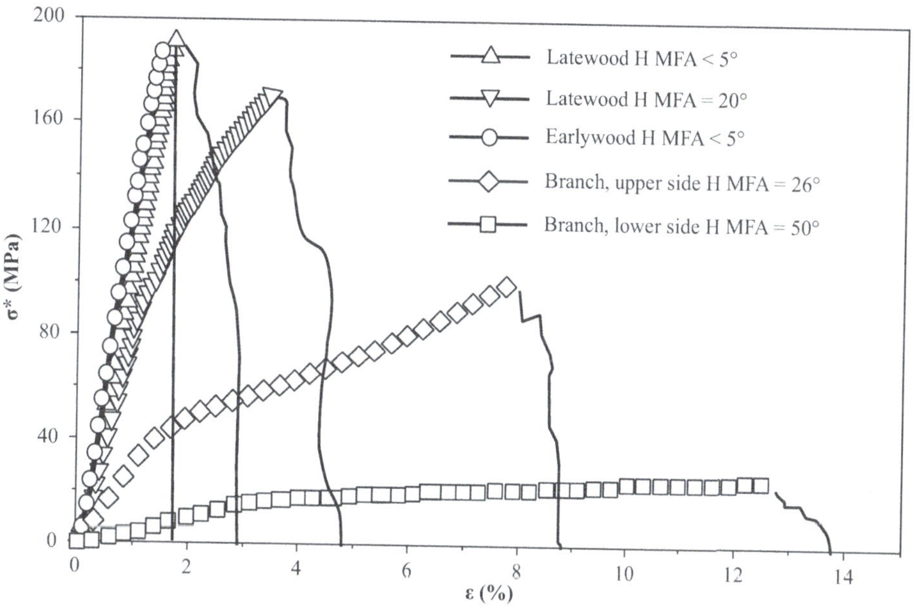

1.2.3 Moisture Content in Green Wood

The moisture content varies within the living tree as well as from year to year (Table 1.7). The moisture content of the heartwood in conifers is typically much lower than that in the sapwood (Fig. 1.10, Table 1.8), but this is not always the case in broad-leaved woods, where the relation between moisture content in heartwood and sapwood depends not only on the species but on the season (Pallardy and Kozlowski 2008). The heartwood content is also dependent on the age of the tree (cf. Fig. 1.3).

Table 1.7 Moisture content of European beech (after Gäumann cited in Mette and Korell 1989)

Material		Moisture content (%)		
		Winter (November–March)	*Summer (May–September)*	*Mean value over the year*
Branches	bark	103	75	82
	wood	73	82	76
Roots	bark	86	106	95
	wood, mainly sapwood	131	98	116
Stem at 0.3 m	bark	66	88	73
	sapwood	100	76	91
	heartwood	72	65	72
Stem at 10 m	bark	64	84	73
	sapwood	104	90	91
	heartwood	75	69	72
Stem at 20 m	bark	59	76	73
	sapwood	94	91	91
	heartwood	79	90	72

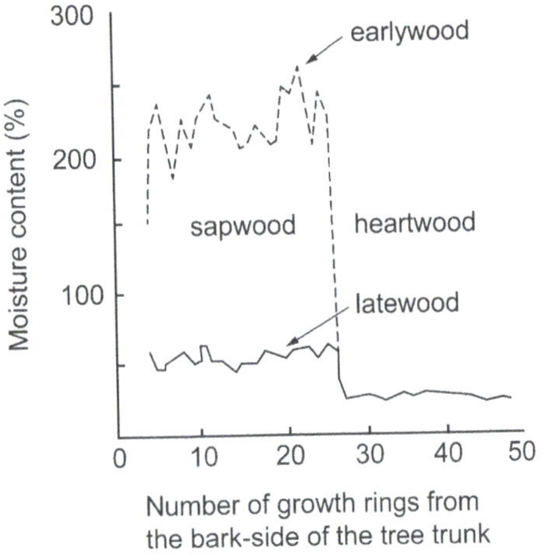

Fig. 1.10 Moisture content of earlywood and latewood in sapwood and heartwood of Scots pine (after Vintila 1939 in Sandberg et al. 2021)

In contrast to heartwood, sapwood in the living tree has a very high moisture content. For example, in Scots pine, there is a significant variation in moisture content between the sapwood between the earlywood and latewood, with the majority of moisture being located in the earlywood. In the heartwood, there is no difference in moisture content between earlywood and latewood (Figs. 1.10, 1.11 and 1.12).

1.3 Moisture of Wood in Service and Climate-Related Data

Moisture is generally omnipresent in wood, as water molecules can be absorbed from the surrounding air. However, the relative humidity of the air may be influenced by season

Table 1.8 The average moisture content of green wood from North American and European species (based on Glass and Zelinka 2010; Mette and Korell 1989; Sandberg et al. 2021). MC – moisture content, HW – heartwood, SW – sapwood.

Species	Broad-leaved trees	MC (%) in:		Species	Conifer trees	MC (%) in:	
		HW	SW			HW	SW
Alder	Red alder		97	Cedar	Eastern red cedar	33	
Apple	Apple	81	74		Incense cedar	40	213
Ash	Black ash	95			Port Orford-cedar	50	98
	Green ash		58		Western red cedar	58	249
	White ash	46	44		Yellow cedar	32	166
Aspen	Aspen (Europe)		110	Cypress	Bald cypress	121	171
	Aspen, (USA)	95	113	Douglas fir	Douglas fir (coast)	37	115
Basswood	American basswood	81	133			29	117
Beech	American beech	55	72	Fir	Balsam fir	88	173
Birch	Paper birch	89	72		Silver fir (Europe)	36	165
	Silver birch		70		Grand fir	91	136
	Sweet birch	75	70		Noble fir	34	115
	Yellow birch	74	72		Pacific silver fir	55	164

(Continued)

Table 1.8 (Continued)

Species	Broad-leaved trees	MC (%) in:	
		HW	SW
Cherry	Black cherry	58	
Chestnut	American chestnut	120	
Cottonwood	Cottonwood	162	146
Elm	American elm	95	92
	Cedar elm	66	61
	Rock elm	44	57
Hackberry	Hackberry	61	65
Hickory	Bitternut hickory	80	54
	Mockernut hickory	70	52
	Pignut hickory	71	49
	Red hickory	69	52
	Sand hickory	68	50
	Water hickory	97	60
Magnolia	Magnolia	80	104
Maple	Silver maple	58	97
	Sugar maple	65	72
Oak	California black oak	76	75
	Northern red oak	80	69
	Southern red oak	83	75
	Water oak	81	81
	White oak	64	78
	Willow oak	82	74
Sweetgum	Sweetgum	79	137
Sycamore	American sycamore	114	130
Tulip tree	Yellow poplar	83	106
Tupelo	Black tupelo	87	115
	Swamp tupelo	101	108
	Water tupelo	150	116
Walnut	Black walnut	90	73

Species	Conifer trees	MC (%) in:	
		HW	SW
	White fir	98	160
Hemlock	Eastern hemlock	97	119
	Western hemlock	85	170
Larch	European	35	100
	Tamarack	49	
	Western larch	54	119
Pine	Loblolly pine	33	110
	Lodgepole pine	41	120
	Longleaf pine	31	106
	Ponderosa pine	40	148
	Red pine	32	134
	Scots pine	35	134
		31	133
	Shortleaf pine	32	122
	Sugar pine	98	219
	Western white pine	62	148
	Weymouth pine	81	219
Redwood	Redwood	86	210
Spruce	Black spruce	52	113
	Engelmann spruce	51	173
	Norway spruce	37	133
		31	133
	Sitka spruce	41	142

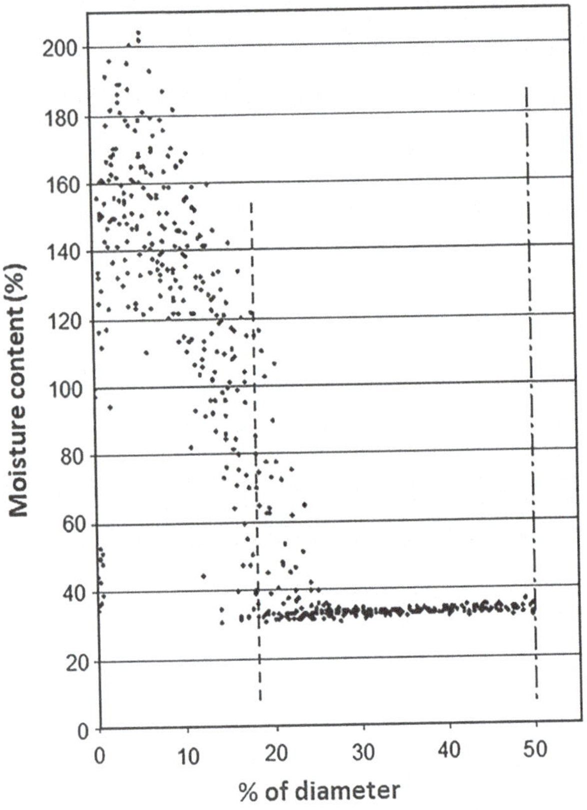

Fig. 1.11 Moisture distribution of cross-sections of Norway spruce logs immediately after felling. Summarised data of four felling dates (December, February, April, July) of trees from the same forest site and age (Fellner and Teischinger 2001). ----- the border between heartwood and sapwood, -.-.- the pith of the log

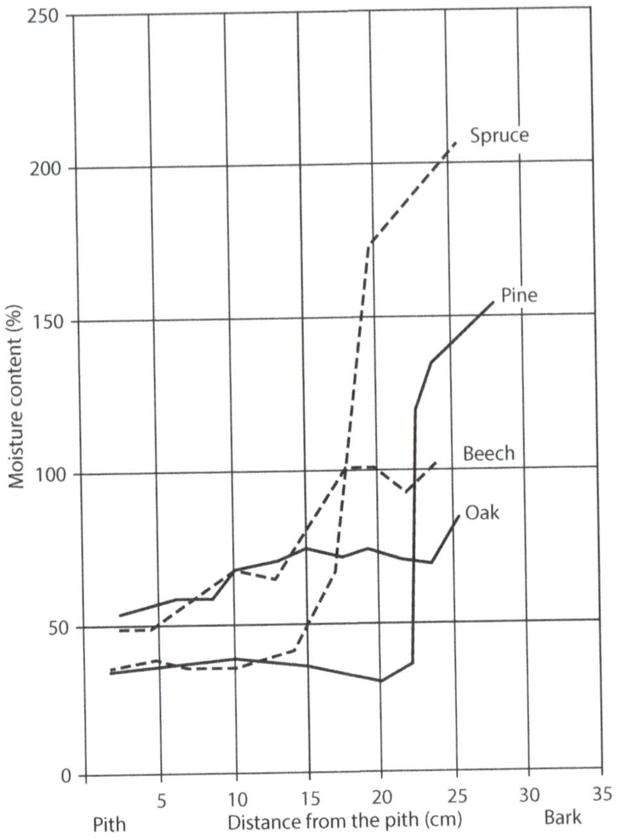

Fig. 1.12 Moisture distribution in standing trees of European beech, European oak, Norway spruce and Scots pine (Trendelenburg and Meyer-Wegelin 1955)

variations, daily temperature fluctuations, heating regimes within buildings, and other factors. Wood in use can also become wet in direct contact with liquid water due to precipitation as rain, hail or snow.

1.3.1 Use Classes and Service Classes – the European System

The European Use Class (UC) system, as described in the EN 335 standard (CEN 2013), does not distinguish between exposure to fresh water and contact with soil, although the Use Classes are defined by moisture conditions and potential degrading organisms in a specific use condition. According to EN 335, the Use Classes are defined as shown in Table 1.9.

Based on the definitions in Table 1.9 and experience in monitoring timber in use, Niklewski et al. (2017) suggested the following moisture content limitations for various Use Classes:

- UC2: the moisture content is occasionally > 20%, with a median of < 17.5%
- UC3.1: the moisture content is frequently > 20%, with a median of 20 ± 2.5%
- UC3.2: the moisture content is frequently > 20%, with a median of 25 ± 2.5%
- UC4: the moisture content is rarely or never < 20%, with a median > 27.5%

The decay risk for wood in a given Use Class can alter the overall performance of the wood component and increase the risk of premature failure. The Use Classes as defined

Table 1.9 Overview of Use Class classifications according to EN 335 (CEN 2013)

Use Class	Definition
1	Situations in which the wood or wood-based product is inside a construction, not exposed to the weather and wetting Attack by disfiguring fungi or wood-destroying fungi is insignificant and always accidental Attack by wood-boring insects, including termites, is possible, although the frequency and importance of such attack depends on the geographical region
2	Situations in which the wood or wood-based product is under cover and not exposed to the weather (particularly rain and driven rain) but where occasional, but not persistent, wetting can occur In this use class, condensation of water on the surface of wood and wood-based products may occur Attack by disfiguring fungi and wood-destroying fungi is possible Attack by wood-boring insects, including termites, is possible although the frequency and importance of such attack depends on the geographical region
3	Situations in which the wood or wood-based product is above ground and exposed to the weather (particularly rain) Attack by disfiguring fungi and wood-destroying fungi is possible Attack by wood-boring insects, including termites, is possible although the frequency and importance of such attack depends on the geographical region 1) A large variety of in-use situations exist and, when relevant, Use Class 3 may be divided into the sub-classes Use Class 3.1 and Use Class 3.2
3.1	In this situation, the wood and wood-based products will not remain wet for long periods. Water will not accumulate
3.2	In this situation, the wood and wood-based products will remain wet for long periods. Water may accumulate
4	A situation in which the wood or wood-based product is in direct contact with ground and/or fresh water Attack by disfiguring fungi and wood-destroying fungi is possible Attack by wood-boring insects, including termites, is possible although the frequency and importance of such attack depends on the geographical region
5	A situation in which the wood or wood-based product is permanently or regularly submerged in salt water (i.e. sea water and brackish water) Attack by invertebrate marine organisms is the principal problem, particularly in the warmer waters where organisms such as *Limnoria* spp., *Teredo* spp. and *Pholads* can cause significant damage. Attack by wood-destroying fungi and the growth of surface moulds and staining fungi is also possible The regions of certain components above water, for example harbour piles, can be exposed to wood-boring insects

Table 1.10 Service Classes according to EN 1995-1-1 (CEN 2010)

Service Class	Description
1	Characterised by a moisture content in the material corresponding to a temperature of 20 °C and a relative humidity in the surrounding air exceeding 65% for only a few weeks per year
2	Characterised by a moisture content in the material corresponding to a temperature of 20 °C and a relative humidity in the surrounding air exceeding 85% for only a few weeks per year
3	Characterised by climatic conditions leading to higher moisture contents than in Service Class 2

by EN 335 refer to the decay risks associated with different wood-destroying organisms, and should be used in conjunction with Service Class definitions given in the EN 1995-1-1 standard (CEN 2010), which define strength values and can be used to calculate the deformations of structural timber members under defined environmental conditions, as shown in Table 1.10.

1.3.2 Scheffer's Climate Index

Early work into a better understanding of how the decay risk is affected by the macro- and micro-climate focused on conditions in the United States (Scheffer 1971), where four different sites were initially evaluated in terms of the effects of temperature and local rainfall on the potential hazard to timber samples:

$$Climate\ index = \frac{\sum_{Dec}^{Jan}[(T-35)(D-3)]}{30} \quad (1.2)$$

where T is the mean day temperature of the month (in Fahrenheit) and D is the number of days with more than 0.1 inches (25.4 mm) of rain per month. The original data showed that the decay hazard ranged from 0.0 in Yuma, Arizona, to 137.5 in West Palm Beach, Florida, and that within the continental USA, three distinct climate zones were noted, indicating three levels of above-ground decay potential (Fig. 1.13).

The concept of the Scheffer Index has been further applied to other regions, including Europe, as shown in Fig. 1.14.

The exposure is expressed as an exposure dose (DEd), for example, as reported by van Niekerk et al. (2022) for Euroep, as shown in Fig. 1.15. The exposure dose is

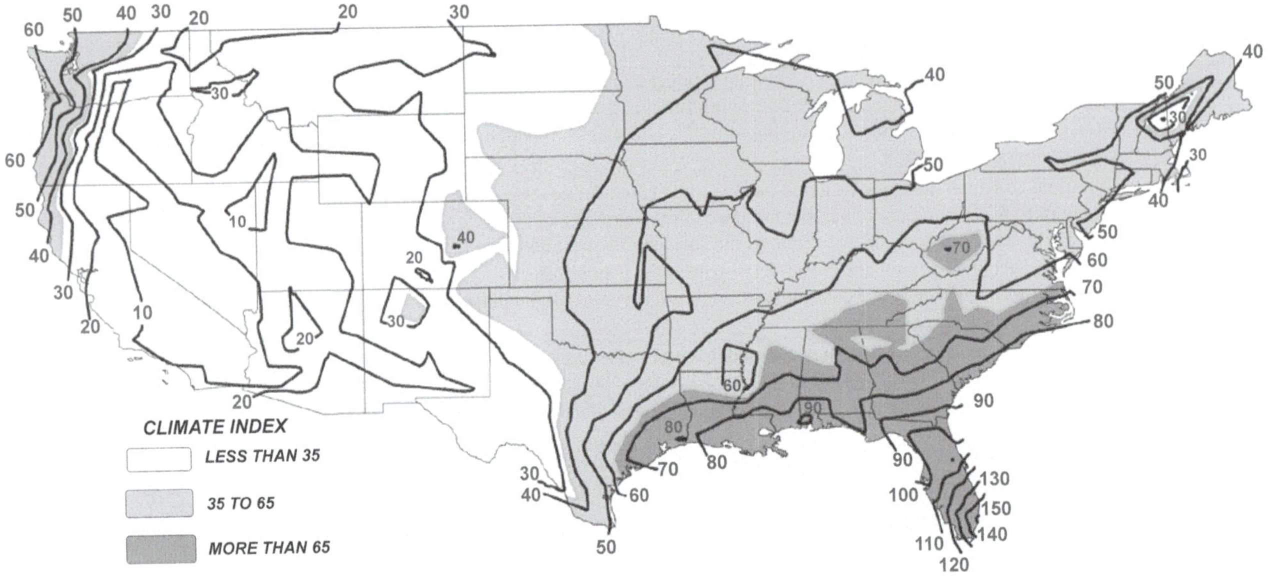

Fig. 1.13 Revised Scheffer´s climate-index map (USA) for decay hazard based on data for 1971–2000. Higher index values indicate a greater decay hazard. Scheffer´s climate-index map of the USA (Carll 2009)

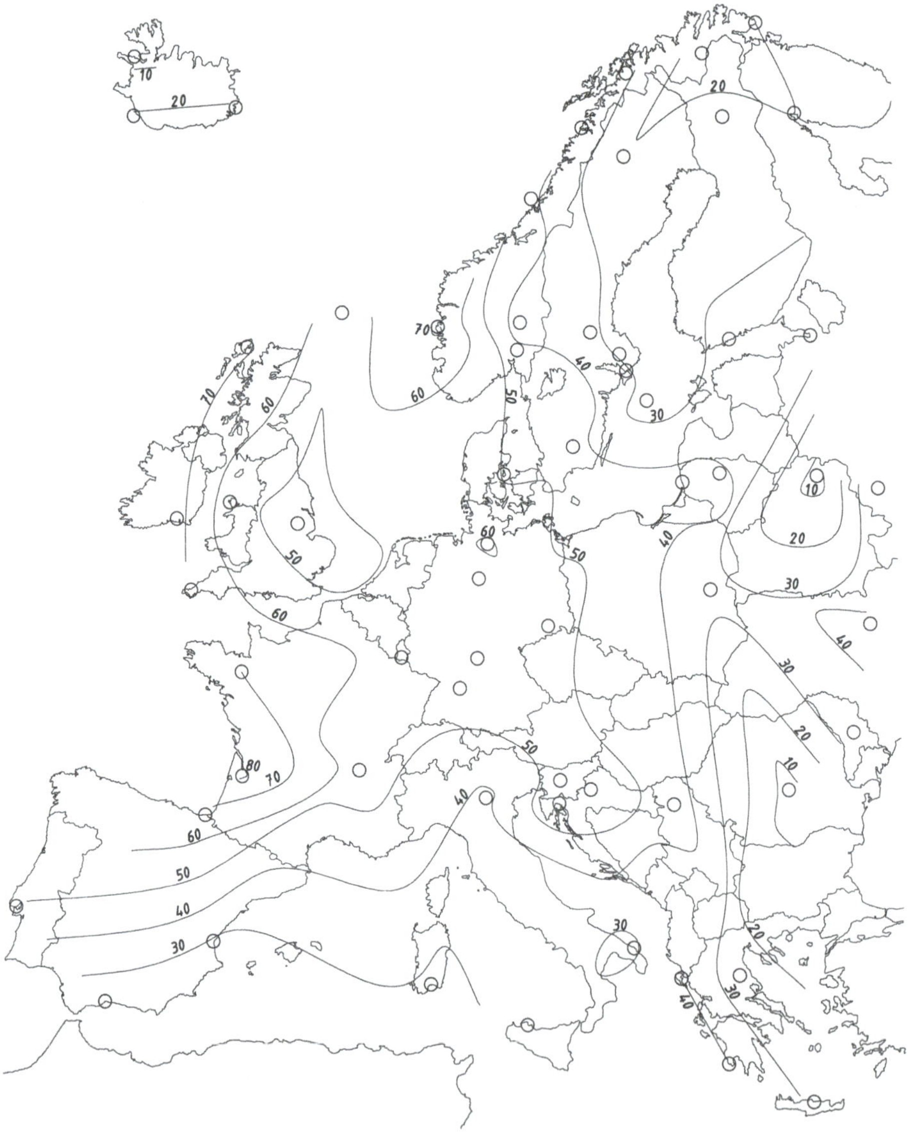

Fig. 1.14 Relative decay potential for Europe in terms of Scheffer's climate index for various European sites (Brischke et al. 2011)

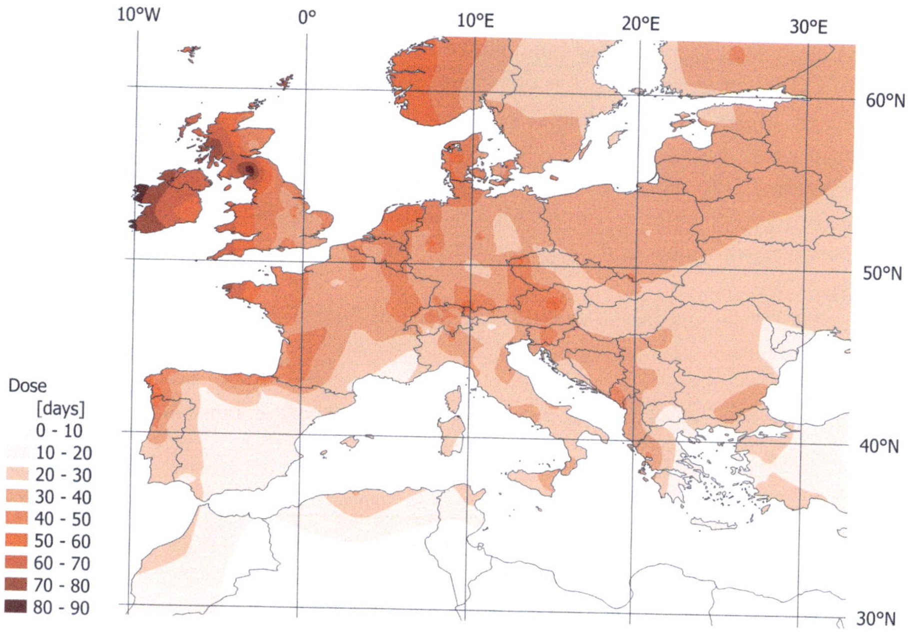

Fig. 1.15 Cumulativ-dose map over a one year period based on the contemporary climate over a major part of Europe (van Niekerk et al. 2022)

determined by a dose–response model based on estimated daily averages of wood temperature and moisture content. The material resistance is expressed as a resistance dose (DRd). Acceptance for a chosen design and material is described as: Exposure dose (DEd) ≤ resistance dose (DRd).

1.3.3 The Moisture Condition in Wood and Its Relation to the Indoor Climate

Wood is a hygroscopic material that constantly strives to remain in equilibrium with the local climate, i.e. the relative humidity (RH) and the temperature. The equilibrium moisture content (EMC) is defined as the moisture content of wood when it is in equilibrium with the surrounding environment. In the hygroscopic range (0% to approximately 30% MC corresponding to 0 to 100% RH), the EMC is described by the moisture isotherm, which is unique but fairly similar for all species (Fig. 1.16).

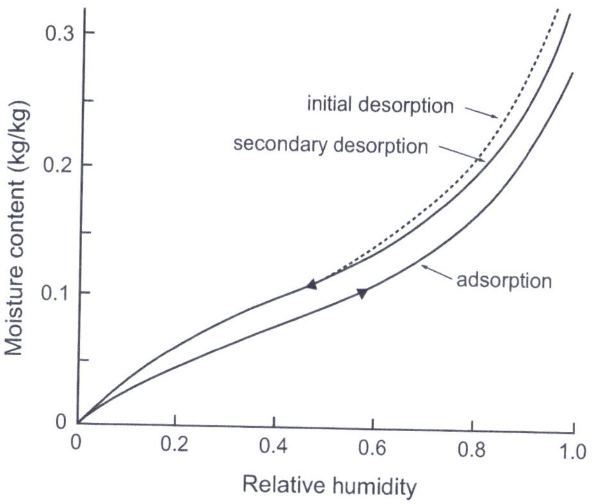

Fig. 1.16 Typical adsorption and desorption isotherms for wood at approximatel 20 °C

The relationship between the moisture content of wood and the temperature and relative humidity of the surrounding air can be approximated by the chart in Fig. 1.17. Figure 1.18 illustrates the potential changes in moisture content in wood during the construction and operation of a timber building. The moisture content at the time of manufacture of a timber building (initial MC) is usually below 15–20% for sawn timber and 5–15% for engineered wood products (EWPs).

Figure 1.19 summarises the estimated vertical movement of each storey of a timber building during the change from the construction phase to the in-service phase of the building. The total movement is divided into shrinkage of the wood material and compression due to the weight of the building. The accumulated shrinkage over four storeys accounts for about 90% of the total vertical movement. This suggests that the vertical deformation caused by changes in the moisture content of the wood is critical and must be considered in the design of timber buildings. The selection of wood-based materials has a considerable influence on the total vertical movement.

The moisture content of wood varies depending on its use, resulting from changes in local temperature and relative humidity throughout the annual cycle. These outdoor climate conditions vary over the world and, depending on how a building is heated or cooled, the climate conditions indoors

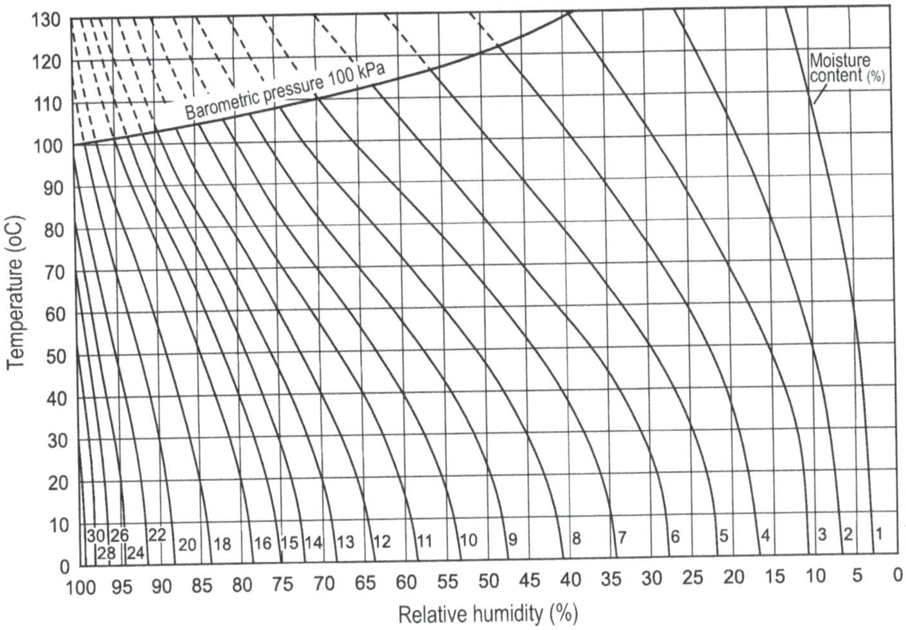

Fig. 1.17 Relationship between the moisture content and the temperature and relative humidity for Sitka spruce (Kollmann and Coté 1968)

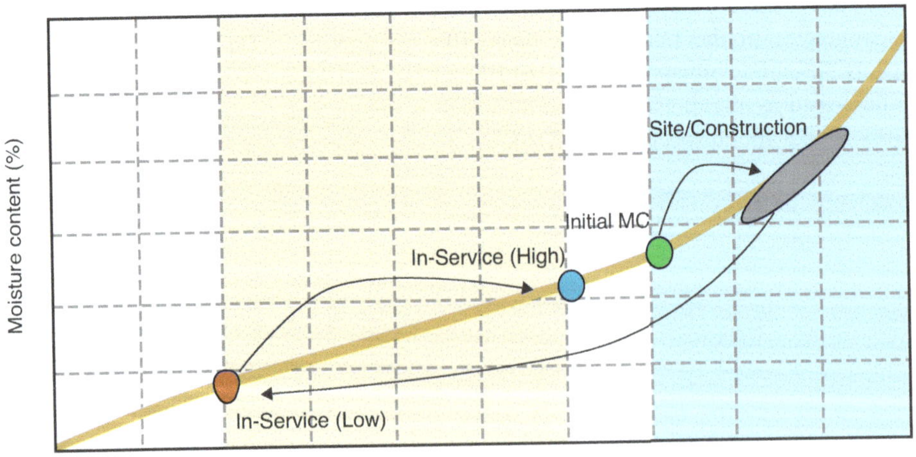

Fig. 1.18 Moisture content in a building under construction (blue region) and in-service (beige region) (Gong 2021)

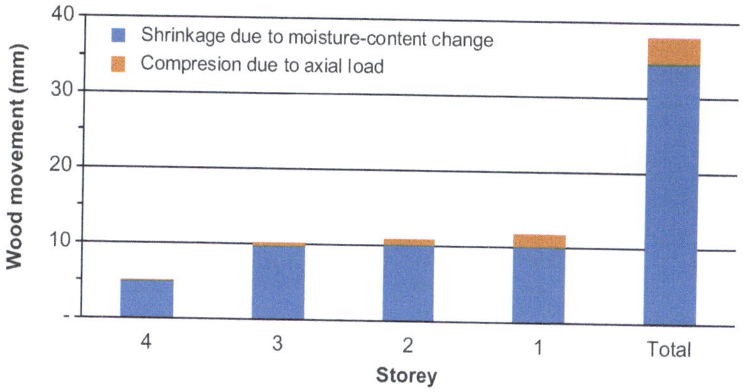

Fig. 1.19 Estimated vertical movement resulting from changes in moisture content and axial compressive load (Gong 2021)

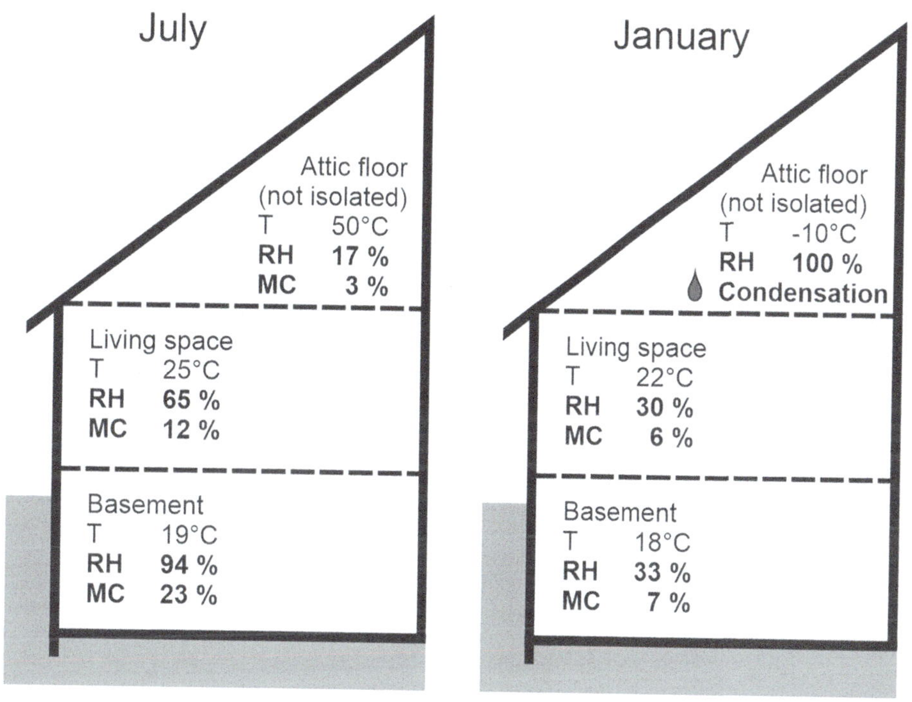

Fig. 1.20 Typical climate conditions in a building in Mid-West USA (Wilcox et al. 1991)

Table 1.11 Required moisture content of wood and wood-based materials for various applications in Central Europe. EMC equilibrium moisture content (Scheer et al. 2004)

Application	EMC (%)
Closed buildings with heating	4–12
Closed buildings without heating	9–15
Components in open or roofed constructions	12–20
Floors, parquet	(5) 7–11 (13)
Interior design	6–10
Living room furniture	8–10
Materials for furniture	10–12
Radiator covers	6–8
Weather-exposed components	>20
Windows and front doors	10–15

also vary (Fig. 1.20), with the result that the moisture content of the wood products has to be adapted (Table 1.11).

References

Abasolo W, Yoshida M, Yamamoto H et al (1999) Internal stress generation in rattan canes. IAWA J 20(1):45–58

Andersson S, Serimaa R, Torkkeli M et al (2000) Microfibril angle of Norway spruce (*Picea abies* (L.) Karst.) compression wood: comparison of measuring techniques. J Wood Sci 46:343–349

Barnett JR, Jeronimidis G (eds) (2003) Wood quality and its biological basis. Blackwell, Oxford

Barnett JR, Bonham VA (2004) Cellulose microfibril angle in the cell wall of wood fibres. Biol Rev 79:4461–4472

Bonham VA, Barnett JR (2001) Fibre length and microfibril angle in silver birch (*Betula pendula* Roth). Holzforschung 55(2):159–162

Brischke C, Frühwald Hansson E, Kavurmaci D et al (2011) Decay hazard mapping for Europe. In: The Proceedings IRG Annual Meeting, Queenstown, New Zealand. International Research Group on Wood Protection IRG/WP 11–20463
Bues C, Fischer S (2012) Rohstoff Holz (Wood as a raw material). TU Dresden, Tharandt
Bues C, Bäuker E (2014) Vertiefende Anatomie des Holzes und der Rinde (In-depth anatomy of wood and bark). TU Dresden, Dresden
Carll CG (2009) Decay hazard (Scheffer) index values calculated from 1971–2000 climate normal data. General Technical Report FPL-GTR-179. Department of Agriculture, Forest Service, Forest Products Laboratory, Madison WI
CEN (2010a) EN 1995-1-1, Eurocode 5: Design of timber structures – Part 1-1: General – Common rules and rules for buildings. CEN-European Committee for Standardization, Brussels
CEN (2013a) EN 335, Durability of wood and wood-based products – Use Classes: Definitions, Application to solid wood and wood-based products. CEN-European Committee for Standardization, Brussels
Donaldson LA (1992) Within- and between-tree variation in microfibril angle in *Pinus radiata*. N Z J For Sci 22:77–86
Donaldson LA, Burdon RD (1995) Clonal variation and repeatability of microfibril angle in *Pinus radiata*. N Z J For Sci 25:164–174
Ekman W, Eneroth O, Hanngren H et al (1922) Handbok i skogsteknologi (Handbook in Forest Technology. Fritzes, Stockholm
Evans R (1998) Rapid scanning of microfibril angle in increment cores by X-ray diffractometry. In: Butterfield, BG (ed) Microfibril angle in wood: Proceedings of the IAWA/IUFRO international workshop on the significance of microfibril angle to wood quality. Westport, N Z University of Canterbury Press, Canterbury
Evans R, Ilic J (2001) Rapid prediction of wood stiffness from microfibril angle and density. For Prod J 51:53–57
Evans R, Stringer S, Kibblewhite RP (2000) Variation of microfibril angle, density and fibre orientation in twenty-nine *Eucalyptus nitens* trees. Appita J 53:450–457
Fellner A, Teischinger A (2001) Alte Holzregeln (Old wooden rules). Österreichischer Kunst- und Kulturverlag, Wien
Fengel D, Wegener G (1984) Wood: Chemistry, ultrastructure, reactions. Walter De Gruyter, Berlin
French J, Conn AB, Batchelor WJ et al (2000) The effect of fibre fibril angle on some handsheet mechanical properties. Appita J 53:210–217
Glass SV, Zelinka SL (2010) Moisture relations and physical properties of wood, Chapter 4. In: Forest Products Laboratory, Wood handbook. Wood as an engineering material (Centential Edition). General Technical Report FPL-GTR-190. Department of Agriculture, Forest Service, Madison WI
Gong M (ed) (2021) Engineered wood products for construction. IntechOpen, London
Gorisek Z, Torelli N (1999) Microfibril angle in juvenile, adult and compression wood of spruce and silver fir. Phyton 39(3):129–132
Herman M, Dutilleul P, Avella-Shaw T (1999) Growth rate effects on intra-ring and inter-ring trajectories of microfibril angle in Norway spruce (*Picea abies*). IAWA J 20:3–21
Kollmann FFP, Côté Jr WA (1968) Principles of wood science and technology (Vol. 1). Springer, Berlin
Lichtenegger K, Reiterer A, Stanzl-Tscheg S et al (1999) Variation of cellulose microfibril angles in softwoods and hardwoods – a possible strategy of mechanical optimisation. J Struct Biol 128:257–269
Lotfy M, El-osta M, Kellogg RM et al (1973) A direct X-ray technique for measuring microfibril angle. Wood and Fiber 5(2), 118–128
McMillin CW (1973) Fibril angle of loblolly pine wood as related to specific gravity, growth rate, and distance from pith. Wood Sci Technol 7:251–255
Mette H, Korell U (1989) Richtzahlen und Tabellen für die Forstwirtschaft (Indicative numbers and tables for forestry). Deutscher Landwirtschaftsverlag, Berlin
Niemz P, Teischinger A, Sandberg D (eds) (2023) Springer Handbook of Wood Science and Technology. Springer Nature, Cham
Nilsson B, Thörnqvist T (2008) Regional amount of forest fuel in Sweden. EC 6th Framework Programme "Clean Hydrogen-rich Synthesis Gas", (Contract No: SES6-CT-2004–502587), Report No. CHRISGAS June 2008_WP5_D37
Niklewski J, Isaksson T, Frühwald Hansson E et al (2017) Moisture conditions of rain-exposed glue-laminated timber members: the effect of different detailing. Wood Mater Sci Eng 13(3):129–140
Reiterer A, Jakob HF, Stanzl-Tschegg SE et al (1998) Spiral angle of elementary cellulose fibrils in cell walls of *Picea abies* determined by small-angle X-ray scattering. Wood Sci Technol 32:335–345
Reiterer A, Lichtenegger H, Tschegg S et al (1999) Experimental evidence for a mechanical function of the cellulose microfibril angle in wood cell walls. Philos Mag A 79(9):2173–2184
Rosenthal M, Bäuker E (2012) Druckholz – Reaktionsholz der Nadelhölzer (Compression wood - reaction wood of softwoods). Holz-Zent.bl 43:1105–1107
Sahlberg U, Salmén L, Oscarsson A (1997) The fibrillar orientation in the S2-layer of wood fibres as determined by X-ray diffraction analysis. Wood Sci Technol 31:77–86
Sandberg D (ed) (2011) Utvändiga träfasader: Inverkan av materialval, konstruktion och ytbehandling på beständigheten hos fasader av gran och tall (External wooden façades: the influence of the choice of material, design and surface treatment on the durability of façades of Norway spruce and Scots pine) Report No. 11, Linnæus University, Växjö
Sandberg D, Kutnar A, Karlsson O et al (2021) Wood modification technologies. Principles, sustainability, and the need for innovation. CRC Press, Boca Raton
Scheer C, Peter M, Stöhr S (2004) Holzbau-Taschenbuch Bd.2. Bemessungsbeispiele nach DIN 1052 (Timber Construction Vol.2. Design examples according to DIN 1052). 10th edn. Ernst & Sohn Verlag fur Architektur und technische Wissenschaften, Berlin
Scheffer TC (1971) A climate index for estimating potential for decay in wood structures above ground. For Prod J 21:25–31
Senft JF, Bendtsen BA (1985) Measuring microfibrillar angles using light microscopy. Wood Fiber Sci 17:564–567
Stuart S, Evans R (1994) X-ray diffraction estimation of the microfibril angle variation in eucalypt wood. Appita 48:197–200
Timell TE (1986) Compression wood in Gymnosperms I-III. Springer-Verlag, Berlin
Trendelenburg R, Meyer-Wegelin (1955) Das Holz als Rohstoff (Wood as a raw material). Carl Hanser Verlag, München
van Niekerk PB, Marais BN, Brischke C et al (2022) Mapping the biotic degradation hazard of wood in Europe – biophysical background, engineering applications, and climate change-induced prospects. Holzforschung 76(2):188–210
Vintila E (1939) Untersuchungen über Raumgewicht und Schwindmass von Früh- und Spätholz bei Nadelhölzern (Studies on density and shrinkage of earlywood and latewood in conifers). Holz Roh- Werkstoff 2(10):345–357
Wardrop AB (1951) Cell wall organisation and the properties of the xylem. 1. Cell wall organisation and the variation of breaking load in tension of the xylem in conifer stems. Aust J Sci Res B 4: 391–414
Watson AJ, Dadswell HE (1964) Influence of fibre morphology on paper properties. 4. Micellar spiral angle. Appita 17:151–156
Wilcox W, Botsai E, Kubler H (1991) Wood as a building material. A guide for designers and builders. John Wiley & Sons, New York
Yamamoto H, Okuyama T, Yoshida M (1993) Method of determining the mean microfibril angle of wood over a wide range by the improved Cave's method. Mokuzai Gakkaishi 39:375–381
Yamashita K, Hirakawa Y, Fujisawa Y et al (2000) Effects of microfibril angle and density on variation of modulus of elasticity of sugi (*Cryptomeria japonica*) logs among eighteen cultivars. Mokuzai Gakkaishi 46:510–522
Zobel BJ, Buijtenen JP v (1989) Wood variation: Its causes and control. Springer-Verlag, Berlin
Zobel BJ, Sprauge JR (1998) Juvenile wood in forest trees. Springer-Verlag, Berlin

Prof.Dr.-Ing.habil. Dr.h.c. Peter Niemz studied at the Faculty of Mechanical Engineering of TU Dresden. From 1972 onwards, he worked at the Institute of Wood Technology Dresden (IHD) and at TU Dresden. From 1993 to 1996, he served as a Full Professor at the Universidad Austral de Chile in Valdivia before joining the Swiss Federal Institute of Technology (ETH Zurich) in 1997. He was Professor of Wood Physics from 2002 until his retirement in 2015.

Prof.Dr. Dick Sandberg received his Doctorate in Mechanical Engineering, Wood Technology and Processing from the Royal Institute of Technology (KTH), Stockholm, Sweden, in 1998. He is a Chaired Professor of Design and Fabrication of Timber Construction at the Norwegian University of Science and Technology (NTNU). He specialises in using new measurement technology, advanced information processing, and wood communication from felling to finished product. He is an Elected Fellow of the International Academy of Wood Science.

Physical and Mechanical Properties of Solid Wood

2

Galina Gorbacheva, Andrade Egas, Raquel Gonçalves, Jiali Jiang, Peter Niemz, Róbert Németh, Robert J. Ross, Dick Sandberg and Christopher Adam Senalik

2.1 Introduction

Peter Niemz, Dick Sandberg

This chapter presents data on the physical and mechanical properties of hundreds of wood species from Africa, China, Europe, Russia, South America, and the USA. Bamboo is also included for China. The extensive data compilation allows for the identification of even very specific wood species worldwide. To create this resource, experts from the respective regions were enlisted to compile the available knowledge. This instruction section discusses various factors that influence wood properties, including wood species, growing conditions, such as climate region, soil, altitude, and social position within the forest, within-stand and interregional variations, as well as tree age and height. Further details can be found in sources such as Kollmann and Côté (1968), Walker (2006), Niemz et al. (2023). The relationship between density and strength is explored for selected species, with a focus on the variation between species, such as the correlation between density and strength. Additionally, the connection between wood structure and selected properties is illustrated using Ashby diagrams. Emphasis is placed on the variability within wood species and the overlap of properties, which is crucial for the practical use of wood and strength grading.

2.2 Variation in Properties

Peter Niemz, Dick Sandberg

Due to the natural variation of wood material, its physical and mechanical properties have in general a large variation between species, withing the same species, and within the same tree or piece of wood (Fig. 2.1).

The original version of the chapter has been revised. A correction to this chapter can be found at https://doi.org/10.1007/978-3-031-38329-8_9

G. Gorbacheva (✉)
Department of Wood Science and Technology, Mytishchi Branch, Bauman Moscow State Technical University, Moscow, Russian Federation
e-mail: gorbacheva@bmstu.ru

A. Egas
Faculty of Agronomy and Forestry Engineering, Wood Science and Technology, Eduardo Mondlane University (UEM), Maputo, Moçambique
e-mail: aegas@uem.mz

R. Gonçalves
School of Agricultural Engineering (FEAGRI), Nondestructive Testing Laboratory (LabEND), University of Campinas (UNICAMP), Sao Paulo, Brazil
e-mail: raquelg@unicamp.br

J. Jiang
Chinese Academy of Forestry, State Key Laboratory of Tree Genetics and Breeding, Research Institute of Wood Industry, Beijing, P. R. China
e-mail: Jialiwood@caf.ac.cn

P. Niemz
Department Wood Physics (today, Wood Materials Science), Institute of Building Materials, ETH Zürich, Zurich, Switzerland
e-mail: niemzp@retired.ethz.ch

R. Németh
Institute of Wood Science, University of Sopron Simonyi Károly Faculty, Sopron, Hungary
e-mail: nemeth.robert@uni-sopron.hu

R. J. Ross · C. A. Senalik
Forest Products Laboratory, USDA Forest Service, Madison, USA
e-mail: robert.j.ross@usda.gov

C. A. Senalik
e-mail: christopher.a.senalik@usda.gov

D. Sandberg
Department of Manufacturing and Civil Engineering, Norwegian University of Science and Technology (NTNU), Gjøvik, Norway
e-mail: dick.sandberg@ntnu.no

P. Niemz et al. (eds.), *Wood Material and Processing Data*, https://doi.org/10.1007/978-3-031-38329-8_2

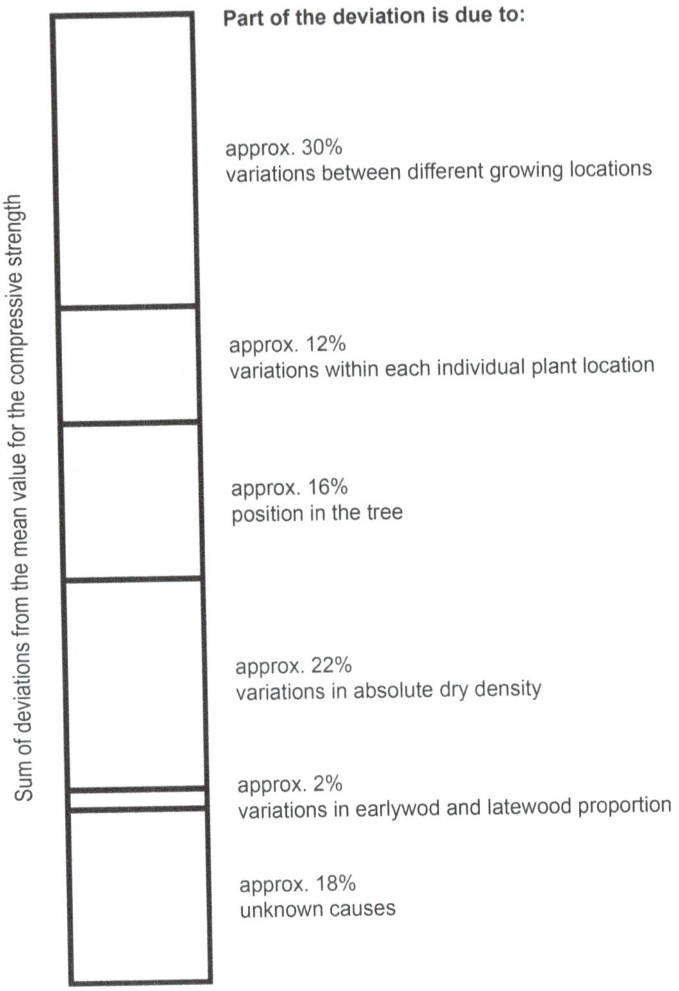

Fig. 2.1 The causes of variations in the compressive strength in Scots pine (after Thunell 1945)

For defect free wood, the density of the material is probably the factor which provides the most reliable guide for the prediction of mechanical properties. The substance of which the wood fibres are formed, and which provides the strength of the material is approximately the same for all species and it is of the order of 1500 kg/m^3. The structural arrangement of the fibres therefore has a considerable influence on the physical and mechanical properties, and the mechanical properties in particular show a pronounced size effect, i.e. the strength of a cellulose chain is approximately 10 times greater than the strength of cell wall in the latewood, which in turn is greater than that of the cell wall in the earlywood. The characteristic strength values used in construction are therefore considerably lower than those which can be achieved when small defect-free specimens are tested on a laboratory scale (Fig. 2.2). Fibre orientation and structural features such as knots have a strong influence on the mechanical properties of timber.

Engineered wood products (EWPs), which are also referred to as reconstituted wood, wood-based products, wood-based composites or man-made wood, include a range of derivative wood products which are manufactured by binding or fixing together fibres, particles, strands, veneers or sawn timber. They have a high load-bearing capacity in relation to their weight, their dimensional stability and their flexibility, and they can be adapted to the required shape and dimensions of the structural element. EWPs have in general more uniform properties in their planar directions and in the cross section of the composite, and the safety margins in strength can be narrower than those required when sawn timber is used (Fig. 2.3), which means that for solid wood, the typical size effect is reduced.

2.2.1 Density of Various wood Species

See Figs. 2.4, 2.5, 2.6 and 2.7

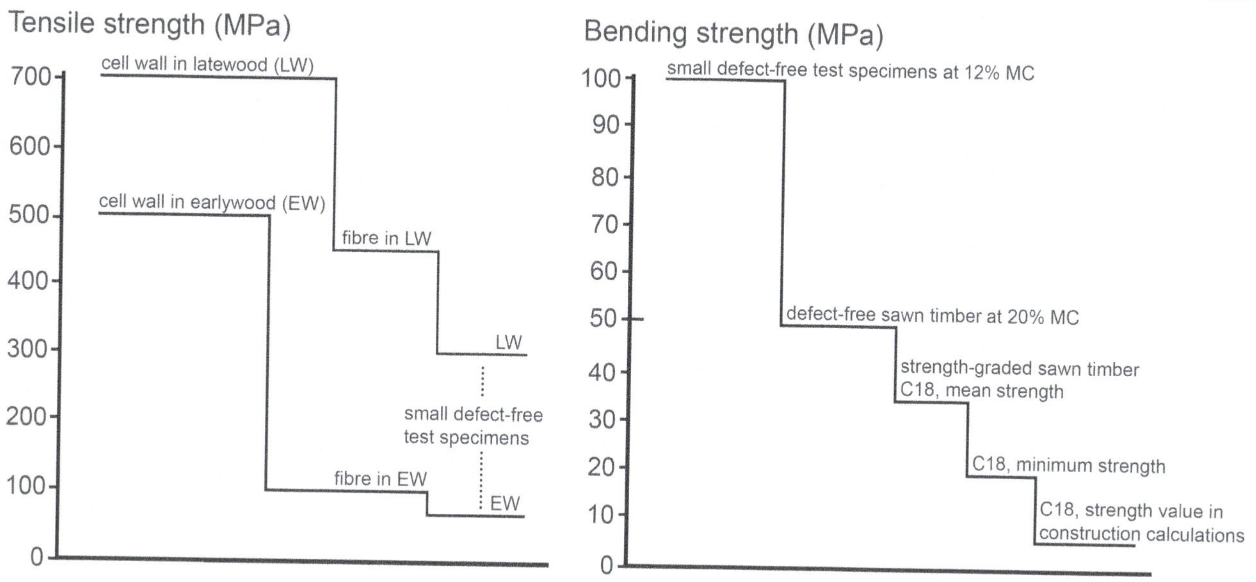

Fig. 2.2 Tensile strength of wood at different structural levels. C18 – Strength Classes, e.g. C14, C18, C24, C30 and C35, are machine stress-graded structural timber according to the EN 14081–1 standard (CEN 2019) intended for load-bearing structures. MC – moisture content

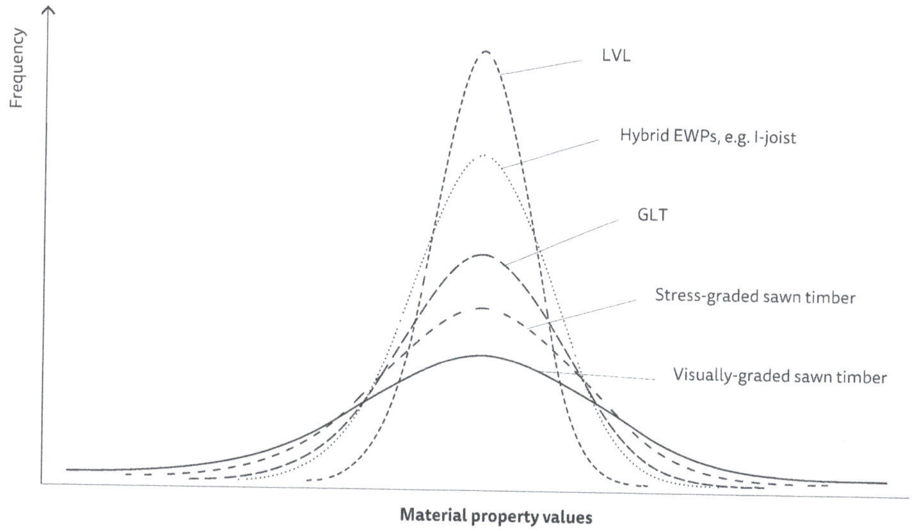

Fig. 2.3 Frequency distribution and variability in mechanical properties of wood and various EWPs. LVL – laminated veneer lumber, GLT – glued-laminated timber (Sandberg et al. 2018)

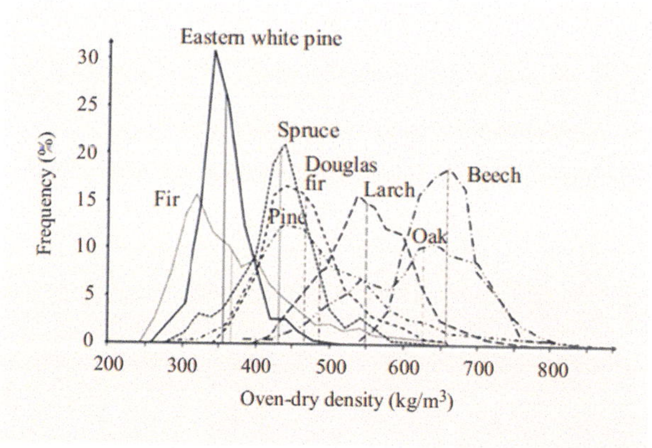

Fig. 2.4 Frequency distributions of the density of different wood species (Knigge and Schulz 1966). Beech – European beech, larch – European larch, oak – European oak, pine – Scots pine, spruce – Norway spruce

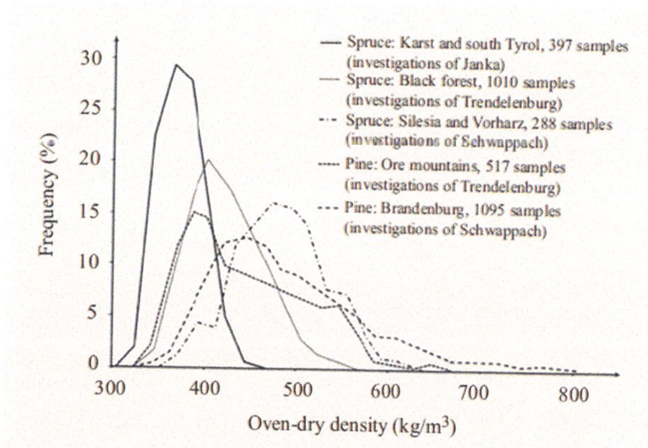

Fig. 2.5 Frequency distributions of the density of Norway spruce and Scots pine from different European sites (Trendelenburg 1935)

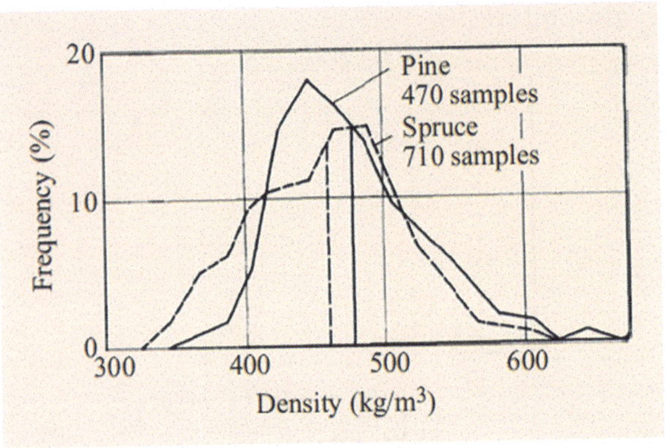

Fig. 2.6 Frequency distributions of the density of Norway spruce and Scots pine grown on the plateau of upper Bavaria, Germany (Trendelenburg 1935). Pine – Scots pine, spruce – Norway spruce

2.2.2 Relationship Between Mechanical Properties and Density

See Figs. 2.8 and 2.9.

2.2.3 Ashby Diagrams for Wood, Wood-Based Materials and Other Materials

An Ashby plot, named after Michael Ashby of Cambridge University, is a scatter plot which displays two or more properties of various classes of materials. Such a plot is useful to compare the ratio between different properties (Figs. 2.10, 2.11, 2.12, 2.13 and 2.14).

Fig. 2.7 Frequency distribution of the density of woods grown in temperate, subtropical and tropical zones (Prütz 1941)

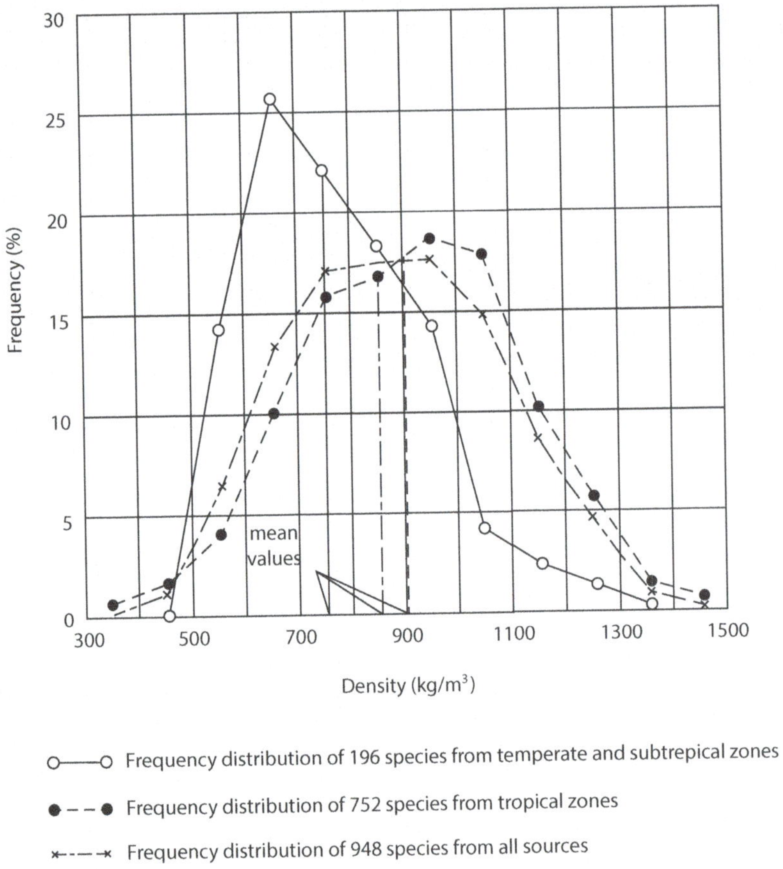

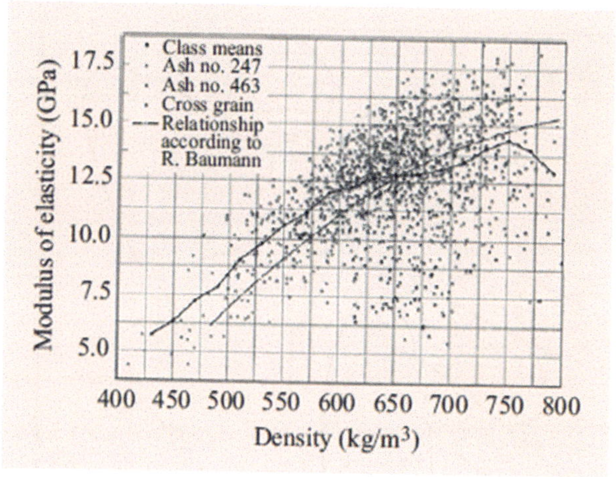

Fig. 2.8 Relationship between modulus of elasticity and density for European ash wood (Kollmann 1941)

2.3 Characterisation of Wood Species from Different Regions of the World

Galina Gorbacheva, Andrada Egas, Jiali Jang, Raquel Gonzalves, Robert J. Ross; Christopher Adam Senalik

Wood is a natural material and a renewable resource that is used in all spheres of human activity. A unique set of properties, a huge variety of wood species, together with traditional and innovative wood processing technologies create a huge potential for using wood as an advanced material.

The number of wood species currently known is 60,065, over 45% (27,203) of which are found in only 10 families (Fig. 2.15) (Beech et al. 2017; Jänicke et al. 2019). The family with the most species is the Leguminosae with 5405 tree species, followed by the Rubiaceae (4827), and Myrtaceae (4330) families.

The country with the most diverse tree flora is Brazil, with 8715 tree species, followed by Colombia (5776 spp.) and Indonesia (5142 spp.), as shown in Fig. 2.16.

Nearly 58% of all tree species (34 575) are endemics to single countries. The countries with the most endemic trees are Brazil (4333 spp.), Madagascar (2991 spp.), Australia (2584 spp.), and China (2149 spp.), as shown in Fig. 2.17.

2.3.1 Species from Africa

Andrada Egas

African florae comprise a huge variety of tree species which have been used for e.g. medicine, food, local construction, energy and the production of high-value timber for trade. There is a huge diversity of timber species in Africa, which have been used both locally and for export. Table. 2.1 show the properties of some of the most frequently used and/or commercialised African hardwood species from the West, Eastern, Central and Southern regions of Africa.

The table includes physical and mechanical properties as well as the chemical composition, the basic density, the density at 12% moisture content and the air-dry density. The basic density is based on the oven-dry weight and the volume of green specimens, while the density at 12% moisture content is based on the weight and volume of samples at 12% moisture content. The moisture contents of the specimens for the calculation of air-dry density are not specified in the references. Rates of shrinkage are reported from the

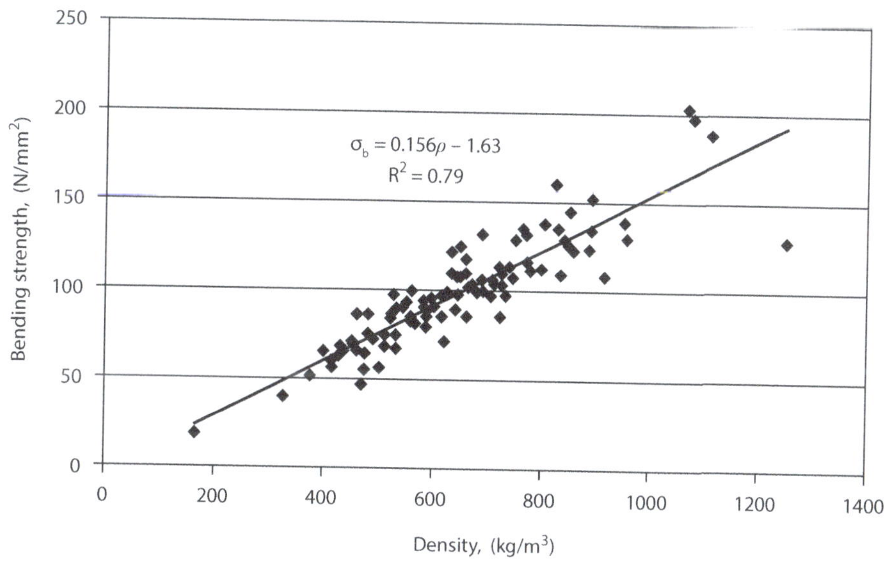

Fig. 2.9 Relationship between bending strength (σ_b) and density (ρ) at 12% moisture content of solid wood using material parameters from the literature for 103 wood species (Niemz and Sonderegger 2017)

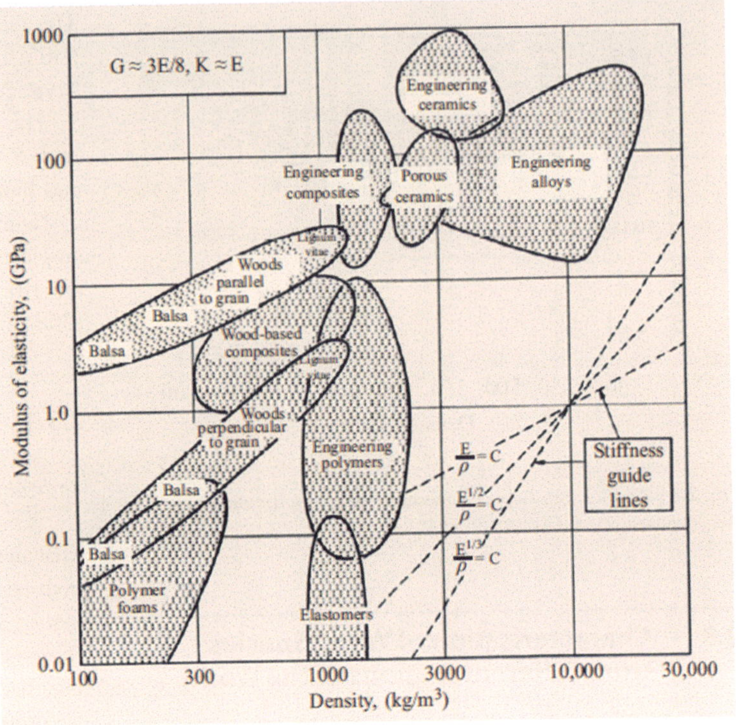

Fig. 2.10 Material-property chart for engineering materials and wood showing the modulus of elasticity (E) plotted against density (ρ). Guidelines show the slopes of three material indices (Wegst and Ashby 2007)

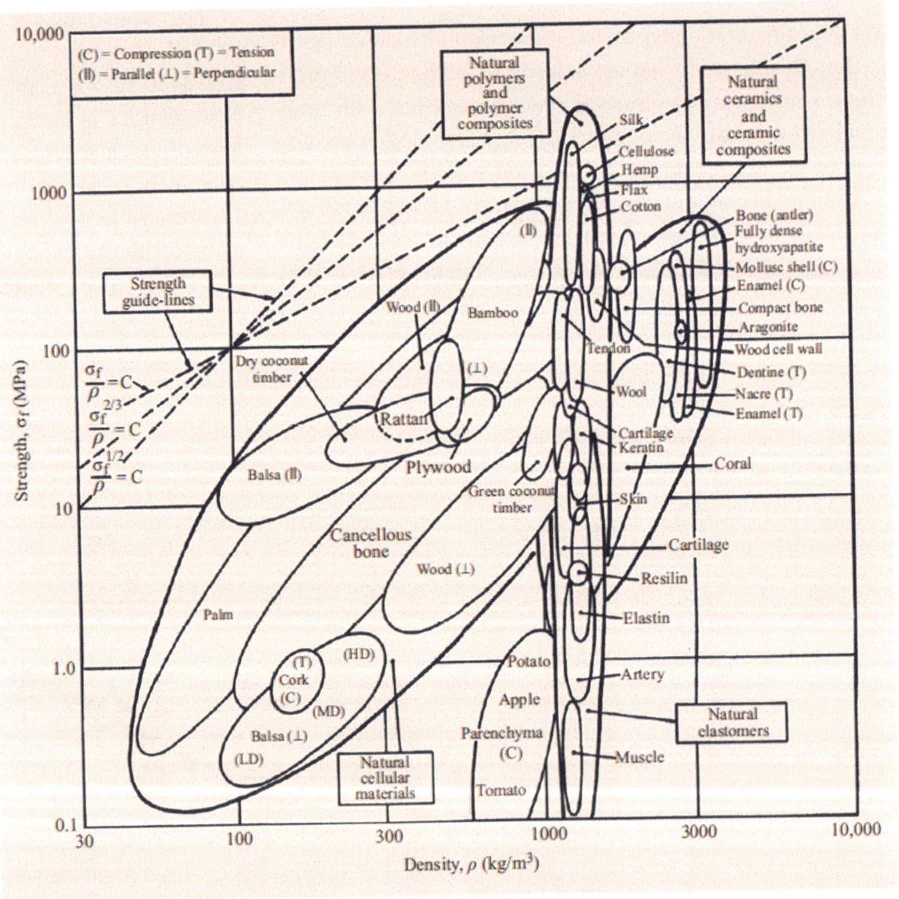

Fig. 2.11 Material-property chart for engineering materials and wood showing strength (σ_f) plotted against density (ρ). Guidelines identify structurally efficient materials that are light and strong (Wegst and Ashby 2007)

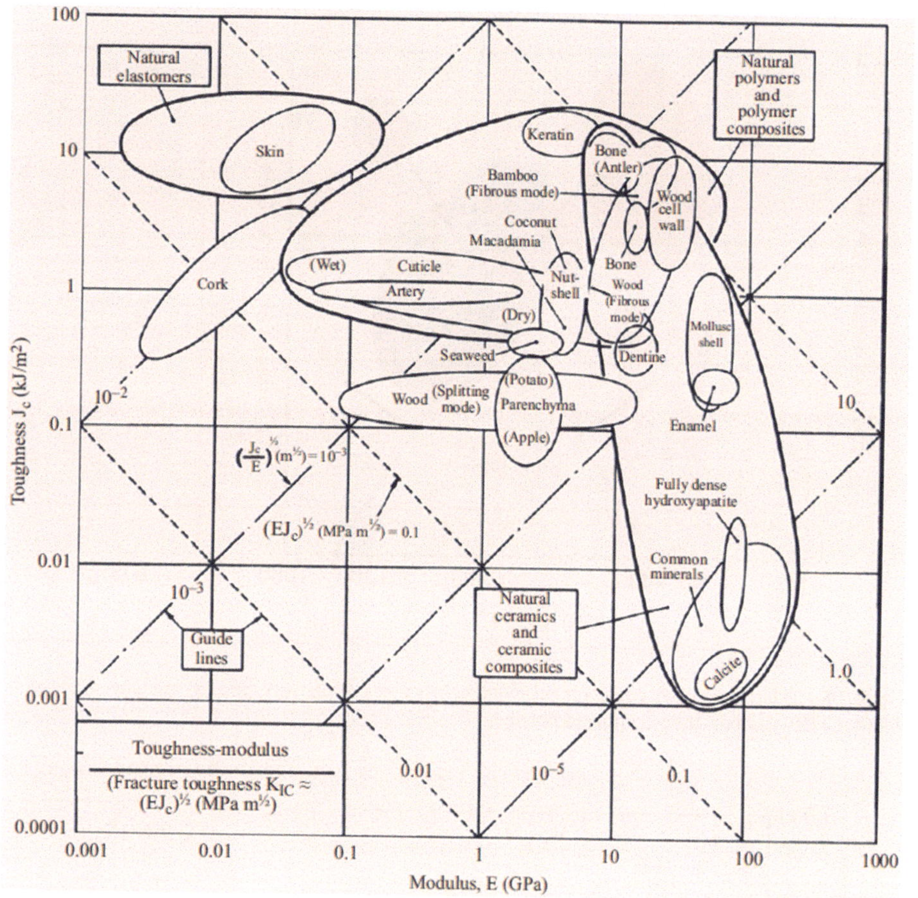

Fig. 2.12 Material-property chart for natural materials, showing toughness (J_c) plotted against modulus of elasticity (E) (Wegst and Ashby 2007)

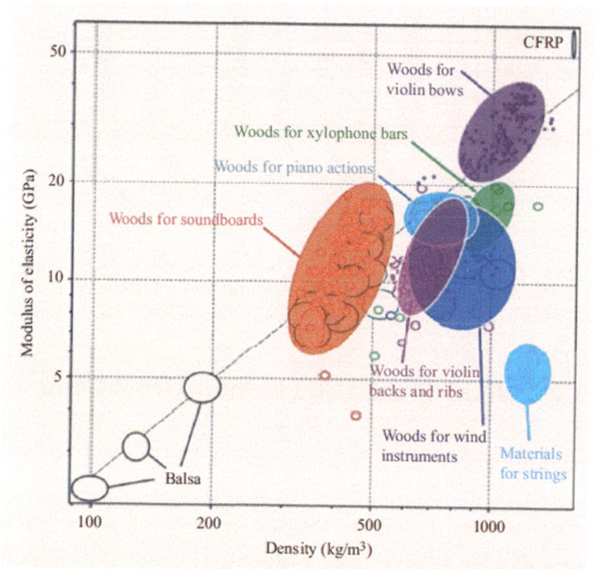

Fig. 2.13 A material-property chart for wood, showing the modulus of elasticity parallel to the fibre direction plotted against density (Wegst 2006). CFRP – carbon-fibre-reinforced polymers

green to the oven-dry states and are presented for both radial and tangential directions; the rate of volumetric shrinkage is also presented. Mechanical properties are based on tests performed at 12% moisture content. The shear strength values from the ITTO database are reported for the radial direction, but in other sources, the directions are not specified.

The information presented in Table. 2.1 is based on some of the major international references which are indicated in the table based on literature published around the world. Due to the diversity of units in the reference sources, the data have been converted to the same units.

For some sources, the information regarding the method used to obtain the results was not found. In other sources a large diversity of methods was used, including American, British, French, Portuguese, European and Brazilian standards or procedures. The information reported in this table thus only provided an indication of the property variation within species regardless of the standards according to which tests were performed. Therefore, due to the broad geographical distribution of the selected species, for timber design and regulatory issues, data based on local tests should be obtained.

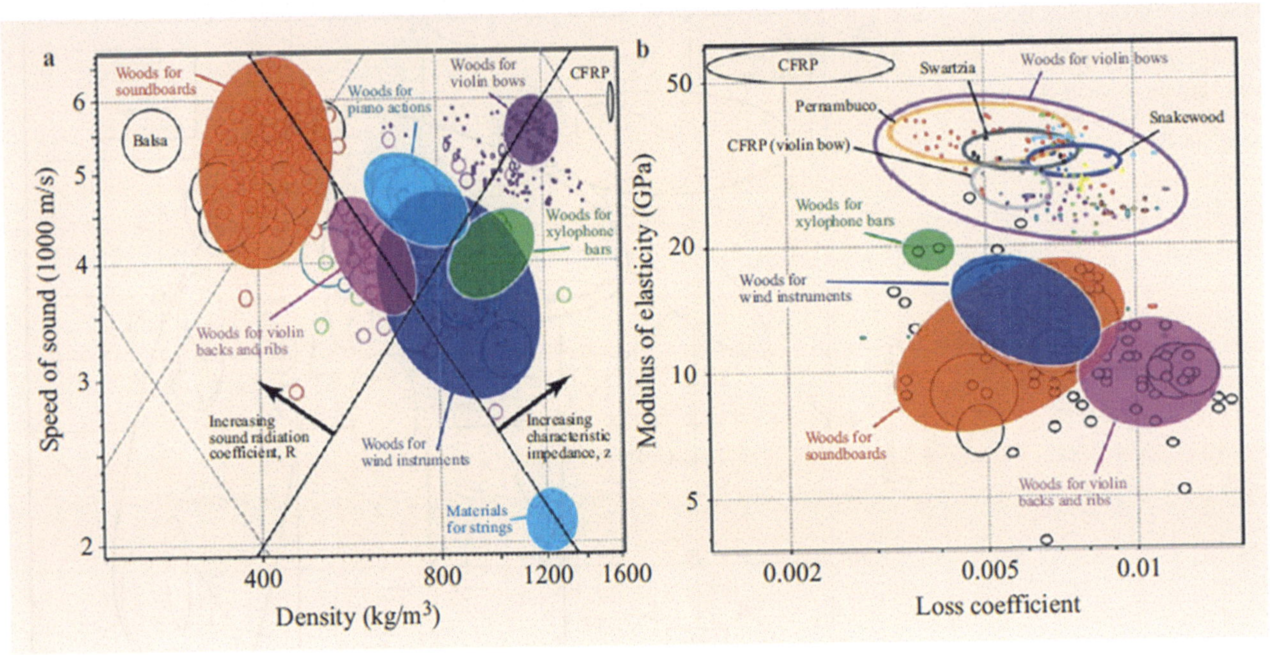

Fig. 2.14 Material-property charts for (a) sound velocity against density, and (b) modulus of elasticity against loss coefficient for wood (Wegst 2006). CFRP – carbon-fibre-reinforced polymers

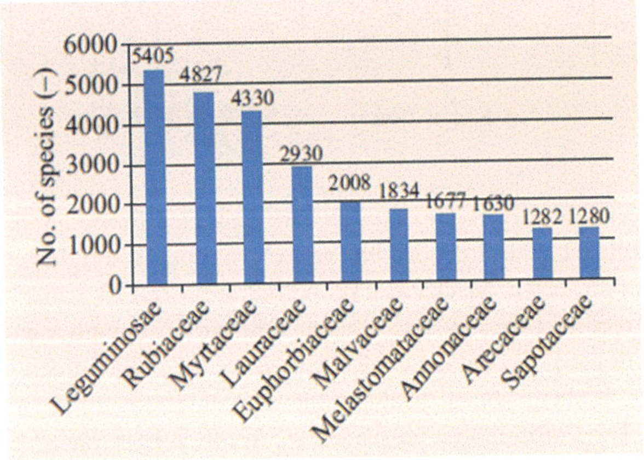

Fig. 2.15 Top ten families with the most tree species (Beech et al. 2017)

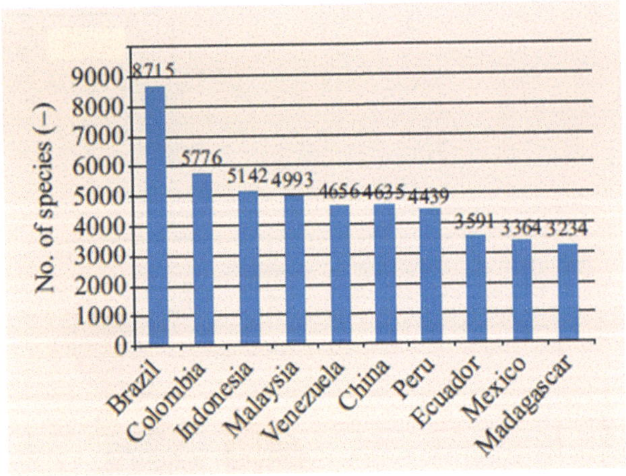

Fig. 2.16 Top ten countries with the most tree species (Beech et al. 2017)

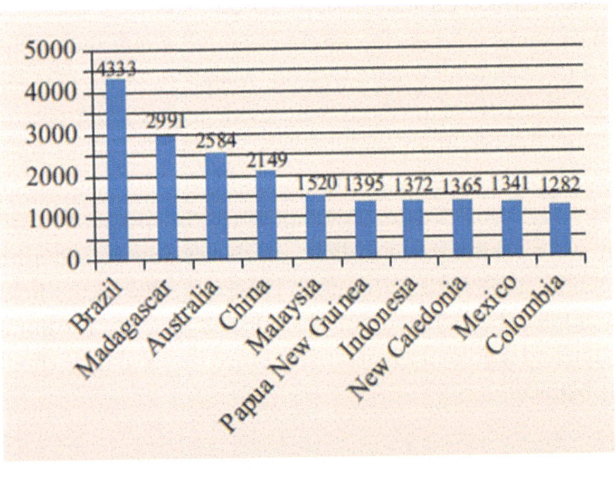

Fig. 2.17 Top ten countries with the largest numbers of endemic species (Beech et al. 2017)

2.3.2 Wood Species and Bamboo from China

Jiali Jang

Native species in China consist mainly of wood (softwoods and hardwoods) and bamboo. They are not only the sources of our commercial products, but they also played a vital role as a turning point in the Chinese cultural revolution. Bamboo and wooden slips were China's main media for documents before the widespread introduction of paper during the first two centuries AD. With the rapid development of technology, remarkable other cellulose-derived products were produced from wood and bamboo, exploiting the physical and mechanical properties of the wood and bamboo structural biopolymers. These physical and mechanical properties are directly linked to the anatomy and to the types of cell-wall polymers.

The existence of vast numbers of native species makes wood and bamboo characterisation difficult. Complete characterisation of wood and bamboo is rare owing to their complex anatomical structures (especially their anisotropy), and the different test methods employed by different institutions (based on national standards). The available information relating to the anatomical structure and chemical composition, and to physical and mechanical properties, has to be collated from individual articles, theses and tables from many research institutions. The majority of these sources contain only partial characterisations, as explained earlier.

As the number of species is too large to cover in this text, only softwoods, hardwoods and bamboo these most commonly used for commercial products are presented.

The data with are presented in Tables 2.2, 2.3, 2.4, 2.5, 2.6 and 2.7. The Tables 2.2, 2.3 and 2.4 provide information on the anatomical structure and chemical composition of various wood and bamboo species. The chemical composition is reported when specimens are oven-dried (absolutely dry wood). For bamboo, these characteristics are based on different positions. Tables 2.12, 2.13 and 2.14 summarise the physical and mechanical properties. For wood (softwoods and hardwoods), the air-dried density, shrinkage coefficients, and thermodynamic constant are reported together wit strength properties in bending, tension, compression, and shear modes. Brinell hardness, modulus of elasticity (MOE), shear modulus, Poisson's ratio, yield stress, and ultimate stress are also reported. In the case of bamboo, the air-dried density and shrinkage coefficients are presented.

2.3.3 European Species and Other Species Used in Europe

Róbert Németh

Wood is the result of a biological optimisation process, which significant differences between hardwoods and softwoods. The former have a more homogeneous cellular structure, a simpler structure, and less differentiated cell types. In hardwood trees, the transporting and solidifying elements (cells) are fundamentally different in structure, but in softwood the same cell type tracheids gives approximately 95% of the wood.

The physical and mechanical properties of wood can vary significantly within a tree species due to habitat, forest management (distance between trees, pruning, fertilisation, etc.) and age. Juvenile wood generally has a lower strength than mature wood. The growth-ring structure (i.e. the width of the ring) and the ratio of earlywood to latewood, affects the strength. In softwoods, the narrow growth-ring structure tends to give a higher density and higher strength, which means that the broader growth-ring tree is stronger. There is no clear correlation in the case of diffuse porous woods (e.g. poplar).

The data in the Table. 2.8 are from standard tests on small, clear specimens, free from irregularities such as knots, cracks, fibre disturbances, etc. The values cannot be interpreted as being valid for load bearing structures e.g. beams with real dimensions, where fibre run-out, knottiness etc. are important strength-reducing factors.

It is important to keep in mind the challenges of the future. The Mediterranean wood species in the tables (e.g. Turkish oak, Calabrian pine), due to their drought-tolerant properties, may in the future spread north to meet the threatening climate change.

In the case of woods with coloured heartwood, the properties of the heartwood are listed. The values refer mainly to mature wood material taken from the breast height diameter of the tree.

The shrinkage and swelling data show the maximum possible changes in the given anatomical direction. Dimensions increase with increasing moisture content (MC) below the fibre-saturation point, but not linearly. At a lower humidity, i.e. a wood moisture content (MC) of 0–5% and near fibre saturation (MC 25–30%), changes are smaller than in the middle of the bound-water range (MC 5–25%).

Strength values are usually the result of static stresses, except for impact bending, which is a dynamic (short-term) test. The greatest standard deviation is in the case of dynamic loads which are extremely sensitive to the fibre direction.

The thermal conductivity depends essentially on the porosity (density). Values perpendicular to the fiber direction are given, as these are of great practical importance (e.g. doors and windows, floorings). It is important to note that calorific values are weight-based quantities. There are significant differences in the volume-related values of lignocellulosic materials.

Basic macromolecular components are also described, supplemented with pH and ash content. Wood is generally acidic in nature. The ash content is given for bark-free wood. It is important to note that the ash content of the bark can be up to 10–20 times that of in the wood itself. Thus, much more residual material is to be expected when burning of branches containing large proportion of bark.

The data for European species (Sect. 2.3.3) partly overlap the data for species from Russia and neighboring countries (Sect. 2.3.4).

Table 2.1 Properties of some of the most used and/or commercialised native species in Africa (Chudnoff 1979; Bunster 2006; Lemmens et al. 2012; ITTO 2019). MC – moisture content

Common name	Scientific name	Reference	Wood Density (kg/m³)			Shrinkage from green to oven dry (%)			Static Bending (MPa)	
			Basic density	Density at 12% MC	Density at air-dry	Volumetric	Radial	Tangential	Modulus of rupture (MOR)	Modulus of elasticity (MOE)
Doussié	*Afzelia africana* Sm. ex Pers.	Lemmens et al. (2012)		720–850			2.5–3.5	3.5–5.0	105–145 (–200)	(9100–) 14 000–17 000
		ITTO (2019)	700	790			3.0	4.2	123	15 117
Doussié	*Afzelia bella* Harms	Lemmens et al. (2012)		700–840			3.1	5.5	90	12 500
		ITTO (2019)	630	700			3.0	5.2	91	12 281
Doussié	*Afzelia bipindensis* Harms	Lemmens et al. (2012)		(750–) 840–950			2.3–3.9	3.5–5.4	150–230	12 150–17 050
		ITTO (2019)	730	830			2.9	4.2	133	18 795
Doussié	*Afzelia pachyloba* Harms	Lemmens et al. (2012)		720-840			2.6–3.6	4.0–5.5	100–120	16 000–18 500
		ITTO (2019)	670	750			3.3	4.9	112	17 550
Chanfuta	*Afzelia quanzensis* Welw.	Bunster (2006)	690	790		6.8	2.9	3.1	110	
		Lemmens et al. (2012)		800–870 (–920)			2.8–3.8	4.0–5.5	83–135	8500–15 000
Mecrusse	*Androstachys johnsonii* Prain	Bunster (2006)	750	880		12.8	5.8	6.5	162	
		Lemmens et al. (2012)		810–960			5.8–6.0	6.2–6.6	92–162	9900–12 150
		Chudnoff (1979)	720		880	12.4	4.0	5.4		
Aniégré	*Aningeria altissima* Aubrev. et Pellegr.	ITTO (2019)	510	560					71	
		Lemmens et al. (2012)		500–580					90	
Aniegre, Asanfina	*Aningeria robusta* Aubrev. et Pellegr.	ITTO (2019)	540	590			3.8	7.1	93	13 740
Aningeria	*Aningeria* spp.	Chudnoff (1979)	400–480		480–580	11.8	3.8	7.0	55	8061
Aniégré	*Aningeria superba* A. Chev.	ITTO (2019)	510	550			3.6	6.8	87	13 618
Okoumé	*Aucoumea klaineana* Pierre	ITTO (2019)	410	440			4.6	6.9	62	9690
		Chudnoff (1979)	370		450	11.3	4.1	6.1		
		Lemmens et al. (2012)		(320–) 350–450 (–580)			3.4–7.0	5.3–11.3	51–109	5500–10 800
Rhodesian-teak, Umgusi	*Baikiaea plurijuga* Harms	ITTO (2019)	790	900					84	8477
		Lemmens et al. (2012)		810–960 (–1000)					84	8500
		Chudnoff (1979)	730		900					
Pau-ferro	*Bobgunnia madagascariensis* (Desv.) J.H.Kirkbr. & Wiersema	Lemmens et al. (2012)		960–1110			3.5	5.8		
		Bunster (2006)	950	1110		11.5	4.6	7.2	189	
Mtundu, Messassa	*Brachystegia spiciformis* Benth.	Bunster (2006)	590	710		13.6	6.0	7.5	126	
		Chudnoff (1979)	670		820					
		Lemmens et al. (2012)		680–920			2.9–4.1	4.3–5.8	88–125	11 100–14 400

(continued)

Table 2.1 (continued)

Compression strength (MPa)		Tensile strength perpendicular to grain (MPa)	Shear strength (MPa)	Hardness (N)		Cleavage (N/mm)	Chemical composition (%)				
Parallel to grain	Perpendicular to grain			Janka side—ry material	Janka end grain—dry material		Silica	Cellulose	Lignin	Pentosan	Ash
57–85			(5–)8–10(–14)	7200–8000		16–25	<0.015	39–44	28–31		
72	10		8	7669	8434						
60–85			7	6000		17	<0.015				
56	7		6	5962	6904						
67–95			7–11			12–22	<0.01	30–35	25–28	14–16	
79	11		9	8453	9110						
70–80			7–10	7500–8500		18–22	<0.015	33	25	15	
72	10		8	6865	7728						
61		2.6									
52–80	14		7–20	6600–9850		17–28	<0.015				
66		3.2									
57–66				11 270–12 580	18 230	19					
45	5			3776	4805						
48							0.3				
52	7		10	4207	5227						
36				3292–5560							
46	6		4	3638	4668						
36	4		6	2354	3295		0.12–0.16				
25–44			4–8	1690–2510	2790–3560	(6–)9–22	<0.3				
66											
66			16	13 500	16 000						
				13 300							
69		3.5									
60		3.3									
				8140							
60–69			11–16	6620–8140	6850–7920	15–16		50–57	27–31	16–18	0.9–1.3

(continued)

Table 2.1 (continued)

Common name	Scientific name	Reference	Wood Density (kg/m³)			Shrinkage from green to oven dry (%)			Static Bending (MPa)	
			Basic density	Density at 12% MC	Density at air-dry	Volumetric	Radial	Tangential	Modulus of rupture (MOR)	Modulus of elasticity (MOE)
Burkea, Mukarati, Mucarala	*Burkea africana* Hook.	Bunster (2006)	760	890		10.0	3.8	5.6	133	
		ITTO (2019)	750	850					134	16 287
		Lemmens et al. (2012)		730–1020			2.9–5.6	4.2–9.2	84–143	12 940
Ceiba, Fromager	*Ceiba pentandra* (L.) Gaertn.	Lemmens et al. (2012)		(200–) 240–380 (–450)			2.5–3.5	3.9–6.7 (–9.1)	26–61	2300–5600
		ITTO (2019)	310	330			2.5	5.2	33	4099
Chanate, Mopane	*Colophospermum mopane* (Benth.) J. Léonard	Bunster (2006)	1060	1250		9.1	3.8	5.3	223	
		Lemmens et al. (2012)		990–1230			4.0	5.2	89–122	10 880–14 000
Monzo	*Combretum imberbe* Wawra	Bunster (2006)	1050	1230		9.9	2.7	6.5	117	
		Lemmens et al. (2012)		1200						
Cordyla, Metondo	*Cordyla africana* Lour.	ITTO (2019)	760	860			4.6	6.2	103	15 807
		Lemmens et al. (2012)		750–910					67–85	8625–12 000
		Chudnoff (1979)	700		860					
Okan	*Cylicodiscus gabunensis* Harms	Lemmens et al. (2012)		770–1100			4.0–7.3	7.2–10.4	(82–) 129–230	14 700–22 600
		ITTO (2019)	800	910			5.9	7.9	134	22 252
		Chudnoff (1979)	800		960	12.6	6.0	8.8	140	16 054
African blackwood, Grenadille Afrique	*Dalbergia melanoxylon* Guill. et Perr.	Bunster (2006)	1150	1330		8.8	3.2	5.6	266	
		ITTO (2019)	1070	1290					162	20 245
		Lemmens et al. (2012)		1180–1330			2.9	4.8	186–267	12 100–20 600
		Chudnoff (1979)	1080		1330	7.6				
Ayan, Movingui	*Distemonanthus benthamianus* Baill.	ITTO (2019)	660	730			3.6	5.8	116	14 737
		Chudnoff (1979)	580		720	10.7	3.1	5.2		
		Lemmens et al. (2012)		570–860			3.0–4.4	5.0–7.2	(103–) 140–207	8400–12 950 (–14 740)
Kosipo	*Entandrophragma candollei* Harms	ITTO (2019)	620	690			4.8	6.7	87	11 186
		Chudnoff (1979)	520–650		640–800					
		Lemmens et al. (2012)		570–810			4.4–5.1	5.7–7.6	82–115 (–150)	7940–11 200
Sapele, Sapelli	*Entandrophragma cylindricum* (Sprague) Sprague (Sapeli)	ITTO (2019)	620	690			5.0	7.2	102	13 957
		Lemmens et al. (2012)		560–750			3.5–7.6	4.3–9.8	(66–) 95–184	8900–13 800
		Chudnoff (1979)	550		670	14.0	4.6	7.4	111	11 713

(continued)

Table 2.1 (continued)

Compression strength (MPa)		Tensile strength perpendicular to grain (MPa)	Shear strength (MPa)	Hardness (N)		Cleavage (N/mm)	Chemical composition (%)				
Parallel to grain	Perpendicular to grain			Janka side—ry material	Janka end grain—dry material		Silica	Cellulose	Lignin	Pentosan	Ash
59		3.0									
71			17								
48–85	12		14–15	6490	7605						
14–26			2–4	1060–1110	1820–1960	4–13		39	24	18	4
19	2		4	1226	1883						
101		2.7									
69–75			17	13 100–21 000	19 400						
66		3.4									
71	10		10	9081	9649						
38–60			12	6225–7025							
				7028							
64–108			8–22	10 600–12 800	11 340–13 740	12–29					
82	12		10	10 268	10 640						
85				12 366							
		3.5									
75	10			20 820	18 662						
71					1785	2.7					
69–75											
64	9		9	6521	7414						
				5471							
45–69(–85)			12–15	5470–6290	7330	13–22	<1.3	37	27	12.5	27
53	7		7	5796	6747						
47–63			6			10–16	<0.5				
62	8		8	5806	6757						
40–75	8		7–18	4180–6730	5650–7450	15–20					
59				6672							

(continued)

Table 2.1 (continued)

Common name	Scientific name	Reference	Wood Density (kg/m^3)			Shrinkage from green to oven dry (%)			Static Bending (MPa)	
			Basic density	Density at 12% MC	Density at air-dry	Volumetric	Radial	Tangential	Modulus of rupture (MOR)	Modulus of elasticity (MOE)
Utile, Sipo	*Entandrophragma utile* (Dawe & Sprague) Sprague	ITTO (2019)	560	620			4.6	6.4	91	13 237
		Chudnoff (1979)	530		660	11.0	4.6	6.4	103	10 748
		Lemmens et al. (2012)		(400–) 550–690 (–740)			(2.9–)3.7–6.4	(4.0–)5.4–9.2	83–152	8830–13 830
Missanda, Tali	*Erythrophleum ivorense* A.Chev.	Lemmens et al. (2012)		900			5.1–5.8	8.4–8.6	99–162	10 550–19 500
		ITTO (2019)	780	890			5.0	8.1	132	20 185
		Chudnoff (1979)	720		900	11.5	5.8	8.6	162	15 434
Missanda, Tali	*Erythrophleum suaveolens* (Guill. & Perr.) Brenan	Bunster (2006)	840	980		8.9	3.7	5.1	127	
		ITTO (2019)	830	960			5.7	10.5	116	13 132
		Lemmens et al. (2012)		900			5.1–5.8	8.4–8.6	99–162	10 550–19 500
Guarea, Bossé clair	*Guarea cedrata* (A.Chev.) Pellegr.	ITTO (2019)	580	630			4.1	6.8	95	12 646
		Lemmens et al. (2012)		540–680			3.5–5.6	5.3–7.9	76–145	9400–12 900
		Chudnoff (1979)	480		580				103	9439
Bushman bean, Large false mopane	*Guibourtia coleosperma* (Benth.) J.Léonard	Lemmens et al. (2012)		(670–) 800–960			2.0–3.7	3.2–5.7	85–142	9210
Muwa, Muimbe	*Julbernardia globiflora* (Benth.) Troupin	Bunster (2006)	710	860		15.1	6.2	9.1	134	
		Lemmens et al. (2012)		820–960			4.8	6.0	97–147	12400–15 600
		Chudnoff (1979)	780		960					
Awoura	*Julbernardia pellegriniana* Troupin	Lemmens et al. (2012)		670–860			3.3–5.7	(6.9–)8.5–10.6	(128–)158–212	(11 500–)13 300–18 900
		ITTO (2019)	690	770			4.3	8.9	128	17 836
African mahogany	*Khaya anthotheca (Welw.)* C. Dc.	ITTO (2019)	550	610			3.7	5.5	83	9242
		Lemmens et al. (2012)		490–660			2.7–4.1	5.8–6.4	50–110	7800–10 300
		Chudnoff (1979)	440		510		3.2	5.6	78-83	8957–9026
Acajou d`Afrique	*Khaya grandifoliola* C. Dc.	ITTO (2019)	600	660			4.3	5.6	89	12 129
		Lemmens et al. (2012)		(560–) 640–730 (–770)					92–119	10 600–11 400
African mahogany	*Khaya grandifoliola* C. Dc. *Khaya senegalensis* (Desr.) A.Juss.	Chudnoff (1979)	550–650		670–800				97	10 611
African mahogany	*Khaya ivorensis* A.Chev.	ITTO (2019)	470	510			2.9	5.5	70	12 240
		Lemmens et al. (2012)		(420–) 460–570			2.2–4.1 (–5.0)	5.0–6.9 (–8.4)	71–126	8700–10 800
		Chudnoff (1979)	440		510		3.2	5.6	78–83	8957–9026
Acajou Cailcedrat	*Khaya senegalensis* (Desr.) A.Juss.	ITTO (2019)	690	780			4.9	5.6	86	11 649
		Lemmens et al. (2012)		(620–) 710–810 (–900)			4.0–5.9	4.3–7.2	82–122	(7200–) 9800–11 650

(continued)

Table 2.1 (continued)

Compression strength (MPa)		Tensile strength perpendicular to grain (MPa)	Shear strength (MPa)	Hardness (N)		Cleavage (N/mm)	Chemical composition (%)				
Parallel to grain	Perpendicular to grain			Janka side—ry material	Janka end grain—dry material		Silica	Cellulose	Lignin	Pentosan	Ash
56	7		7	4599	5619						
60				5605							
45–72			5–17	3330–5610	5250–7960	11–23					
56–97				13 000							
79	12		10	9699	10 170						
97				13 033							
82		3.3									
79	11		9	11 258	11 444						
56–97				13 000							
55	7		8	4864	5864						
47–60			10–15	4000	5420	11–24	<1.0	54	30	13	
53				4448							
51–58	14		14–16	8820–9065	9800–11150	20					
65		4.2									
57–81			14–20	1100–7700	9600						
				10 987							
61–74			10–17			19–34					
68	9		11	7316	8130						
49	6		8	4452	5472						
24–53			8–14	3250–5120	3600–6360						
44–46				3692-3825							
55	7		8	5256	6247						
53–74			15–17	6090		14					
55				6005							
41	5		6	3148	4148						
37–48	4–8		8–12	3210–3700	4810	10–17					
44-46				3692–3825							
54	7		9	7424	8218						
45–54(–72)						15–28					

(continued)

Table 2.1 (continued)

Common name	Scientific name	Reference	Wood Density (kg/m³)			Shrinkage from green to oven dry (%)			Static Bending (MPa)	
			Basic density	Density at 12% MC	Density at air-dry	Volumetric	Radial	Tangential	Modulus of rupture (MOR)	Modulus of elasticity (MOE)
Ekki, Azobé	*Lophira alata* Banks ex P.Gaertn.	Lemmens et al. (2012)		1010–1150			(5.6–) 7.1–9.2	(8.3–) 10.7–13.2	(159–) 201–287	13 200–22 000 (–23 500)
		ITTO (2019)	910	1060			7.3	10.3	162	21 410
		Chudnoff (1979)	900		1120	17.0	8.4	11.0	178	16 881
Iroko, Tule	*Milicia excelsa* (Welw.) C.C.Berg	Bunster (2006)	540	630		9.1	3.7	5.8	98	
		ITTO (2019)	590	650			3.6	5.5	88	13 309
		Lemmens et al. (2012)		550–750			1.7–4.1 (–5.6)	2.4–6.3 (–9.8)	75–156	8300–13 300
Wengé	*Millettia laurentii* De Wild.	ITTO (2019)	770	870			5.9	9.1	144	21 036
		Lemmens et al. (2012)		750–960			4.5–6.2	8.6–10.0	123–246	16 500–22 400
Wengé, Jambire	*Millettia stuhlmannii* Taub.	Bunster (2006)	710	830		9.9	3.6	5.8	152	
		Lemmens et al. (2012)		720–990			2.2–3.1	3.7–5.8	112	13 600
Afrormosia	*Pericopsis elata* (Harms) Meeuwen	Lemmens et al. (2012)		(620–) 700–800			3.0–3.7	5.8–7.0	93–155	9450–13 300
		ITTO (2019)	660	740			3.2	5.9	93	13 132
		Chudnoff (1979)	570		690	10.7	3.0	6.4	134	12 471
Muninga, Umbila	*Pterocarpus angolensis* DC.	Bunster (2006)	560	640		6.4	2.2	3.6	135	
		ITTO (2019)	560	620					86	11 541
		Chudnoff (1979)	590		660				94	8406
		Lemmens et al. (2012)		400–700			1.7	2.5	82–94	8200–9200
Ven	*Pterocarpus erinaceus* Poir.	Lemmens et al. (2012)		(560–) 800–890 (–940)			3.0–3.5	5.2–7.4	(76–) 137–198	11 500–15 700
		ITTO (2019)	780	890			3.2	5.9	130	15 564
Padouk, African	*Pterocarpus tinctorius* Welw.	ITTO (2019)	800	920			3.3	5.2	104	18 725
		Lemmens et al. (2012)		450–900			3.3	5.5	91–147	9100–15 000
Tambooti, (African cocobolo, sandalo)	*Spirostachys africana* Sond.	Bunster (2006)	860	1020		12.3	3.9	7.8	124	
		Lemmens et al. (2012)		910–1090			2.1–3.5	4.0–6.7	102–108	8600–9210
Afara, Limba, Frake	*Terminalia superba* Engl. et Diels	ITTO (2019)	500	540			4.3	6.1	80	11 741
		Lemmens et al. (2012)		(370–) 430–730			2.0–6.4	4.4–8.7	50–133(–157)	(3625–) 4700–14 300 (–16 660)
		Chudnoff (1979)	450		540	10.8	4.5	6.2		
Obeche, Ayous, Samba	*Triplochiton scleroxylon* K.Schum.	Lemmens et al. (2012)		320–440 (–490)			2.5–4.1	4.2–6.6	52–110	4800–9200
		ITTO (2019)	360	380			2.9	5.0	52	7261
		Chudnoff (1979)	320		380	9.2	3.0	5.4	54	5512

(continued)

Table 2.1 (continued)

Compression strength (MPa)		Tensile strength perpendicular to grain (MPa)	Shear strength (MPa)	Hardness (N)		Cleavage (N/mm)	Chemical composition (%)				
Parallel to grain	Perpendicular to grain			Janka side—ry material	Janka end grain—dry material		Silica	Cellulose	Lignin	Pentosan	Ash
(75–)85–109			(10–)14–20	9990–17540	12 300–21 200	20–33					
96	15		14.51	14013	13 621						
90				14902							
39		2.2									
55	7		7	5119	6110						
42–65			5–14	4400–5610	5360–6640	10–21					
85	13		12	9307	9836						
69–100			11–12			13–22					
69		2.3									
69			16	7250		73					
63–71			9–16	6940–7115		17–24					
64	9		9	6620	7502						
71				6940							
47		2.2									
50			12								
57				6583							
50–57			9–16	4450–6580	5380–7420						
62–80	3		7–10			20					
76	11		9	9699	10 170						
78	11		5	10424	10 768						
45–77						8					
57		2.8									
57–60			16	8940							
47	6		6	3560	4580						
26–67	8–12		4–10	3020	3420	8–24		40–45	28–35	14–18	0.9–3.4
24–43			3–8	1910–2100	3070–3600	5–15					
30	3		4	1706	2560						
28				1913							

Table 2.2 Anatomical structure and chemical composition of Chinese softwood species, average values in parentheses (Cheng 1985; Cheng et al. 1985; Jiang et al. 2010)

General characteristics		Chinese fir (*Cunninghamia lanceolata*)	Dahurian larch (*Larix gmelini*)	Armand pine (*Pinus armandii*)	Red pine (*Pinus koraiensis*)	Masson pine (*Pinus massoniana*)
Tracheids	Diameter[a] (μm)	Earlywood: 25–40 (31)	Earlywood: 25–51 (40)	Earlywood: 37–53 (44)	Earlywood: 29–42 (34)	Earlywood: 43–62 (53)
		Latewood: 25–38 (30)	Latewood: 28–42 (36)	Latewood: 31–46 (37)	Latewood: 26–36 (30)	Latewood: 28–53 (44)
	Length (mm)	Earlywood: 3.58–5.54 (4.41)	Earlywood: 2.66–5.53 (4.25)	Earlywood: 3.17–4.61 (3.86)	Earlywood: 3.09–4.52 (3.75)	Earlywood: 3.48–5.61 (4.40)
		Latewood: 3.96–6.35 (4.86)	Latewood: 3.94–5.94 (4.90)	Latewood: 3.33–5.00 (4.23)	Latewood: 3.37–4.57 (3.94)	Latewood: 3.72–6.46 (4.96)
	Wall thickness (μm)	Earlywood: 2.2–3.5 (2.7)	Earlywood: 2.1–3.4 (2.8)	Earlywood: 2.0–3.7 (2.8)	Earlywood: 1.6–4.4 (2.9)	Earlywood: 3.7–5.3 (3.7)
		Latewood: 4.2–7.1 (5.1)	Latewood: 5.5–10.6 (8.1)	Latewood: 3.3–4.4 (3.9)	Latewood: 3.3–5.2 (3.9)	Latewood: 4.7–9.3 (6.8)
	Arrangement	Gradual	Abrupt	Gradual	Gradual	Abrupt
Longitudinal parenchyma		Diffuse and tangential banded	Absent	Absent	Absent	Absent
Wood rays	Height (μm)	30–348 consist of 1–21 cells	52–742 consist of 1–34 cells	25–452 consist of 1–18 cells	52–577 consist of 1–22 cells	32–500 consist of 1–20 cells
	Density	3–6/mm	6–9/mm	3–6/mm	4–8/mm	3–6/mm
	Type	Uniseriate, rarely 2 row	Uniseriate and fusiform	Uniseriate and fusiform	Uniseriate and fusiform	Uniseriate and fusiform
Cross-field pitting		Taxodioid	Piceoid, few taxodioid	Fenestriform	Fenestriform	Fenestriform
Chemical composition	α-cellulose (%)	42.47	40.17	56.43	37.68	43.47
	Hemicelluloses (%)	13.41	12.46	13.78	16.30	12.67
	Lignin (%)	32.44	26.46	30.28	25.56	24.69
	Extractives[b] (%)	2.87	2.58	5.56	7.54	9.93
	Ash (%)	0.26	0.38	0.15	0.30	0.27

General characteristics		Khingan fir (*Abies nephrolepis*)	Chinese cryptomeria (*Cryptomeria fortunei*)	Little-seed spruce (*Picea jezoensis*)	Chinese hemlock (*Tsuga longibracteata*)	Java podocarpus (*Podocarpus imbricatus*)
Tracheids	Diameter[a] (μm)	Earlywood: 25–95 (34)	Earlywood: 27–48 (32)	Earlywood: 32–48 (37)	Earlywood: 33–48 (40)	Earlywood: 21–34 (27)
		Latewood: 23–42 (31)	Latewood: 26–41 (34)	Latewood: 30–47 (38)	Latewood: 24–40 (33)	Latewood: 22–30 (26)
	Length (mm)	Earlywood: 2.59–3.72 (3.04)	Earlywood: 2.27–4.13 (3.26)	Earlywood: 2.74–4.15 (3.45)	Earlywood: 2.47–3.78 (3.17)	Earlywood: 2.78–4.39 (3.54)
		Latewood: 2.22–4.3 (3.28)	Latewood: 2.75–4.63 (3.60)	Latewood: 3.07–4.70 (3.97)	Latewood: 2.56–4.53 (3.42)	Latewood: 3.21–5.37 (3.99)
	Wall thickness (μm)	Earlywood: 1.9–4.7 (3.2)	Earlywood: 2.0–4.8 (3.3)	Earlywood: 2.2–5.1 (3.5)	Earlywood: 2.0–3.6 (2.9)	Earlywood: 2.5–4.3 (3.1)
		Latewood: 2.4–5.4 (3.8)	Latewood: 2.9–4.4 (4.1)	Latewood: 4.0–6.0 (4.2)	Latewood: 3.9–7.7 (6.6)	Latewood: 3.6–4.1 (3.2)
	Arrangement	Gradual	Gradual	Gradual	Abrupt	Gradual
Longitudinal parenchyma		Very few, diffuse	Diffuse and tangential banded	Absent	Very few, terminal	Diffuse, short tangential lines
Wood rays	Height (μm)	30–410 consist of 1–18 cells	25–360 consist of 1–17 cells	48–710 consist of 1–35 cells	28–415 consist of 1–20 cells	26–515 consist of 1–30 cells
	Density	3–8/mm	3–6/mm	5–9/mm	3–6/mm	5–8/mm
	Type	Uniseriate	Uniseriate, rarely 2 row	Uniseriate and fusiform	2 seriate or partial 2 seriate	Uniseriate
Cross-field pitting		Taxodioid	Taxodioid	Piceoid	Cupressoid	Podocarpodioid
Chemical composition	α-cellulose (%)	41.34	43.03	42.48	44.96	42.47
	Hemicelluloses (%)	17.87	12.24	17.37	10.83	14.41
	Lignin (%)	28.96	34.24	28.58	31.13	31.54
	Extractives[b] (%)	3.37	2.47	1.63	3.47	2.11
	Ash (%)	0.50	0.66	0.29	0.18	0.43

Note: (a) tangential diameter; (b) extracted with ethanol–benzene.

Table 2.3 Anatomical structure and chemical composition of hardwood species from China (Cheng 1985; Cheng et al. 1985)

General characteristics		Tree of heaven (Ailanthus altissima)	Asian white birch (Betule platyphylla)	Camphor tree (Cinnamomum camphora)	Longpetiole beech (Fagus longipetiolata)	Manchurian ash (Fraxinus mandshurica)
Vessels	Diameter[a] (μm)	50–255	55–80	70–145	35–55	180–245
	Length (μm)	170–529	600–1220	326–661	350–880	160–320
	Wall thickness (μm)	3–4	3.7	5.8	2.5	4.4
	Density		65 in 1 mm^2 cross section	16 in 1 mm^2 cross section	>100 in 1 mm^2 cross section	
	Arrangement	Ring porous	Diffuse porous	Diffuse porous to semi-ring porous	Semi-ring porous	Ring porous
Longitudinal parenchyma		Vasicentric, aliform, paratracheal, boundary	Boundary	Vasicentric, boundary	Diffuse-in-aggregate, apotracheal banded	Vasicentric, circumferential
Fibres	Lumen diameter (μm)	20–30	18–24	15–25	15	17–25
	Length (μm)	530–1280	980 ~ 2000	916–1521	810–1750	840–1520
Wood rays	Height (μm)	150–900 consist of 5–55 cells	160–340 consist of 7–20 cells	42–649 consist of 1–32 cells	34–680 consist of 1–20 cells	34–403 consist of 1–24 cells
	Density	1–4/mm	7–9/mm	4 ~ 8/mm	5–11/mm	4–7/mm
	Width	9–78 consist of 1–8 cells	9–31 consist of 1–3 cells	13–46 consist of 1–3 cells	11–20	10–35 consist of 1–3 cells
Chemical composition	α-cellulose (%)	45.64	41.82	43.00	43.77	46.20
	Hemicelluloses (%)	13.34	18.18	13.64	12.02	11.61
	Lignin (%)	21.22	20.37	24.52	27.34	21.57
	Extractives[b] (%)	2.47	3.08	4.92	1.65	2.36
	Ash (%)	0.12	0.33	0.12	0.53	0.72

General characteristics		Manchurian walnut (Juglans mandshurica)	Amur cork-tree (Phellodendron amurense)	Chinese white poplar (Populus tomentosa)	Mongolian oak (Quercus mongolica)	Black locust (Robinia pseudoacacia)
Vessels	Diameter[a] (μm)	165–195	25–260	60–83	230–290	70–310
	Length (μm)	320–670	250–610	500–1000	340–600	130–220
	Wall thickness (μm)	3.0	4.2	3.3	3.4	5.0
	Density	7 in 1 mm^2 cross section		80 in 1 mm^2 cross section		
	Arrangement	Semi-ring porous, or diffuse porous	Ring porous	Diffuse porous to semi-ring porous	Ring porous	Ring porous to semi-ring porous
Longitudinal parenchyma		Apotracheal banded, diffuse-in-aggregate	Vasicentric, boundary	Boundary	Diffuse-in-aggregate, apotracheal banded	Vasicentric, boundary
Fibers	Lumen diameter (μm)	20–26	16–28	15–21	14–20	15–20
	Length (μm)	980–1820	880–1520	1030–2050	1000–1620	600–1350
Wood rays	Height (μm)	31–810 consist of 2–53 cells	50–680 consist of 2–32 cells	180–350 consist of 8–18 cells	25–465 consist of 2–29 cells	34–825 consist of 2–52 cells
	Density	6–8/mm	2–3/mm	9–11/mm	11–12/mm	4–7/mm
	Width	11–37 consist of 1–4 cells	14–56 consist of 1–5 cells	8–18 consist of 1 cell	11–22 consist of 1–2 cells	9–85 consist of 1–6 cells
Chemical composition	α-cellulose (%)	46.06	43.67	45.33	44.91	35.86
	Hemicelluloses (%)	13.59	11.61	14.03	13.05	14.10
	Lignin (%)	18.61	20.89	21.38	21.72	21.80
	Extractives[b] (%)	5.39	4.55	4.45	4.04	8.16
	Ash (%)	0.50	0.49	0.54	0.55	1.30

Note: (a) tangential diameter; (b) extracted with ethanol–benzene.

Table 2.4 Anatomical structure and chemical composition of bamboo species from China (Cheng 1985; Yao et al. 1992, 2002)

General characteristics			*Phyllostachys bambusoides*	*Phyllostachys lithophila*	*Phyllostachys glauca*	*Phyllostachys pubescens*	*Pleioblastus amarus*	*Pseudosasa amabilis*	*Sinocalamus latiflorus*
Epidermis cell	Elongated cells	Length (μm)	40	30	20	55	80	70	40
		Width (μm)	10	9	10	11	11	10	10
	Short cells	Cork cells	Saddle	Square	Rectangle	Rectangle	Rectangle	Rectangle	Square
		Silica cells	Circle	Circle	Oval	Oval	Rhombus	Circle	Rhomboid
Exodermis cells			1 layer	1 layer	1 layer	1 layer	1 layer	1 layer	5–6 layers
Cortex cells			3–4 layers	3–4 layers	3 layers	5–6 layers	3 layers	3–4 layers	5–6 layers
VBMCB			III	III	III	III	III	III	I
CM	VBM	Type	IV	III	III	III	III	III	II
		Length (mm)	0.25	0.31	0.23	0.54	0.25	0.54	0.80
		Width (mm)	0.28	0.28	0.25	0.51	0.37	0.46	0.70
		RRTVB	0.90	1.10	0.90	1.10	0.80	1.20	1.10
	VBI		III	III	III	III	III	III	II, III
	TCPC		6–8 layers, rectangle	4–5 layers, long oval	4–5 layers, rectangle	10 layers, rectangle	4–5 layers, rectangle	4–5 layers, rhomboid	6–8 layers, rectangle
VBMCT			IV	III	III	III	III, IV	IV	II
Pitting types in metaxylem			Alternate reticulation	Alternate reticulation	Alternate reticulation	Alternate reticulation	Opposite scalariform	Scalariform	Opposite reticulation
Tylosoids in primary phloem			Present	Present	Present	Present	Present	Present	Absent
Rhizome			Monopodial	Monopodial	Monopodial	Monopodial	Amphipodial	Amphipodial	Sympodial
Chemical composition[(a)]	α-cellulose (%)		43.50	42.59	42.82	42.35	47.69	43.28	39.44
	Hemicelluloses (%)		10.93	11.13	12.86	11.27	10.12	11.42	9.78
	Lignin (%)		23.62	24.86	24.57	24.49	26.26	21.88	21.69
	Extractives[(b)] (%)		7.45	7.00	6.88	7.35	2.97	7.95	8.84
	Ash (%)		1.30	1.28	1.30	1.20	1.30	1.26	1.26

Note: (a) middle of bamboo; (b) extracted with ethanol–benzene. CM: culm middle; RRTVB: the ratio of the radial diameter and tangential diameter of vascualr bundle in transverse section; TCPC: tissue cells around pith cavity; VBI: vascular bundles in inner part; VBM: vascular bundles in middle part; VBMCB: vascular bundles in middle part of culm base; VBMCT: vascular bundles in middle part of culm top. I: double-broken-waist type; II: broken-waist type; III: open-type; IV: semi-open type.

2.3.4 Species from Russia and Neighboring Countries

Galina Gorbacheva

Russia has the largest amount of forest area in the world (20.4% in 2015 FAO 2015; Keenan 2015), and 15.4% of the world's wood reserves (Köhl et al. 2015). The physical and mechanical properties of the wood are therefore very important for the effective development of the wood industry. The first standard methods for the physical and mechanical testing of wood (OST VKS-7653) were developed by Perelygin and approved in 1935, and there are now more than 40 standard methods for the determination of the physical and mechanical properties of wood.

The continuously growing demand for wood, widely used as a structural and decorative material and as a raw material for chemical and chemical–mechanical processing made it necessary to provide specialists in various sectors of the national economy with reliable information about the characteristic physical and mechanical properties of small clear specimens of wood and of the structural wood used on building sites.

As a result of a joint decision of Gosstandart and the Ministry of Forestry of the USSR, an automated data centre „Wood" of the State service of standard reference data

(GSSSD) was created, and the management of the subsystem of the centre for the properties of clear wood and sawn timber was entrusted to the Central Research Institute of mechanical processing of wood (CNIIMOD), which has summarised an extensive amount of wood property data.

The Wood Handbook (Borovikov and Ugolev 1989), was the first attempt at a concentrated presentation of the results of numerous studies of wood properties, conducted by various research organisations and universities in USSR. It contained summary tables of experimental data that had passed the examination in the all-Union research center of substances and materials of the state standard, and which were assigned categories as standard and recommended reference data. In the Wood Handbook, these tables are preceded by basic information on wood such as the structure of wood, physical and mechanical properties, standardisation of test methods, classification of defects and their impact on the properties of wood and the characteristics of the most common and rare wood species.

Small clear wood specimens were used for testing. They had a sufficiently small cross-section to avoid the influence of the curvature of the growth rings and included a sufficiently large number of characteristic anatomical elements of this species, i.e. at least 4–5 growth rings, with a basic cross section of 20×20 mm. The growth rings on the end surfaces of the specimens were parallel to one pair of opposite faces and perpendicular to the other, and the direction of the wood fibres coincided with the longitudinal axis of the specimens. Such specimens made it possible to determine the properties of wood as an orthotropic material.

The physical and mechanical properties of small clear wood specimens of 91 wood species from different regions of Russia and neighbouring countries (former USSR) are presented by Borovikov and Ugolev (1989) in Table. 2.9. Data are given for different regions due primarily to a significant variability in growing conditions. They were prepared on the basis of the Technical guidance materials (Anon. 1962) taking into account the results of tests performed in 1962–1981 by CNIIMOD and other organisations using standardised methods.

The data for the wood species in Table. 2.9 are related mainly to a forest stand of exploitable age and for site index (bonitet) II and III. The differences in wood properties due to differences in forest productivity (site index I–IV) and over an age difference of 80 to 300 years did not exceed 10%. Coefficients of variation and average values for all wood species are given by Borovikov and Ugolev (1989), calculated under the assumption of a normal statistical distribution (Table. 2.10). Values of physical and mechanical properties in Table. 2.9 relate to a moisture content of 12% and a temperature of 20 °C.

The data for species in Russia and neighboring countries (Sect. 2.3.4) partly overlap the data for species in Europe (Sect. 2.3.3).

The averages of the main physical and mechanical properties (density, swelling, ultimate strength, MOE) for small pure specimens of 35 common wood species of Russia and neighboring countries are given in Table. 2.11 (Borovikov and Ugolev 1989). The average main physical and mechanical properties of wood are presented at moisture contents of 12% and of 30% or more. In Tables. 2.12 and 2.13 the data on chemical compositions of some wood species are presented (Borovikov and Ugolev 1989). The chemical compositions of 62 hardwoods and 18 softwoods, collected and published by Nikitin (1962) are presented in Tables. 2.14 and 2.15. Physical and mechanical properties, workability and serviceability of common wood species in Russia are presented in Tables. 2.16, 2.17, 2.18, 2.19, 2.20, 2.21, 2.22, 2.23, 2.24, 2.25, 2.26, 2.27, 2.28, 2.29 and 2.30 (Borovikov and Ugolev 1989).

2.3.5 Species from South America

Raquel Gonzalves

Hardwoods predominate in tropical regions. In Brazil, for example, the only native softwood species is the *Araucaria angustifolia*.

The large number of native species makes wood characterisation difficult, and in general are property data related only to the longitudinal direction are available. Complete characterisation of tropical species is rare, because the test methods require complex removal of specimens (on and off axes) and the tests are expensive with strain measurement systems in different directions. However, with the growing use of nondestructive testing means that more complete information can be inferred by characterisation, using wave propagation methods.

Information has here been collected from various scientific journals, thesis and documents from research institutions linked to wood. Most of these sources however provide only a partial characterisation (Tables. 2.31, 2.32 and 2.33).

Table 2.5 Chemical, physical, and mechanical properties of softwood species from China (Cheng 1985[1], Cheng et al. 1985[2], Rong 2008[3], Li 2014[4], Peng et al. 2016[5], Jiang et al. 2017[6]). UTS – ultimate tension strength parallel to the fibre direction. References in [ ]

Chinese softwood species																		
Wood species	Chemical composition (oven-dry specimens)						Shrinkage coefficient (%/%)			Bending strength (MPa)	MOE (GPa)	UTS (MPa)	UCS (MPa)	Impact toughness (KJ/m²)	Brinell hardness (HB) (MPa)			
	α-cellulose (%)	Hemicelluloses (%)	Lignin(%)	Extractives[(a)] (%)	Ash (%)	Air-dried density (kg/m³)	Radial	Tangential	Volumetric						Ovendried density (kg/m³)	Cross section	Radial surface	Tangential surface
Chinese fir (*Cunninghamia lanceolata*) [1–5]	42.47	13.41	32.44	2.87	0.26	371	0.123	0.277	0.42	62.52	9.41	77.52	37.04	26.12	369 (30)	52.60 (16.80)	11.40 (3.5)	12.60 (3.3)
Dahurian larch (*Larix gmelini*) [1–4]	40.17	12.46	26.46	2.28	0.38	641	0.169	0.398	0.588	111.03	14.21		56.44	50.00		40.67	31.75	31.46
Armand pine (*Pinus armandii*) [1–2]	46.43	13.78	30.28	5.56	0.15	430	0.151	0.330	0.498	58.02	9.02		36.85	31.73		24.70	20.38	19.11
Red pine (*Pinus koraiensis*) [1–2, 4–5]	37.68	16.3	25.56	7.54	0.3	440	0.122	0.321	0.459	63.99	9.80		32.73	35.71	481 (11)	56.60 (7.00)	13.3 (2.0)	11.80 (1.9)
Masson pine (*Pinus massoniana*) [1–2, 4]	43.47	12.67	24.69	9.93	0.27	592	0.187	0.327	0.543	77.81	11.76	97.02	36.16	46.22		41.36	31.56	35.28
Khingan fir (*Abies nephrolepis*) [1-2, 4]	41.34	17.87	28.96	3.37	0.50	384	0.129	0.366	0.472	63.80	9.41		32.83	32.14		21.56	16.07	14.01
Chinese cryptomeria (*Cryptomeria fortunei*) [1–2]	43.03	12.24	34.24	2.47	0.66	368	0.108	0.28	0.410	52.14	8.13		27.93	22.55		22.74	12.05	14.99
Little-seed spruce (*Picea jezoensis*) [1–2, 4–5]	42.48	17.37	28.58	1.63	0.29	451	0.171	0.349	0.528	73.60	10.39		41.55	49.39		24.50	17.25	15.78
Chinese hemlock (*Tsuga longibracteata*) [1–2, 4]	44.96	10.83	31.13	3.47	0.18	511	0.149	0.273	0.439	89.67	11.07		45.37	40.82		39.98	26.56	28.62
Java podocarpus (*Podocarpus imbricatus*) [1–2, 4]	42.47	14.41	31.54	2.11	0.43	510	0.139	0.240	0.402	80.85	9.31		34.01	32.96		41.94	28.32	28.52

(continued)

Table 2.5 (continued)

Thermodynamic constants (Air-dried wood)					Shear strength parallel to grain (MPa)		Felling shake (N/mm)		MOE (GPa) and CoV (%)											
Thermal conductivity (W/(m K)		Thermal diffusivity (10^{-4} m^2/s)		specific heat capacity (kJ/(kg K))	Radial surface	Tangential surface	Radial surface	Tangential surface	Moisture content (%)	Density (kg/m^3)	Air-dried density (kg/m^3)	Electrometric method			Tension			Compression		
Radial	Tangential	Radial	Tangential									E_L	E_R	E_T	E_L	E_R	E_T	E_L	E_R	E_T
0.1082	0.1070	0.001336	0.001308	1.762	6.6	5.4	5.19	6.27	10.3	314 (5.4)					7.37 (9.8)	0.48 (12.5)	0.35 (8.1)	6.24 (6.7)	0.44 (15.6)	0.33 (17.5)
									12.2	341 (5.7)					6.88 (6.5)	0.42 (11.6)	0.30 (6.0)	5.71 (12.7)	0.39 (15.1)	0.28 (15.1)
									14.6	351 (6.7)					6.04 (9.5)	0.34 (7.2)	0.25 (4.6)	5.19 (11.9)	0.32 (15.3)	0.22 (18.5)
									16.7	332 (5.2)					5.55 (7.2)	0.30 (6.6)	0.21 (4.5)	4.69 (12.3)	0.29 (15.1)	0.18 (16.8)
0.1605		0.001683		1.720							600	12.06	1.04	0.69						
0.0991	0.1039	0.001456	0.001511	1.708																
0.1163	0.0989	0.001536	0.001334	1.666																
0.1256	0.1221	0.001386	0.001406	1.737																
0.0868	0.1087	0.002078	0.001719	1.700																
0.0955	0.0997	0.001378	0.001419	1.624							440	11.73	1.20	1.04						
0.1523	0.1616	0.001586	0.001867	1.607																
0.1500		0.001683		1.687																

(continued)

Table 2.5 (continued)

Chinese softwood species																					
Wood species	Shear moduli (MPa). R^2								Poisson ration and CoV (%)												
	Moisture content (%)	Air-dried density (kg/m)^3	Electrometric method			$G_{LR/RL}$	$G_{LT/TL}$	$G_{RT/TR}$	Moisture content (%)	Electrometric method						Tension					
			G_{RL}	G_{TL}	G_{RT}					υ_{LR}	υ_{LT}	υ_{RL}	υ_{RT}	υ_{TL}	υ_{TR}	υ_{LR}	υ_{LT}	υ_{RL}	υ_{RT}	υ_{TL}	υ_{TR}
Chinese fir (*Cunninghamia lanceolata*) [1–5]	10.3					0.62 (R^2= 0.89)	0.49 (R^2= 0.94)	0.13 (R^2= 0.96)	10.3							0.05 (24.3)	0.07 (15.3)	0.44 (21.5)	0.32 (18.0)	0.54 (19.7)	0.76 (13.8)
	12.2					0.54 (R^2= 0.93)	0.45 (R^2= 0.88)	0.11 (R^2= 0.92)	12.2							0.06 (17.1)	0.06 (25.2)	0.41 (16.7)	0.37 (17.6)	0.52 (11.6)	0.78 (11.2)
	14.6					0.45 (R^2= 0.94)	0.36 (R^2= 0.96)	0.10 (R^2= 0.92)	14.6							0.04 (27.4)	0.05 (25.9)	0.40 (19.0)	0.35 (13.5)	0.57 (17.0)	0.82 (14.5)
	16.7					0.42 (R^2= 0.87)	0.32 (R^2= 0.91)	0.09 (R^2= 0.88)	16.7							0.05 (24.0)	0.07 (26.8)	0.45 (18.9)	0.34 (17.4)	0.58 (19.7)	0.86 (11.2)
Dahurian larch (*Larix gmelini*) [1–4]		600	1084.00	1045.33	275.33					0.35	0.46	0.07	0.47	0.04	0.36						
Armand pine (*Pinus armandii*) [1–2]																					
Red pine (*Pinus koraiensis*) [1–2, 4–5]																					
Masson pine (*Pinus massoniana*) [1–2, 4]																					
Khingan fir (*Abies nephrolepis*) [1-2, 4]																					
Chinese cryptomeria (*Cryptomeria fortunei*) [1–2]																					
Little-seed spruce (*Picea jezoensis*) [1–2, 4–5]		440	495.30	209	84.67					0.34	0.44	0.04	0.46	0.07	0.41						
Chinese hemlock (*Tsuga longibracteata*) [1–2, 4]																					
Java podocarpus (*Podocarpus imbricatus*) [1–2, 4]																					

(continued)

Table 2.5 (continued)

Compression						Yield stress (MPa) and CoV (%)					Ultimate stress (MPa), strain (%), and CoV (%)							
						Moisture content (%)	Tension		Compression		Moisture content (%)	Tension					Compression	
υ_{LR}	υ_{LT}	υ_{RL}	υ_{RT}	υ_{LR}	υ_{LR}		σ_R	σ_T	σ_R	σ_T		UTS_L	UTS_R	ε_R	UTS_T	ε_T	UCS_L	ε_L
0.15 (19.3)	0.10 (22.8)	0.32 (26.7)	0.26 (23.1)	0.40 (16.9)	0.63 (12.4)	10.3	3.28 (15.8)	2.87 (12.3)	3.24 (5.2)	2.64 (18.7)	10.3	/	3.68 (13.3)	0.84 (14.1)	3.23 (10.7)	0.94 (10.4)	28.50 (8.1)	0.93 (22.1)
0.14 (18.0)	0.11 (24.0)	0.33 (19.4)	0.34 (16.5)	0.38 (24.4)	0.69 (12.9)	12.2	2.99 (13.6)	2.58 (9.7)	2.96 (5.8)	2.30 (9.9)	12.2	75.1	3.44 (12.5)	0.92 (8.4)	2.81 (8.4)	1.07 (7.1)	26.22 (2.9)	0.81 (18.6)
0.16 (18.9)	0.09 (31.7)	0.30 (19.6)	0.31 (22.4)	0.43 (25.9)	0.66 (15.5)	14.6	2.54 (12.0)	2.09 (10.5)	2.44 (10.7)	2.02 (2.3)	14.6		2.98 (9.7)	1.05 (12.4)	2.38 (11.9)	1.20 (12.6)	24.70 (5.8)	0.68 (9.8)
0.14 (25.5)	0.12 (32.8)	0.34 (23.1)	0.27 (9.7)	0.46 (28.8)	0.69 (14.5)	16.7	2.18 (7.9)	1.80 (12.6)	2.15 (10.6)	1.87 (12.1)	16.7		2.44 (8.0)	1.20 (15.3)	2.07 (13.2)	1.44 (19.4)	23.08 (8.4)	0.53 (13.1)

Note: (a) extracted with ethanol-benzene, MOE - the modulus of elasticity in static bending of wood, UTS - ultimate tension strength parallel to grain, and UCS - ultimate

Note: (a) extracted with ethanol-benzene, MOE - the modulus of elasticity in static bending of wood, UTS - ultimate tension strength parallel to grain, and UCS - ultimate compressive strength parallel to grain.

Table 2.6 Chemical, physical, and mechanical properties of hardwood species from China (Cheng 1985[1], Cheng et al. 1985[2], Xu 1988[3], Rong 2008[4], Li 2014[5], Peng et al. 2016[6], Jiang et al. 2017[7]). References in [ ]

Chinese hardwood species																	
Wood species	Chemical composition (oven-dry specimens)					Shrinkage coefficient (%/%)				Bending strength (MPa)	MOE (GPa)	UTS (MPa)	UCS (MPa)	Impact toughness (kJ/m²)	Brinell hardness (HB) (MPa)		
	α-cellulose (%)	Hemicelluloses (%)	Lignin (%)	Extractives[(a)] (%)	Ash (%)	Air-dried density (kg/m³)	Radial	Tangential	volumetric						Cross section	Radial surface	Tangential surface
Tree of heaven (*Ailanthus altissima*) [1–2]	45.64	13.34	21.22	2.47	0.12	672	0.182	0.289	0.495	81.24	10.49		37.53	56.12	53.70	43.90	48.31
Asian white birch (*Betule platyphylla*) [1–2, 5]	41.82	18.18	20.37	3.08	0.33	615	0.188	0.258	0.466	85.75	9.02		41.75	81.43	37.53	37.73	36.46
Camphor tree (*Cinnamomum camphora*) [1–2]	43	13.64	24.52	4.92	0.12	580	0.154	0.245	0.412	73.60	9.02		40.77	40.20	39.69	33.91	31.36
Longpetiole beech (*Fagus longipetiolata*) [1–2]	43.77	12.02	27.34	1.65	0.53	793	0.204	0.387	0.617	113.19	13.43		51.55	138.37	62.13	61.05	59.78
Manchurian ash (*Fraxinus mandshurica*) [1–2, 5]	46.2	11.61	21.57	2.36	0.72	643	0.171	0.322	0.519	105.94	12.64		48.80	66.63	58.70	45.37	44.88
Manchurian walnut (*Juglans mandshurica*) [1–2, 5]	46.06	13.59	18.61	5.39	0.50	528	0.190	0.300	0.516	85.26	11.47		43.32	53.37	32.34	27.34	26.36
Amur cork-tree (*Phellodendron amurense*) [1–2]	43.67	11.61	20.89	4.55	0.49	449	0.128	0.242	0.368	74.58	8.82		33.03		32.44	23.42	21.07
Chinese white poplar (*Populus tomentosa*) [2–4]	45.33	14.03	21.38	4.45	0.54	525	0.142	0.289	0.458	77.03	10.19	91.63	38.22	81.73	38.42	31.46	32.14
Mongolian oak [*Quercus mongolica*) [4, 6–7]	44.91	13.05	21.72	4.04	0.55	748	0.181	0.318	0.520	116.23	12.94		53.41	120.00	71.44	59.09	58.80
Black locust (*Robinia pseudoacacia*) [1–2]	35.86	14.10	21.80	8.16	1.30	792	0.210	0.327	0.548	124.26	12.74		52.82	177.45	67.13	65.07	67.23

(continued)

Table 2.6 (continued)

Thermodynamic constants(Air-dried wood)					Shear strength parallel to grain (MPa)		Felling shake (N/mm)		MOE (GPa)						Poisson ration						Shear modulus (MPa)					
Thermal conductivity (W/(m K))		Thermal diffusivity (10^{-4} m^2/s)		Specific heat capacity (kJ/(kg K))	Radial surface	Tangential surface	Radial surface	Tangential surface	Moisture content (%)	density (kg/m^3)	Air-dried density (kg/m^3)	E_L	E_R	E_T	υ_{LR}	υ_{LT}	υ_{RL}	υ_{RT}	υ_{TL}	υ_{TR}	G_{LT}	G_{LR}	G_{TR}	G_{RL}	G_{TL}	G_{RT}
Radial	**Tangential**	**Radial**	**Tangential**																							
0.1233	0.1430	0.001283	0.001481	1.633																						
									11	750		13.70	0.224	0.114	0.45	0.51		0.75			1060	1610	460			
0.1768	0.1524	0.001436	0.001247	1.771																						
0.1233	0.1233	0.001494	0.001272	1.875																						
					7.35	9.41	11.56	13.92			430	5.35	1.200	1.040	0.64	0.66	0.04	0.45	0.03	0.40				621	466	94
0.1942	0.171	0.001583	0.001342	1.846			11.47	21.46																		

Note: (a) extracted with ethanol-benzene, MOE - the modulus of elasticity in static bending of wood, UTS - ultimate tension strength parallel to grain, and UCS - ultimate

Note: (a) extracted with ethanol-benzene, MOE - the modulus of elasticity in static bending of wood, UTS - ultimate tension strength parallel to grain, and UCS - ultimate compressive strength parallel to grain.

Table 2.7 Chemical, physical, and mechanical properties of bamboo species from China (Yao et al. 1992[1], Lin et al. 1999[2], Yu et al. 2005[3], Li et al. 2006[4], Qiu 2016[5]). References in [ ]

Bamboo species		Chemical composition (oven-dry specimens)					Shrinkage coefficient				Bending strength (MPa)		MOE (GPa) Bending	Shear strength parallel to grain (MPa)	UTS (MPa)	UCS (MPa)
		α-cellulose (%)	Hemicelluloses (%)	Lignin (%)	Extractives[(a)] (%)	Ash (%)	Air-dried density (kg/m^3)	Radial	Tangential	Volumetric	Radial	Tangential				
Phyllostachys bambusoides [1, 4]	top part	42.48	10.48	24.86	5.47	1.12	772	0.167	0.251		130.7	155.8	13.2	17.1	224.3	69.2
	middle part	43.50	10.93	23.62	7.45	1.30										
	base part	43.87	10.16	23.69	8.08	1.80										
Phyllostachys lithophila [1, 4]	top part	39.72	12.53	25.08	8.46	0.80	562				228.5					81.4
	middle part	42.59	11.11	24.86	7	1.28										
	base part	43.4	10.11	26.03	6.40	2.04										
Phyllostachys migra var. Henonis [1, 4]	top part	41.64	10.42	25.68	6.59	2.06	818	0.178	0.281		136.4	186.9	16.5	15.7	271.4	
	middle part	42.82	12.86	24.57	6.88	1.30										
	base part	42.24	11.05	25.65	6.15	1.10										
Phyllostachys pubescens [1, 4]	top part	41.22	12.92	24.73	5.99	1.22	740	0.236	0.252		162.3	162.4	12.5	18.6	202.9	60.7
	middle part	42.35	11.27	24.49	7.35	1.20										
	base part	42.89	11.50	23.97	7.39	1.10										
Pleioblastus amarus [1, 3, 4]	top part	41.91	10.77	27.81	3.00	1.38	880	0.560	0.590	0.740			11.1		211.8	72.2
	middle part	47.69	10.12	26.26	2.97	1.30	810	0.590	0.600	0.690			9.2		222.3	67.9
	base part	48.32	9.99	25.69	3.01	3.20	710	0.710	0.810	0.840			6.5		182.3	51.7
Pseudosasa amabilis [1, 4]	top part	10.14	12.57	22.79	7.41	1.75										
	middle part	43.28	11.42	21.88	7.95	1.26										
	base part	41.52	12.54	22.11	8.63	1.19										
Sinocalamus latiflorus [1, 2, 4]	top part	40.89	10.06	22.85	9.01	1.71	592				100.0			13.2		48.8
	middle part	39.44	9.78	21.69	8.84	1.26	546				87.8			10.4		42.2
	base part	40.67	10.08	22.17	8.88	1.37	475				68.4			8.5		28.4
Dendrocalamus giganteus [5]	top part						869	0.510	0.470	0.960	479.5		55.2	8.9		84.3
	middle part						748	0.500	0.460	0.940	258.1		19.4	7.9		80.2
	base part						680	0.480	0.440	0.920	187.3		10.1	7.6		73.1

Note: (a) extracted with ethanol-benzene, MOE - the modulus of elasticity in static bending of wood, UTS – ultimate tension strength parallel to the fibre direction, UCS – ultimate compressive strength parallel to the fibre direction. The physical and mechanical parameters (basic density, shrinkage coefficient, MOR, Shear strength, MOE, UTS and UCS) were obtained for a 3-year-old bamboo.

As the number of species is so large, it is impossible to cover them in this section, and only the most common tropical species used in civil construction (timber, ceilings, panels, floors, frames) have been selected. Some of the species presented are not traditionally used but have recently been studied to replace traditional species which are now restricted or prohibited due to the risk of extinction.

The data relating to tropical woods are presented in Tables. 2.31, 2.32 and 2.33. Table. 2.31 presents static test results obtained in compression (longitudinal direction only) and in bending. Table. 2.32 presents values of stiffness coefficients obtained using ultrasound tests. The stiffness coefficients (C_{11}, C_{22}, C_{33}) are related to the longitudinal modulus, and C_{44}, C_{55} and C_{66} to the shear modulus, but it is only possible with the complete matrix (including stiffness coefficients obtained for the axes C_{12}, C_{13} and C_{23}) is it possible to calculate the twelve elements of the compliance matrix and to obtain a characterisation (12 elastic constants). Table. 2.33 presents the complete elastic characterization of some species, obtained using ultrasound tests.

2.3.6 Species from USA

Robert J. Ross; Christopher Adam Senalik

The USDA Forest Service Forest Products Laboratory (FPL) was founded in 1910 in Madison, Wisconsin to serve as a centralised wood research laboratory for the United States. One of the first tasks initiated at FPL was an extensive study to obtain fundamental physical and mechanical property information for wood from many species of trees grown in the United States (Markwardt and Wilson 1935). It involved testing several hundred thousand specimens from 164 tree species of clear wood free from defects. Table. 2.34 presents a summary of the study for hardwoods. It was developed (1) to compare properties of different species, (2) to calculate the strength of wood members, (3) to establish safe working stresses, and (4) to group species into classes with similar properties for a variety of end-uses. The experimental design recommended the use of five trees of each species from a single location, and the following data were recorded for each tree harvested for the study.

1. Common and botanical names of the species.
2. Place of growth, specifically the county and state in the United States.
3. Both green and dry materials were tested. All tests on green wood were made at approximately the moisture content of the living tree.
4. The number of trees from which the specimens were obtained.
5. The number of growth rings per cm, to indicate the rate of growth in diameter of the trees from which the test specimens were cut. Rings per cm were counted along a radial line on the cross-section of each specimen.
6. Percentage of latewood.

The fundamental information generated from this study still provides a broad basis for the establishment of design values for structural timber and sawn-timber products, and it is an important reference for the physical and mechanical properties of wood used by individuals around the world. It also serves as a baseline for investigations of the effects of global climate change on the growth of trees, and for studies that focus on the effect timber that harvesting practices have on tree growth rates and subsequent wood quality.

These data, and other fundamental information on wood properties, are summarised and presented in Wood Handbook—Wood as an Engineering Material (Forest Products Laboratory 2021), which is perhaps the most well-known technical document produced by the USDA Forest Service. It was first issued in 1935, and slightly revised in 1939, as an unnumbered publication. Further revisions in 1955, 1974, and 1987 were published by the U.S. Department of Agriculture as Agriculture Handbook No. 72. The 1999 revision was published by the FPL as General Technical Report FPL–GTR–113 and reprinted for broader distribution by the Forest Products Society. In 2010 a special edition, the Centennial Edition was published (Forest Products Laboratory 2010).

Tables. 2.35, 2.36, 2.37, 2.38, 2.39 and 2.40 present fundamental information on the physical and mechanical properties of woods native to the United States. This information was taken directly from the 2010 and 2021 editions of Wood Handbook.

Table 2.8 Characterisation of European wood species (Jenkins and Ebeling 1985, Niemz 1993, Kucera and Gfeller 1994, Voulgaridis and Passialis 1995, Rousodimos 1997, Matsunaga and Minato 1998, Babiak et al. 2000, Németh et al. 2000, Molnár and Bariska, 2002 Németh 2002, Molnár 2004, Molnár and Komán 2006, Wagenführ 2007, Günther et al. 2012, Le CIRAD 2012, Carrasco et al. 2013, Ravenshorst et al. 2013, Stangerlin al. 2013, García et al. 2014, Jankowska and Kozakiewicz 2014, Mazlan et al. 2015, Vasubsbu et al. 2015, de Oliveira Araújo et al. 2016, Limama et al. 2016, Molnár et al 2016, Fodor et al. 2017, Grabner 2017, Kamperidou and Barboutis 2017, Moreira et al. 2017, Komán 2018, ITTO 2023). FSP – fibre-saturation point, L – longitudinal, R – radial, T – tangential, // - parallel, ⊥ - perpendicular, MOE–modulus of elasticity (bending modulus), HW – heartwood, SW – sapwood

Wood species	Botanical name	Density		Shrinkage (maximum FSP → 0%)			strength //	Tensile strength		Compression strength	
		Absolutely dry	Air dry	Longitudinal	Radial	Tangential		//	⊥	//	⊥
		(kg/m³)	(kg/m³)	(%)	(%)	(%)	(MPa)	(MPa)	(MPa)	(MPa)	(MPa)
Softwoods											
Aleppo pine	*Pinus halepensis*		580	0.2	4.9	7.1	119			55	11.8
Black pine	*Pinus nigra*	560	600	0.4	4.3	6.9	132			62	7.0
Calabrian pine	*Pinus brutia*	517		0.3	4.3	6.0	90			44	9.8
Cembra	*Pinus cembra*	430	460		3.2	5.9	82	115		45	
Douglas fir	*Pseudotsuga*	470-530	510-580	0.3	4.0-4.8	7.0-7.7	70-100	82-105	2.4	42-68	
Juniper	*Juniperus communis*	450	500	0.2	3.8	6.9	79	34		43	
Norway spruce	*Picea abies*	300-430-640	330-470-680	0.3	3.5-3.7	7.8-8.0	49-78-136	21-90-245	1.6-2.7-4.0	33-59-79	
Scots pine	*Pinus sylvestris*	300-490-860	330-510-890	0.2-0.4	3.3-4.5	7.5-8.7	41-80-205	35-104-196	1.0-3.0-4.4	33-55-94	
Silver fir	*Abies alba*	320-410-710	350-450-750	0.1	2.9-3.8	7.2-7.6	47-73-118	48-84-120		31-47-59	
White pine	*Pinus strobus*	310-370-470	340-400-500	0.2	2.1-2.7	5.8-6.8	41-61-71	53-76-136	2.1-2.6	28-34-43	
Yew	*Taxus baccata*	610-640-740	641-670-810	0.5	3.7	5.3	85-92	108		58	
Hardwoods											
Aspen	*Populus tremula*	350-450-560	400-490-600		3.5	6.7-8.5	52-60	75	1.7	25-40	
Beech	*Fagus sylvatica*	490-680-880	540-720-910	0.3	5.8	11.8	74-123-210	57-135-180	7.0-10.7	41-62-99	
Black locust	*Robinia pseudoacacia*	540-740-870	580-770-900	0.1	3.2-4.6	5.4-7.2	103-136-169	132-195	5.6	79-93	13.8-19.0
Black poplar	*Populus nigra*	370-410-520	410-450-560	0.3-0.6	3.3-5.2	7.1-9.8	43-60-94	43-77-110	1.7-2.8	26-35-56	
Black walnut	*Juglans nigra*	560-610	580-640-810	0.4	4.8-5.5	7.1-7.8	90-103		4.7	44-53	
Cherry	*Prunus avium*	490-550-670	600-630-690		3.5-5.0	6.5-8.7	85-110	98		45-54	
Chestnut	*Castanea Sativa*	530-590	590-620-680	0.6	3.4-4.3	6.4-6.7	64-77-91	115-142		40-52-57	
Common alder	*Alnus glutinosa*	450-510-600	490-550-640	0.5	4.4-4.8	7.7-9.3	44-92-172	55-94-140	6.9-7.3-7.9	31-55-77	
Common ash	*Fraxinus excelsior*	410-650-820	450-720-860	0.2	4.6-5.0	8.0-8.4	58-105-210	75-165-293	7.0-11.2	23-52-80	
Common olive	*Olea europaea*	797	990	0.9	3.6	4.9	70			56	
Common pear	*Pyrus communis*	650-700-760	690-740-800	0.4	4.6	9.1	77-98-112	100	5.3	41-54-60	
Common walnut	*Juglans regia*	450-640-750	570-680-810	0.5	5.4	7.5	99-147-178	99-100-125	3.5	42-72-89	

(continued)

Table 2.8 (continued)

Shear strength	Impact bending	Brinell Hardness		MOE	Thermal	Calorific	Lignin	Sugars	Cellulose	Pentosane	Ash	pH-value
//		//	⊥	//	conductivity ⊥	value						
(MPa)	(J/cm²)	(MPa)	(MPa)	(GPa)	(W/mK)	(MJ/kg)	(%)	(%)	(%)	(%)	(%)	(-)
15.1	4.6			8.3-13.5	0.20 (MC = 5%)		24.9-29.3		52.6-57.4		0.2-0.3	
	4.6	54		15.3			27.9-28.2		55.7-57.9		0.3-0.4	
14.0	3.7						28.0-29.0		52.6-57.9		0.3-0.5	
	2.5		12	9.8		20.9	24.8					
7.0-10.2	3.8-6.0	39-49	31	12	0.12	19.8	25.4-34.8	67.0-71.4	43.7-55.6	6.0-11.9	0.6-1.1	3.3-4.2 (HW) 5.8 (SW)
12.9	4.8			7.6		20.5	33.9-34.3		40.3-45.9		0.3-0.4	5.0
4.0-6.7-12.0	1.0-4.6-11.0	32	12	7.3-11-21.4	0.10-0.12	16.3	19.0-29.0	67.9-80.9	38.1-46.0	6.8-12.0	0.3-0.8	4.0-5.3
6.1-10.0-14.6	1.5-4.0-13.0	35-40-95	13-19-24	6.9-12-20.1	0.14	15.9	26.3-31.4	64.3-72.3	41.9-52.2	8.2-13.4	0.2-0.5	5.1
3.7-5.0-6.3	3.0-4.2-12	18-30-57	13-16	6.6-11-17			28.0-29.9		40.6-44.1	11.5	0.42-1.10	5.5-6.1
5.5-6.4-7.0	3.5-4.9	20-27	8-11	8.3-10-11.4			25.6-27.7	66.6-70.6	41.4-49.4	5.5-5.8	0.2-0.5	3.4
	15	65-71	31	12-15		20.8	28.8	62.3-70.8	32.6	11.9	0.2-0.3	
6.8	3.0-4.0	20-23	10-11	7.5-7.8	0.12-0.13		18.2-26.4	80.3	47.1-62.8	17.6-27.5	0.3-0.4	5.8
6.5-8.0-19.0	3.0-10.0-19.0	72	34	10.0-16.0-18.0		18.4	11.6-22.7		33.7-46.4	17.8-25.5	0.3-1.2	5.1-5.4
11.0-13.0-16.0	12.0-14.0-18.0	67-78-88	35	9.0-11.3-13.6	0.15	17.7	20.6-29.1	81.7	39.1-50.1	20.8-23.7	0.1-0.6	4.1-5.3
4.0-5.0-8.0	3.0-5.0-7.0	24-37	10-15	4.0-8.8-11.7	0.12-0.13	19.1	13.8-24.5	75.0-80.0	31.0-60.0	15.0-23.0	0.3-0.8	5.8-6.7
8.8-9.6	5.8-6.8	50	26	11.3-13.5		19.8	29.1		40.8-42.6	12.6	0.4-0.8	4.7
14.8	5.6	51-59	28-31	10-11			18.2	81.5	46.3	23.5	0.2	4.5
8.0-9.5	5.5-5.9	27-38-47	15-23	6.4-9.0		17.4	25.9-34.7		42.0	52.6	0.3-0.4	2.8
3.0-4.5-6.5	2.5-5.4-10.8	33-38	7-17	7.7-11.7		19.6	22.6-25.8	76.0-79.0	39.6-46.5	18.8-25.1	0.5-0.6	4.5-4.9
9.0-12.0-14.6	1.0-6.8-8.0	36-65-100	37-41	4.4-13.4-18.1	0.15	14.2	21.1-30.4	76.1-82.1	40.9-46.8	21.2-26.7	0.4-0.6	5.8
	20.7	138	130									
							25.9		32.5-41.7	22.4-23.8	0.4	4.8
7.0-9.0	9.5-19.0	50-70	28-52	12.5-13.0		19.6	29.1		40.8-42.6	12.6	0.4-0.8	4.7

(continued)

Table 2.8 (continued)

Wood species	Botanical name	Density		Shrinkage (maximum FSP→ 0%)				Tensile strength		Compression strength	
		Absolutely dry	Air dry	Longitudinal	Radial	Tangential	strength //	//	⊥	//	⊥
		(kg/m³)	(kg/m³)	(%)	(%)	(%)	(MPa)	(MPa)	(MPa)	(MPa)	(MPa)
Softwoods											
Crab apple	*Malus sylvestris*	700	700-740		7.0	13.7	96	154		44	
Elm	*Ulmus* spp	440-640-820	480-680-850	0.3	4.6-4.8	6.9-8.3	65-72-111	65-80-210	4.0	33-46-56	
European oak	*Quercus robur*	390-650-930	430-690-960	0.4	3.5-4.6	7.8-10.0	74-88-105	50-90-180	2.4-4.0-9.6	54-61-67	
Hornbeam	*Carpinus betulus*	689	761	0.4	6.5	10.9	144	47-135-200	24.0	59	
Horse-chestnut	*Aesculus hippocastanum*	460-510-550	500-550-590	0.9	3.3	6.8	54-64-70	56-81-100		28-31-38	
Lime spp	*Tilia cordata*	320-490-560	350-530-630	0.2-0.3	5.5-6.6	9.1-10.7	46-106-147	23-85-145	5.0-5.8-8.6	26-52-78	
Pannonia poplar	*Populus euramericana*		410				67	56		33	
Red oak	*Quercus rubra*	480-660-870	550-700-980	0.5-0.7	3.7-4.3	8.2-10.0	89-130	163	5.5	36-61	
Sessile oak	*Quercus petrea*	390-650-930	430-690-960	0.4	3.5-4.6	7.8-10.0	78-110-117	50-90-180	2.4-4.0-9.6	54-61-67	
Silver birch	*Betula pendula*	650-700-760	690-740-880	0.4	4.6	9.1	77-98-112	100	53	41-54-60	
Sycamore	*Platanus acerifolia*	380-580-650	420-620-680	0.5	4.5-5.0	8.4-8.7	70-99-107	98	5.3	38-46-53	
Sycamor maple	*Acer pseudoplatanus*	380-580-650	420-620-680	0.5	4.5-5.0	8.4-8.7	70-99-107	98	4.3	38-46-53	
Turkish oak	*Quercus cerris*	570-690-850	720-750-810	0.3-0.4	4.4-4.9	8.5-9.8	94-128-136	100-136-139		44-62-71	17.8
White willow	*Salix alba*	270-330-380	290-350-420	0.5-0.9-1.1	1.9-2.4-3.9	5.4-6.3-8.7	30-47-71	33-46-85		18-24-34	
Imported species											
Bangkirai (Balau)	*Shorea optusa*	800-900	900-1100		2.7	5.3	125-140			68-80	
Cumaru	*Dipteryx odorata*	860	1070	0.2	4.4	6.1	145-192			95-111	
Massaranduba	*Manilkara bidentata*		900-1000		7.1	9.4	200			92	
Meranti	*Shorea albida*	560-670-810	590-710-890	0.2-0.3-0.4	3.4-4.1-4.6	7.1-9.7-11.0	77-119-158	66-146-222	2.7	53-63-74	
Teak	*Tectona grandis*	440-630-720	520-660-700	0.4-0.6	2.1-3.0	4.2-5.8	58-109	95-120-155	2.3-5.4	42-59	

(continued)

Table 2.8 (continued)

Shear strength	Impact bending	Brinell Hardness		MOE	Thermal	Calorific	Lignin	Sugars	Cellulose	Pentosane	Ash	pH-value
//		//	⊥	//	conductivity ⊥	value						
(MPa)	(J/cm²)	(MPa)	(MPa)	(GPa)	(W/mK)	(MJ/kg)	(%)	(%)	(%)	(%)	(%)	(-)
	5.4		31	7.2		19.9						
6.9-11.0	2.0-6.0-10.0	64	30	5.9-11.0-16.0	0.14		20.5-27.3	73.0-80.3	35.9-55.2	12.2-23.7	0.3-0.9	6.4
6.0-11.0-13.0	1.0-6.0-16.0	50-66	25-34	10.0-11.7-13.2	0.13-0.20	14.5	24.9-34.3		37.6-42.8	19.0-25.5	0.3-0.6	3.9
8.5-16.0	8.4	67	29	15.4		19.3	17.8-22.5	79.4-86.7	42.0-46.4	20.1-27.0	0.3-0.4	4.8-5.2
	1.3-3.5-5.9	33	4.6-5.4-6.5				21.3-26.2	79.4	41.6-47.5	18.1	22.1	4.5
4.0-4.5-6.0	4.8-13.0	27-38-47	13-16-20	5.8-7.4-17.2			16.5-20.0	76.7-81.9	35.2-48.2	16.1-23.1	0.5-1.5	4.5-5.2
8.3		21	11	6.5								
12.0	7.4-8.6-13.8	53-66	29-36	12.0-13.0		17.7	19.0-24.5		40.4-49.9	29.0-24.8	0.1-0.2	3.4-4.2
6.0-11.0-13.0	1.0-6.0-16.0	50-66	25-34	9.2-13.0-13.5	0.13-0.20		24.9-34.3		37.6-42.8	19.0-25.5	0.3-0.6	3.9
	1.8-3.1-4.3	60-97	32	6-8-10	0.26-0.31	20.1	19.3-24.7	70.9-82.0	45.3-56.5	23.6-27.1	0.3-0.4	4.8
10.0-12.5	3.8-7.0-12.7	44-46-52	17-21-24	8.6-10.5-12.7			27.7-29.1		41.1-50.7	17.9-24.9	0.6-0.8	
10.0-12.5	3.8-7.0-12.7	44-46-52	17-21-24	8.6-10.5-12.7		19.6	27.7-29.1		41.1-50.7	17.9-24.9	0.6-0.8	5.3
9.0	10.0	57				16.7-19.2	24.2-25.8					5.7
5.7-6.4-8.0	1.7-3.5-7.2	16-23-29	8-13-16	4.4-7.2-10.1		19.4	22.7-27.2		34.2-50.0	16.3-26.7	0.3-0.6	5.0
10.0-15.3	8.1		37	12.6-15.4 (20)								
	5.7			23.4-29.8			28.3	62.4			0.5	
23.9				12.9-23.5	0.21							4.9-5.2
7.1-9.0-10.6	4.0-6.0-9.0	35-48-60	17-24-34	11.7-14.5-19.4	0.19	18.8-20.4	36.0		52.0	10.8	0.1-0.5	4.7-6.5
8.3-9.5	3.7-8.6	63-71	28-39	9.5-13.2	0.13	18.9	30.1-38.2	61.1	40.3-43.1	12.6-19.4	1.1-1.4	5.1

Table 2.9 Physical and mechanical properties of wood from different regions in Russia and neighbouring countries (Borovikov and Ugolev 1989). MoR – modulus of rupture, R – radial, T – tangential, V – volumetric, // – parallel, RF – Russian Federation

Scientific name	Common name	Growing area	Rate of growth (number of growth rings per cm)	Percentage of late-wood (%)	Density (kg/m^3)	Swelling coefficient (%/% unit MC)			MOR (MPa)	Compression // to grain (MPa)
						R	T	V		
Abies alba	European silver fir	Ukraine	4.2	25	426	0.15	0.29	0.46	74.6	44.0
Abies alba	European silver fir (Cultures)	Latvia	2.6	26	346				43.8	34.8
Abies holophylla	Manchurian fir	Primorsky Krai (RF)	5.6	19	396	0.15	0.35	0.52	70.2	37.8
Abies nephrolepis	Khingam fir	Primorsky and Khabarovsk Krai (RF)	6.0	18	396	0.13	0.34	0.48	75.6	41.7
Abies nordmaniana	Nordmann fir	Krasnodar Krai (RF) and Abkhazia	5.0	31	436	0.18	0.34	0.54	80.7	45.1
Abies nordmaniana	Nordmann fir (cultures)	Latvia	3.0	28	406				57.7	41.6
Abies sibirica	Siberian fir	Ural (RF)	7.5	22	386	0.13	0.25	0.40	65.5	38.2
		Krasnoyarsk Krai (RF)	7.3	13	386				68.7	40.9
		West Siberia (RF)	4.2	31	356	0.12	0.36	0.50	64.0	36.7
		Altai (RF)			406				82.8	45.2
		East Siberia (RF)	4.5	24	346	0.09	0.33	0.44	58.3	39.0
Acer campestre	Field maple	Ukraine and Ural (RF)	6.2		693	0.21	0.34	0.57	115.0	58.4
Acer mandschuricum	Manchurian maple	Primorsky Krai (RF)			683	0.19	0.32	0.53	102.0	48.9
Acer mono Maxim	Japanese maple	Primorsky Krai (RF)			713	0.20	0.31	0.53	129.0	57.6
Acer pictum	Shir-daar maple	Azerbaijan	3.0						104.0	58.5
Acer platanoides	Norway maple	Ukraine and Belarus	6.4		703	0.21	0.34	0.57	119.0	60.6
Acer velutinum	Velvet maple	Azerbaijan			683	0.20	0.27	0.49	123.0	66.4
Albizia julibrissin	Lankaran albitia (silk acacia)	Azerbaijan			584	0.15	0.23	0.40		57.5
Alnus glutinosa	Black alder	Ukraine			535	0.15	0.29	0.46	91.2	43.6
		Belarus			525	0.20	0.30	0.52	73.8	43.9
		Lithuania	7.0		525	0.20	0.30	0.52	79.1	45.7
		Latvia	5.8		525	0.18	0.39	0.59	79.1	46.3
Alnus subcordata	Caucasian alder	Azerbaijan			535	0.14	0.21	0.37	71.7	42.5
Betula costata	Mongolian birch	Primorsky Krai (RF) Khabarovsk Krai (RF)	5.7		685				119.0	60.0
Betula dahurica	Asian black birch	Primorsky Krai (RF) Khabarovsk Krai (RF)	4.5		725				116	48.9
Betula pendula	Silver birch	Estonia			645				116	56.4
		Republic of Karelia (RF)	7.8		655	0.31	0.37	0.70	98.0	58.1

(continued)

Table 2.9 (continued)

Scientific name	Common name	Growing area	Rate of growth (number of growth rings per cm)	Percentage of late-wood (%)	Density (kg/m³)	Swelling coefficient (%/% unit MC)			MOR (MPa)	Compression // to grain (MPa)
						R	T	V		
Betula pubescens	European white birch	Latvia			665	0.30	0.35	0.63	121.0	58.7
		Belarus	4.1		616	0.26	0.34	0.62	116.0	45.0
		Central areas of the European part of RF	5.5		616	0.27	0.32	0.61	98.6	55.4
		Ukraine			655				118.0	47.2
		Ural (RF)	9.0		645	0.28	0.34	0.64	108.0	47.2
		Western Siberia (RF)	6.2		645	0.30	0.34	0.64	108.0	59.6
Betula schmidtii	Schmidt's birch	Primorsky Krai (RF)			937				136.0	81.7
Buxus hyrcana	Boxwood	Azerbaijan			960				108.0	80.6
Carpinus betulus	European hornbeam	Caucasus	7.0		794				137.0	61.5
		Armenia	9.0		774	0.26	0.36	0.64	120.0	63.8
		Ukraine			804				120.0	55.3
		Belarus			804	0.27	0.35	0.64	137.0	55.8
Carpinus betulus	European hornbeam (cultures)	Latvia	3.3		715				107.0	54.3
Carpinus caucasica	Caucasian hornbeam	Azerbaijan	6.1		854	0.20	0.34	0.56	127.0	55.4
		Armenia	6.0		745	0.29	0.39	0.70	125.0	73.5
Castanea sativa	Sweet chestnut	Caucasus	4.0		485				72.8	45.7
Cerasus avium	Prunus avium	Armenia	2.9		594	0.14	0.27	0.43	89.0	55.9
Chosenia arbutifolia	Chosenia	Khabarovsk Krai (RF)	3.9		465				73.5	40.8
Diospyros lotus	Caucasian persimmon	Azerbaijan			733	0.16	0.23	0.41	92.2	53.8
Fagus orientalis	Oriental beech	Krasnodar region (RF)	6.5		645	0.15	0.33	0.50	103.0	52.1
		Azerbaijan	6.2		715	0.20	0.35	0.57	112.0	55.1
		Armenia			665	0.20	0.38	0.60	101.0	53.5
Fagus sylvatica	European beech	Ukraine. Carpathians			675				101.0	52.9
Fagus sylvatica	European beech (cultures)	Latvia	4.8		725				108.0	57.0
Fraxinus coriariifolia	Sumach ash	Azerbaijan			762	0.24	0.30	0.56		80.4
Fraxinus excelsior	*European ash*	Tatarstan Republic Chuvash Republic (RF)	5.0	58	663				115.0	57.4
		Latvia	6.5		673	0.17	0.31	0.50	105.0	50.5
		Belarus	5.0	59	673	0.21	0.30	0.53	120.0	57
		Ukraine	4.7	70	723				130.0	55.9
		Georgia			683				123.0	59.7
Fraxinus mandshurica	Japanese ash	Primorsky and Khabarovsk Krai (RF)	6.7	51	653	0.2	0.32	0.54	108.0	50.8
Fraxinus oxycarpa	Raywood ash	Armenia	4.4		802	0.25	0.38	0.65	137.0	67.8
Fraxinus pennsylvanica	Red ash (cultures)	European part of RF	5.0	71	742	0.22	0.32	0.56	111.0	52.6
		Ukraine	3.9	85	733	0.18	0.23	0.43	114.0	61.7
Gleditsia caspia	Caspian locust	Azerbaijan			782	0.2	0.29	0.51	93.1	64.3
Juglans regia	European walnut	Ukraine. Tajikistan and Caucasus			594	0.18	0.28	0.48	108.0	55.4

(continued)

Table 2.9 (continued)

Scientific name	Common name	Growing area	Rate of growth (number of growth rings per cm)	Percentage of late-wood (%)	Density (kg/m³)	Swelling coefficient (%/% unit MC)			MOR (MPa)	Compression // to grain (MPa)
						R	T	V		
Juniperus communis	Common juniper	The Central European part of the RF	13.0		485	0.11	0.19	0.32	86.0	54.7
Juniperus foetidissima	Stinking juniper	Armenia	11.1		534	0.13	0.17	0.32	76.7	45
Juniperus polycarpos	Persian juniper	Armenia	12.1		554	0.11	0.17	0.32	85.7	48.4
Larix decidua	European larch (cultures)	Ukraine		29	516	0.17	0.32	0.51	93.6	47.4
		Belarus	3.4	34	506	0.14	0.31	0.47	82.9	46.6
		Lithuania	3.5		536				69.6	48.9
Larix gmelinii	Dahurian larch	Primorsky and Khabarovsk Krai (RF)	7.0	32	645	0.18	0.37	0.57	117.0	64.8
		Republic of Sakha (RF)	17.5	33	616	0.18	0.34	0.54	103.0	59.4
Larix leptolepis	Japanese larch	Latvia	4.0	36	536				76.8	57.4
		Belarus	3.6	33	496	0.14	0.31	0.47	81.8	46.6
Larix sibirica	Siberian larch	The North European part of the RF	9.5		675	0.25	0.39	0.66	107.0	58.0
		Ural (RF)	10.8	30	675	0.22	0.40	0.64	107.0	57.6
		West Siberia	5.5	34	655	0.18	0.43	0.63	108.0	69.5
		Altai (RF)	12.0	292	725	0.20	0.40	0.62	111.0	61.5
		East Siberia (RF)	13.5	29	636	0.18	0.37	0.57	106.0	62.7
		Krasnoyarsk Krai (RF)	7.4	31	645	0.18	0.38	0.58	111.0	60.9
Larix sibirica	Siberian larch (cultures)	Belarus	4.6	39	546	0.16	0.32	0.50	91.2	51.8
Parrotia persica	Persian iron wood	Azerbaijan			822	0.22	0.37	0.61	106.0	58.8
Picea abies	Norway spruce	The North of the European part of the RF	12.1	21	475	0.11	0.24	0.37	83.4	49.3
		Central regions of the European part of RF	9.0	26	455	0.14	0.24	0.40	80.7	44.4
		Ukraine. Carpathians	5.6	25	426	0.17	0.30	0.49	80.7	44.0
		Belarus			465	0.22	0.37	0.61	74.5	43.2
		Lithuania	4.4	21	465	0.2	0.36	0.58	81.6	47.5
Picea ajanensis	Ayan spruce	Primorsky and Khabarovsk Krai (RF)	6.8	23	446	0.19	0.36	0.57	83.9	45.1
Picea obovata	Siberian spruce	Ural (RF)	10.1	20	436				71.6	40.8
		West Siberia (RF)	6.5	25	386	0.15	0.30	0.47	67.6	40.8
		East Siberia (RF)	9.0	25	436	0.12	0.26	0.40	81.5	49.8
		Primorsky Krai Khabarovsk Krai (RF)	6.6	26	455				80.5	44.9
Pinus funebris	Chinese pine	Primorsky Krai (RF)		19	485				81.2	39.9
Pinus koraiensis	Korean pine	Primorsky and Khabarovsk Krai (RF)	6.8	22	436	0.12	0.29	0.43	71.5	38.9
Pinus nigra	European black pine	Latvia	3.6	26	475				54.7	47.3

(continued)

Table 2.9 (continued)

Scientific name	Common name	Growing area	Rate of growth (number of growth rings per cm)	Percentage of late-wood (%)	Density (kg/m³)	Swelling coefficient (%/% unit MC)			MOR (MPa)	Compression // to grain (MPa)
						R	T	V		
Pinus sibirica	Siberian pine	Ural (RF)	10.0	25	426				67.5	43.4
		Krasnoyarsk Krai (RF)	9.1	34	436	0.13	0.24	0.39	54.1	33.7
		West Siberia (RF)	4.9	23	436	0.12	0.27	0.14	72.1	40.6
		Altai (RF)	10.3	20	455				79.9	40.0
		East Siberia (RF)	7.0	29	446	0.13	0.28	0.43	70.2	43.6
Pinus strobus	Weymouth pine (cultures)	Belarus	4.0	18	376	0.14	0.29	0.45	66.6	38.1
		Lithuania	5.3	21	356				356	33.6
		Latvia	2.4	19	346				45.0	31.3
Pinus sylvestris	Scots pine	Kola Peninsula (RF)	14.0	25	505	0.15	0.27	0.44	89.1	48.0
		The North European part of the RF	11.8	26	546				95.1	53.0
		Komi Republic (RF)	8.5	26	465	0.17	0.31	0.50	75.5	41.7
		The Central European part of the RF	6.4	26	525	0.18	0.33	0.53	88.3	50.5
		Latvia	6.8	38	535	0.22	0.35	0.59	91.3	54.7
		Lithuania	7.8	26	505	0.12	0.30	0.48	89.0	51.0
		Belarus	8.8	28	515	0.21	0.32	0.55	84.8	44.3
		Ukraine	5.1	22	535	0.18	0.31	0.51	81.5	44.1
		Ural (RF)	7.0	27	525				79.8	49.2
		West Siberia (RF)	6.9	29	475	0.17	0.32	0.51	82.1	49.2
		East Siberia (RF)	11.2	27	465	0.17	0.30	0.49	80.2	45.7
		Khabarovsk Krai (RF)	7.0	21	475				77.5	40.1
Pinus sylvestris L.	Scots pine (cultures)	Belarus	4.1	27	465	0.16	0.27	0.45	84.5	45.4
Platanus orientalis	Oriental plane	Armenia	5.2		644	0.17	0.24	0.43	75.1	38.4
Populus alba	White poplar	European part of the RF	2.9		416	0.12	0.23	0.37	59.7	35.6
Populus bolleana	Silver poplar	Armenia			426	0.11	0.25	0.38	61.8	41.8
Populus canescens	Grey poplar	The European part of the RF	2.3		465	0.16	0.31	0.49	67.0	40.5
Populus deltoides	Eastern cottonwood	Caucasus	2.6		396	0.14	0.25	0.41	62.2	41.4
Populus italica	Lombardy poplar	Armenia			475	0.13	0.27	0.42	74.8	45.5
		Ukraine	1.3		337				58.9	31.4
Populus Maxim	Maximowicz' poplar	Primorsky Krai and Khabarovsk Krai (RF)	4.2		406				62.8	34.8
Populus nigra	Black poplar	Ukraine			535	0.16	0.30	0.48	94.0	48.7
		Republic of Bashkortostan (RF)	2.9		396	0.15	0.28	0.45		35.1
		Uzbekistan			554	0.15	0.26	0.43	55.7	35.2
Populus Schischkinii	Schischkin's poplar	Armenia			584	0.14	0.36	0.52	83.1	50.3

(continued)

Table 2.9 (continued)

Scientific name	Common name	Growing area	Rate of growth (number of growth rings per cm)	Percentage of late-wood (%)	Density (kg/m³)	Swelling coefficient (%/% unit MC)			MOR (MPa)	Compression // to grain (MPa)
						R	T	V		
Populus tremula	Aspen	The Central European part of the RF	5.4		485	0.20	0.32	0.54	77.4	44.7
		Latvia	3.9		475	0.13	0.30	0.47	71.0	42.8
		Belarus			495	0.18	0.32	0.50	71.7	42.3
		Ukraine			525	0.13	0.32	0.47	91.0	46.1
		Armenia	3.4		515	0.11	0.26	0.39	71.4	39.5
		Primorsky and Khabarovsk Krai (RF)			416				65.0	35.7
Pseudotsuga taxifolia	Douglas fir. Columbian pine	Latvia	2.6	35	465				68.1	46.1
Pterocarya pterocarpa	Wingnuts	Azerbaijan			455	0.14	0.20	0.36	64.2	44.5
Pyrus caucasica	Caucasian pear	Armenia			693	0.17	0.35	0.54	81.8	54.0
Pyrus communis	Common pear	Caucasus	6.5		733	0.21	0.28	0.51	130	63.1
Pyrus hyrcana	Hirkan pear	Azerbaijan			703	0.19	0.22	0.43		56.2
Quercus araxina	Arax oak	Armenia			802	0.16	0.25	0.43	83.3	53.3
Quercus castaneifilia	Chestnut oak	Azerbaijan	5.2	74	782	0.24	0.36	0.62	123.0	61.2
Quercus iberica	Georgian oak	Armenia			693	0.13	0.21	0.36	75.3	51.3
		Azerbaijan			871	0.20	0.30	0.52	98.4	60.1
Quercus macranthera	Caucasian oak	Armenia			703	0.14	0.23	0.39	82.4	52.0
Quercus robur	English oak	Central regions of the European part of RF	5.5	65	703	0.20	0.30	0.52	105.0	58.2
		Ukraine	6	68	723	0.19	0.30	0.51	99.2	59.8
		Belarus			703	0.20	0.30	0.52	118.0	56.7
		Latvia	6.2		663				97.8	51.9
Quercus rubra	Red oak (cultures)	Latvia	3.0		713				86.9	61.6
Robinia pseudacacia	Black locust (cultures)	Georgia	1.9	84	824	0.23	0.33	0.58	160.0	73.3
		Ukraine	3.6	73	774	0.26	0.34	0.62	135.0	72.9
Salix alba	White willow	The European part of the RF	3.0		416	0.11	0.27	0.40	63.4	35.3
Salix fragilis	Brittle willow	Ukraine			485	0.10	0.32	0.44	78.1	40.9
Sorbus aucuparia	Rowan tree	The Central European part of the RF	6.0		594	0.19	0.26	0.47	97.9	56.5
Taxus baccata	Common yew	Armenia			584	0.18	0.31	0.51	65.8	45.9
		Western regions of the Ukraine	9.4	24	782	0.15	0.20	0.37	109.0	64.6
Tilia amurensis	Amur linden	Primorsky Krai (RF)			475				70.4	38.5
Tilia cordata	Small-leaved linden	European part of RF	6.8		485	0.22	0.31	0.55	73.0	42.5
Tilia platyphyllos	Large-leaved linden	Azerbaijan	2.5		505	0.24	0.34	0.60	99.9	49.0
Ulmus carpinifolia	Common elm	Azerbaijan	10.7	53					81.0	41.8
		Uzbekistan			752	0.18	0.32	0.52		47.5

(continued)

Table 2.9 (continued)

Scientific name	Common name	Growing area	Rate of growth (number of growth rings per cm)	Percentage of late-wood (%)	Density (kg/m³)	Swelling coefficient (%/% unit MC)			MOR (MPa)	Compression // to grain (MPa)
						R	T	V		
Ulmus glabra	Wych elm	Ukraine	4.6	83	683				116.0	54.7
		Republic of Bashkortostan (RF)	6.0	60	614	0.22	0.40	0.64	86.4	43.3
		Azerbaijan	6.6	70	742	0.16	0.31	0.49	113.0	50.0
Ulmus glabra	Wych elm (elliptic)	Armenia			653	0.16	0.28	0.46	74.8	39.7
Ulmus japonica	Japanese elm	Primorsky Krai (RF)	7.6	47	584				82.2	40.9
		Khabarovsk Krai (RF)								
Ulmus laevis	European elm	European part of RF	5.0	80	544	0.15	0.32	0.49	92.6	46.4
Zelkova carpinifolia	Zelkowa	Azerbaijan	4.1	85	782	0.18	0.32	0.52	125.0	60.1

Table 2.10 Average coefficients of variation (CV) of wood properties (Borovikov and Ugolev 1989)

Wood properties	CV (%)
Latewood	28
Density	10
Swelling:	
in radial and tangential directions	28
volumetric	16
Ultimate strength in:	
compression parallel to grain	13
tension parallel to grain	20
static bending	15
shear parallel to grain	20
Modulus of elasticity	20
Impact bending	32
Static hardness	17

Table 2.11. The main average physical and mechanical properties of common wood species in Russia and neighboring countries (Borovikov and Ugolev 1989). x/y=property value at 12% MC/ property value at 30% MC or more, MC—moisture content, MOE - modulus of elasticity

Wood species	Density (kg/m³)			Swelling coefficient (%/% MC)			Ultimate strength (MPa)					Impact bending (kJ/m²)	Hardness (MPa)			MOE at static bending (GPa)
	At 12% MC	At oven-dry	Basic	Radial	Tangential	Volumetric	Static bending	Compression parallel to grain	Tension parallel to grain	Shear parallel to grain in: Radial	Shear parallel to grain in: Tangential		End	Radial	Tangential	
Alder	525	495	430	0.17	0.30	0.49	78.9/48.4	44.5/23.1	97.3/74.3	8.0/5.1	9.8/6.2	51.6/43.0	39.2/23.5	26.5/15.9	28.2/16.8	9.33/6.00
Ash, European	680	645	560	0.19	0.31	0.52	118.0/72.8	56.2/31.8	140.0/107.0	13.4/9.2	13.0/8.6	88.7/74.0	78.3/46.8	57.1/34.1	65.1/38.9	
Ash, Japanese	655	620	535	0.20	0.32	0.54	108.0/65.8	50.8/28.7	146.0/111.0	13.0/8.9	12.2/8.0	61.9/53.0	65.4/39.1			13.7/10.3
Ash, Raywood	805	755	645	0.25	0.38	0.65	137.0/87.0	67.8/39.4		12.0/8.4	13.1/8.8		72.5/43.3	70.3/42.0	64.3/38.4	
Ash, red (cultures)	740	700	610	0.20	0.28	0.50	112.0/70.2	57.0/33.1		13.7/9.5	13.9/9.2	163.0/139.0	74.7/44.7	65.5/39.2	65.5/39.2	
Ash, Sumach	760	725	620	0.24	0.30	0.56		80.4/46.2		12.3/8.6	11.8/7.8	94.9/80.0				
Aspen	495	465	410	0.15	0.30	0.47	76.5/44.5	43.1/18.8	121.0/92.7	6.2/3.5	8.4/4.9	84.6/72.0	25.8/15.4	18.7/11.2	19.6/11.7	11.2/7.7
Beech	680	650	560	0.19	0.35	0.56	104.0/62.9	52.9/26.8	124.0/92.6	12.1/7.3	14.0/8.7	76.3/69.0	65.1/40.5	53.2/29.8	49.5/30.5	12.4/10.1
Birch, European	640	620	520	0.29	0.34	0.65	109.5/64.5	54.0/26.3	136.5/102.0	9.0/5.8	10.9/7.0	92.9/78.0	46.3/27.5	35.9/21.5	32.1/19.2	14.2/10.8
Birch, Mongolian	685	650	530				119.0/65.6	60.0/	212.0/162.0	13.5/7.4	14.1/7.5	111.0/94.0				12.6/9.2
Birch, Red	725	690	570				116.0/64.9	48.9/20.6	196.0/149.0	12.2/6.7	14.9/8.0	84.6/72.0				
Birch, Schmidt's	975	920	760				135.0/81.0	81.7/36.6			15.0/8.2	86.7/74.0	110.0/65.8	110.0/65.5	108.0/64.6	16.7/13.5
Elm	650	620	535	0.17	0.33	0.52	92.4/57.9	45.6/24.7	84.5	8.9/6.4	9.9/7.2	92.8/78.0	54.7/32.7	41.2/24.6	41.1/24.6	10.1/6.9
Fir, Caucasian	435	415	355	0.18	0.34	0.54	80.7/47.4	45.1/19.5	113.0/86.0	8.4/5.8	8.9/5.7	41.3/35.0	36.4/17.2			10.6/7.2
Fir, Khingam	395	375	325	0.12	3.34	0.48	75.6/44.5	41.7/18.0		4.4/3.0	4.8/3.0	33.0/27.0				11.3/7.7
Fir, Manchurian	395	375	325	0.15	0.35	0.52	70.2/41.2	37.8/16.3	98.2/74.9	6.3/4.4	6.2/3.9	28.9/25.0				9.7/6.3
Fir, Siberian	375	350	310	0.11	0.10	0.44	67.9/39.6	40.0/17.2	66.3/50.5	5.9/3.7	5.7/3.6	31.8/25.0	27.4/12.9	15.1/7.1	14.2/—	9.0/5.6
Fir, Silver	425	400	350	0.15	0.29	0.46	74.6/43.8	44.0/18.9		6.5/4.5	6.9/4.4	35.1/29.0	27.8/13.1	17.9/8.4	17.5/8.3	10.6/7.2
Hornbeam	795	760	640	0.26	0.36	0.64	127.7/93.8	60.9/36.4	128.5/96.3	14.7/8.5	18.5/10.7	99.0/84.0	88.4/52.5	75.9/45.2	78.1/46.4	13.2/11.7
Larch	665	635	540	0.20	0.38	0.6	108.8/60.5	61.5/24.8	124.0/94.5	9.8/6.2	9.1/5.7	53.1/43.0	42.0/20.1	31.5/14.8	33.4/15.1	14.3/10.8
Linden	495	475	400	0.23	0.33	0.58	86.4/53.1	45.8/23.7	117.0/89.4	8.4/5.5	8.0/4.9	57.8/49.0	25.0/15.0	16.7/10.0	17.4/10.4	8.9/5.6
Black locust	800	770	650	0.25	0.34	0.61	148.0/95.6	73.1/40.8	171.0/107.0	13.2/9.7	14.7/10.8	190.0/161.0	94.2/56.2	66.2/39.6	75.9/45.4	16.3/12.9
Maple	690	655	570	0.20	0.38	0.60	115.0/66.0	58.5/28.1		12.0/7.7	13.7/8.5	76.3/59.0	73.8/44.1	54.1/32.3	57.4/34.3	11.9/8.7
Oak, Arax	800	750	665	0.16	0.25	0.43	83.3/55.1	53.3/29.1		13.7/10.4	15.4/11.8		89.8/53.7	72.0/43.0	82.5/49.4	
Oak, Caucasian	700	655	585	0.14	0.23	0.39	82.4/53.3	52.0/28.7		11.8/8.9	12.6/9.5		69.3/41.4	54.5/32.6	60.0/35.9	

(continued)

Table 2.11 (continued)

Wood species	Density (kg/m³)			Swelling coefficient (%/% MC)			Ultimate strength (MPa)					Impact bending (kJ/m²)	Hardness (MPa)			MOE at static bending (GPa)
	At 12% MC	At oven-dry	Basic	Radial	Tangential	Volumetric	Static bending	Compression parallel to grain	Tension parallel to grain	Shear parallel to grain in: Radial	Shear parallel to grain in: Tangential		End	Radial	Tangential	
Oak, Chestnut	785	750	635	0.24	0.36	0.62	123.0/81.2	61.2/33.2		12.1/9.2	13.2/10.0	111.0/94.0				
Oak, Georgian	780	735	645	0.17	0.26	0.45	87.3/57.6	55.9/30.3		10.7/8.1	12.7/9.7		57.3/34.3	48.2/28.8	52.8/31.6	
Oak, English	690	655	570	0.20	0.30	0.52	103.0/66.4	57.1/30.5	105.0	9.9/7.4	11.8/8.8	76.3/65.0	65.5/39.2	54.5/32.6	47.5/28.4	10.2/7.0
Pear	710	670	585	0.19	0.28	0.49	106.0/62.1	57.7/26.2		8.6/5.5	13.3/7.9	116.0/98.0	77.0/46.1	57.7/34.5	58.9/36.2	11.9/8.7
Pine, Scots	505	480	415	0.18	0.31	0.51	84.5/48.5	46.3/20.8	102.0/77.6	7.4/4.2	7.2/4.4	41.3/35.0	28.4/13.2	22.5/10.6	23.2/10.9	12.2/8.7
Pine, Siberian stone, S. yellow	435	405	360	0.12	0.27	0.41	69.2/36.3	40.0/16.4	89.2/68.0	6.4/3.8	6.4/4.0	30.9/25.0	21.6/11.6	14.8/—	15.4/—	9.1/7.1
Poplar	455	425	375	0.14	0.28	0.44	68.0/39.5	40.0/17.4	87.8/67.0	6.0/3.3	7.2/4.1	39.2/33.0	26.7/15.4	18.5/11.1		10.3/6.9
Spruce, Norway	445	420	365	0.17	0.31	0.50	78.6/43.0	45.0/19.2	101.0/77.2	6.8/4.0	6.7/4.3	39.2/33.0	25.3/12.0	17.5/8.2	17.8/8.5	9.6/6.2
Walnut, European	590	560	490	0.18	0.28	0.43	108.0/59.5	55.4/23.3		10.7/5.8	11.4/6.0	74.3/63.0	62.0/38.0			11.7/8.3
Willow	455	425	380	0.11	0.30	0.43	70.7/40.8	38.2/16.5	99.1/75.8	7.3/4.0	10.3/6.0	55.7/47.0	27.4/16.5	20.9/12.4	20.7/12.4	9.0/5.6

Table 2.12 The elemental chemical composition of ash given as percentage of oven-dry mass (Borovikov and Ugolev 1989)

Wood species	Ash	K_2O	Na_2O	MgO	CaO	P_2O_5	O	SiO_2
Beech	0.55	0.09	0.02	0.06	0.31	0.03	0.01	0.03
Birch	0.26	0.03	0.02	0.02	0.15	0.02	0.01	0.01
Larch	0.25	0.04	0.02	0.07	0.07	0.03	0.01	0.01
Oak	0.51	0.05	0.02	0.02	0.37	0.03	0.01	0.01
Pine	0.29	0.04	0.01	0.03	0.14	0.02	0.01	0.04

Table 2.13 The chemical composition of some wood species given in as percentage of the total weight (total mass) (Borovikov and Ugolev 1989)

Wood species	Cellulose	Lignin	Pentosans	Hexozans	Ash	Extractive substances soluble in:	
						ester	water
Birch	34.1	21.2	22.0	20	0.4	1.3	
Oak	38.9	23.8	22.8	6.3	0.3	4.3	1.8
Pine	51.9	20.5	11.2	9.3	0.2	2.6	0.6
Poplar	39.8	18.8	23.5	8.3	0.3	6.8	2.1
Siberian larch	45.8	29.5	9.3		1.0	3.6	5.1
Spruce	45.2	29.0	9.3	9.2	0.3	1.5	1.4

Table 2.14 Chemical compositions of softwood (%) in Russia and in neighboring countries (Nikitin 1962)

Common name	Scientific name	Growing area	Lignin	Kürschner cellulose	Pentosans in cellulose	Cellulose without hemicelluloses	Hemicelluloses	Pentosans	Uronic acids	Ethereal extract	Ethanolic extract	Water extract	Ash
Ailanto	*Ailanthus glandulosa* Desf	Caucasus	14.0	56.1	2.3	52.6	31.4	18.1	3.7	6.0	3.2	2.7	0.90
Amur cork tree	*Phellodendron amurense* Rupr	Far East region (RF)	21.5	48.4		25.9	25.9	20.1		0.8	1.5	1.9	0.39
Apricot tree	*Prunus armeniaca* L	Central Asia	18.3	39.4	2.0	29.5	33.7	23.9		3.0	11.1	1.1	0.51
Ash	*Fraxinus excelsior* L	Central Black Earth region (RF)	25.2	44.1	1.1	38.5	31.3	25.4		1.2	3.4	0.7	0.51
Aspen	*Populus tremula* L	Leningrad region (RF)	21.3			52.1[a]		22.8		1.5		2.4	0.26
Beech	*Fagus* L. spp.	Caucasus	20.8	46.0	3.8	42.2	36.9	29.3	4.8	0.5	1.2	0.6	0.50
Black alder	*Alnus glutinosa* (L.) Gaertn	Leningrad region (RF)	22.5	48.3	2.9	42.7	30.8	24.3		0.9	3.0	0.4	0.29
Black poplar	*Populus nigra* L	Central Black Earth region (RF)	18.8	48.0	2.8	44.6	28.8	23.4		1.8	5.2	0.6	0.35
Black saxaul	*Haloxylon aphyllum* (Minkw.) Iljin	Central Asia	27.8	32.2			37.1	21.2		0.7	1.3	3.4	2.90
Castor aralia	*Kalopanax ricinifolium* Mig	Far East region (RF)	21.8	54.3			20.8	21.5		0.8	1.5	2.8	0.47
Caucasian persimmon	*Diospyros lotus* L	Caucasus	19.2	45.3	3.2	40.2	37.5	24.2	3.5	2.3	4.1	2.3	0.80
Checker tree	*Sorbus torminalis* (L.) Crantz	Caucasus	25.6	41.8	3.0	39.2	37.0	27.4	6.0	0.4	1.4	0.3	0.67
Cherry laurel	*Prunus laurocerasus* L	Caucasus	26.6	44.8	1.8	38.3	37.3	25.6		0.5	1.3	1.1	0.54
Chosenia	*Chosenia macrolepis* (Turcz.) Kom	Far Eastern	22.2	51.4	3.9	47.5[b]		16.1	5.0	0.9		2.5	0.20
Common cotoneaster	*Cotoneaster vulgaris* Hook.f	Leningrad region (RF)	22.4	44.3	3.6	36.9	43.0	31.3		0.5	1.4	0.4	0.43
Common elm	*Ulmus foliaceae* Gilib	Central Black Earth region (RF)	26.2	47.6	2.7	43.6	24.1	18.6		0.5	2.3	0.9	1.14
Common hazel	*Corylus avellana* L	Central Black Earth region (RF)	22.0	47.4	3.8	40.9	33.5	29.4		0.6	2.6	0.9	0.42
Common pear	*Pyrus communis* L	Caucasus	24.4	43.7	3.4	39.0	31.1	25.8	6.0	0.7	2.0	0.5	0.37
Crabapple	*Malus prunifolia* (Willd.) Borkh	Caucasus	24.6	44.5	3.2	40.8	30.8	23.8	5.6	0.8	1.0	0.6	0.48
Crabapple	*Malus prunifolia* (Willd.) Borkh	Central Asia	19.5	40.9	3.4	38.5	37.3	26.2		1.1	7.1	0.5	0.41
Dogwoods	*Cornus* L. spp.	Caucasus	23.6	37.0	4.9	36.2	20.4	16.2	4.0	7.2	10.7	2.6	0.58
European box	*Buxus sempervirens* L	Caucasus	30.0	40.0		36.4	32.5	26.4		0.8	2.8	1.3	0.46

(continued)

Table 2.14 (continued)

Common name	Scientific name	Growing area	Lignin	Kürschner cellulose	Pentosans in cellulose	Cellulose without hemicelluloses	Hemicelluloses	Pentosans	Uronic acids	Ethereal extract	Ethanolic extract	Water extract	Ash
European chestnut	*Castanea sativa* Mill	Caucasus	23.2	42.5	3.1	40.4	25.0	19.6	5.2	1.4	8.1	2.7	0.35
European elder	*Sambucus nigra* L	Caucasus	30.0	47.8	4.3	43.2	25.6	25.4		0.4	1.6	0.7	0.61
European hackberry	*Celtis australis* L	Crimea (RF)	20.7	41.9	2.1	37.3	35.6	28.5		0.8	6.5	2.2	1.26
European plane	*Platanus orientalis* L	Caucasus	21.2	44.1	1.5	41.2	31.6	20.6	2.6	1.2	3.2	1.4	1.28
European white birch	*Betula pubescens* Ehrh	Leningrad region (RF)	21.2			34.1[e]	42.0	22.0	4.0	1.3			0.35
European white birch	*Betula pubescens* Ehrh	Republic of Karelia (RF)	20.3	45.8				28.7	4.9				
French tamarisk	*Tamarix gallica* L	Crimea (RF)	18.3	34.5	1.5	30.9	33.8	20.5		0.7	7.7	8.5	5.43
Greek straw-berry tree	*Arbutus andrachne* L	Crimea (RF)	23.9	37.6	1.5	32.1	33.5	25.7		0.7	10.5	1.4	0.83
Hackberry	*Prunus padus* L	Leningrad region (RF)	20.2	46.5	2.5	41.2	38.1	27.6		0.5	0.7	0.6	0.19
Hawthorn	*Crataegus* L	Central Black Earth region (RF)	22.6	42.7	1.0	39.5	38.6	29.1		0.6	3.6	1.0	0.44
Hophornbeam	*Ostrya carpinfolia* Scop	Caucasus	21.1	48.5	3.4	44.4	28.4	24	5.8	0.8	2.0	1.3	0.61
Hornbeam	*Carpinus betulus* L	Caucasus	19.4	46.7	5.7	41.8	34.2	26.0	3.6	0.9	1.2	1.2	0.46
Ivy	*Hedera helix* L	Crimea (RF)	24.6	39.3	0.5	34.5	35.4	28.9		1.0	2.1	2.0	0.83
Laurel	*Laurus nobilis* L	Crimea (RF)	20.9	43.2	2.7	37.8	35.1	28.5		0.7	4.8	2.5	0.73
Manchurian birch	*Betula mandschurica* (Reg.) Nak	Primorsky Krai (RF)	20.4	42.9			24.1			1.5	1.3	2.5	0.25
Manchurian walnut	*Juglans mandshurica* Maxim	Far East region (RF)	19.7	51.2			22.2	15.9		2.2	4.4	1.9	0.38
Mongolian oak	*Quercus mongolica* Fisch. ex Ledeb	Far East region (RF)	22.1	46.6			31.9	24.1		0.9	1.5	1.5	0.18
Norway maple24:524:53	*Acer platanoides* L	Central Black Earth region (RF)	23.1	48.0	4.0	41.5	33.3	25.6		0.5	2.0	0.5	0.28
Olive tree	*Olea europaea* L	Crimea (RF)	19.8	32.5	0.2	29.2	35.1	23.9		2.4	14.2	1.0	1.11
Osage orange	*Maclura aurantiaca* Nutt	Caucasus	19.4	40.2	2.4	36.9	30.3	21.2	2.8	3.0	8.9	2.0	0.64
Pedunculate oak	*Quercus pedunculata* Ehrh	Moscow region (RF)	22.5			37.1[b]		22.6	4.7	0.5		7.0	0.27
Persian ironwood	*Parrotia persica* (DC.) C.A.Mey	Caucasus	20.4	46.2	3.8	42.3	30.3	25.7	7.9	1.4	2.4	0.7	0.46
Persian walnut	*Juglans regia* L	Caucasus	22.3	48.8	3.3	47.8	25.2	20.2	5.9	2.2	5.1	1.0	0.45
Pistachio tree	*Pistacia vera* L	Caucasus	21.5	33.8	4.3	29.8	30.7	23.2	4.2	3.3	8.8	3.7	0.22

(continued)

Table 2.14 (continued)

Common name	Scientific name	Growing area	Lignin	Kürschner cellulose	Pentosans in cellulose	Cellulose without hemicelluloses	Hemicelluloses	Pentosans	Uronic acids	Ethereal extract	Ethanolic extract	Water extract	Ash
Pomegranate tree	*Punica granatum* L	Crimea (RF)	21.1	39.3	0.5	34.9	37.8	24.6		0.8	3.8	3.3	1.21
Princesstree	*Paulownia tomentosa* (Thunb.) Steud	Caucasus	19.7	45.7		39.8	31.8	23.5		1.2	5.6	2.1	0.29
Red birch	*Betula dahurica* Pall	Far East region (RF)	19.4	50.0			30.0	26.9		1.6	2.0	0.3	0.22
Rhododendron	*Rhododendron ponticum* L	Caucasus	28.8	41.7	2.7	35.7	33.9	25.1		0.7	4.4	0.5	0.33
Rowan	*Sorbus aucuparia* L	Leningrad region (RF)	22.4	46.4	4.0	39.3	33.5	29.5		0.9	2.6	0.7	0.63
Sand acacias	*Ammodendron conollyi* Boiss	Central Asia	25.5	42.6			22.5	21.0		4.2	2.9	1.7	0.38
Schmidt's birch	*Betula schmidtii* Regel	Far East region (RF)	17.7				29.2	24.6		1.2	8.9	0.4	0.20
Sessile oak	*Quercus sessiliflora* Salisb	Central Black Earth region (RF)	28.8	43.6	3.0	38.9	29.1	22.8		0.9	3.4	1.8	0.26
Small-leaved linden	*Tilia cordata* Mill	Central Black Earth region (RF)	18.3	49.8	1.2	45.6	27.7	23.3		5.7	1.6	1.2	0.60
Tianshan birch	*Betula tianschanica* Rupr	Central Asia	18.6	43.1	0.9	36.8	41.2	32.4		2.2	2.3	0.6	0.32
Tung-oil tree	*Aleurites* J.R.Forst. & G.Forst	Caucasus	23.5	46.3	2.8	40.9	32.2	27.1		1.0	2.4	0.8	0.50
White birch	*Betula alba* L	Germany	19.6	64.2[d]		45.3[f]		27.1		0.7	1.1		0.30
White willow	*Salix alba* L	Central Black Earth region (RF)	28.4	46.3	1.1	43.7	24.3	24.7		1.2	2.2	1.4	0.48
Wild Cherry	*Prunus avium* L	Caucasus	18.1	44.8	3.8	40.7	28.4	24.2	5.8	2.8	7.1	0.8	0.26
Zelkova	*Zelkova carpinifolia* Dipp	Caucasus	20.4	33.4	2.7	31.7	18.9	20.6	5.2	1.7	15.4	1.4	0.77

a - cellulose isolated by chlorine method, free from pentosans.
b - Kürschner cellulose, free from pentosans.
c - ethanol benzol extract.
d - cellulose isolated by chlorine method.
e - cellulose isolated by Kizel's and Semiganovskij method.
f - cellulose isolated by chlorine method without pentosans.

Table 2.15 Chemical compositions of softwood (%) in Russia and in neighboring countries (Nikitin 1962)

Common name	Scientific name	Growing Area	Lignin	Kürschner cellulose	Pentosans in cellulose	Cellulose without hemicelluloses	Hemicelluloses	Pentosans	Mannan	Galactan	Uronic acids	Ethereal extract	Ethanolic extract	Water extract	Ash
Ayan spruce	*Picea ajanensis* (Lindl. et Gord.) Fisch. ex Carr.	Sakhalin (RF)	28.7	46.9	1.9	45.0[b]	13.3	6.9	9.1	0.7	3.3	3.1		4.1	0.20
Common yew	*Taxus baccata* L.	Caucasus	29.4	42.6	2.4	37.5	21.8	11.7	-			2.3	3.0	0.5	0.39
Dahurian larch	*Larix dahurica* Turcz. ex Trautv.	Far East region (RF)	28.6	58.8[d]				9.2	4.9	7.4					0.41
Dahurian larch	*Larix dahurica* Turcz. ex Trautv.	Far East region (RF)	26.6	52.1			20.4	11.6				1.3	1.4	2.2	0.21
Dahurian larch	*Larix dahurica* Turcz. ex Trautv.	Sakhalin (RF)	25.3	39.5	3.3	36.2		8.8	6.1	14.4	2.9	1.4	6.3[c]	22.6	0.30
Finnish (Karelian) spruce	*Picea fennica* (Regel) Kom.	Republic of Karelia (RF)	28.7			48.0[b]		10.0	7.4		3.2	1.4		1.2	0.32
Japanese yew	*Taxus cuspidata* Siebold et Zucc. ex Endl.	Far East region (RF)		61.5[d]				9.6	7.0	0.1					0.18
Juniper, common	*Juniperus communis* L.	Leningrad region (RF)	21.7	48.3	1.9	40.8	29.1	15.5				6.3	2.6	0.5	0.22
Korean pine	*Pinus koraiensis* Siebold et Zucc.	Far East region (RF)	25.0	46.6			21.0	12.3				2.3	2.2	5.9	0.14
Korean pine	*Pinus koraiensis* Siebold et Zucc.	Far East region (RF)	29.0	60.3[d]				6.5	4.9	0.8					0.19
Korean pine	*Pinus koraiensis* Siebold et Zucc.	Primorski Krai (RF)	25.7	44.0			29.8					6.7	2.7	7.8	0.21
Manchurian fir	*Abies holophylla* Maxim.	Primorski Krai (RF)	30.3	42.8			14.3					1.4	2.1	2.5	0.56
Manchurian fir	*Abies holophylla* Maxim.	Far East region (RF)		63.6[d]				9.6	7.0	0.2					1.45
Manchurian fir	*Abies holophylla* Maxim.	Far East region (RF)	33.8	63.0[d]				5.5	8.0	0.1					1.14
Manchurian fir	*Abies holophylla* Maxim.	Far East region (RF)	28.0	55.7	3.6	52.1[b]	8.8	5.4	10.5	traces	3.8	0.7		3.4	0.40
Manchurian fir	*Abies holophylla* Maxim	Primorski Krai (RF)	30.9	45.6			21.7					2.1	2.4	3	0.83
Nordmann fir	*Abies nordmanniana* (Stev) Spach.	Caucasus	29.0	46.2	2.3	44.8	22.2	10.3				2.5	3.8	0.4	0.43
Norway spruce	*Picea abies* (L.) H. Karst.	Republic of Karelia (RF)	29.0			45.2[b]		9.5	9.2		3.3	1.5		1.4	0.26

(continued)

Table 2.15 (continued)

Common name	Scientific name	Growing Area	Lignin	Kürschner cellulose	Pentosans in cellulose	Cellulose without hemicel-luloses	Hemicelluloses	Pentosans	Mannan	Galactan	Uronic acids	Ethereal extract	Ethanolic extract	Water extract	Ash
Norway spruce	*Picea abies* (L.) H. Karst.	Leningrad region (RF)	28.1			59.1[a]		10.0	7.6	2.6		1.1		1.8	0.20
Sakhalin fir	*Abies sachalinensis* F. Schmidt	Sakhalin (RF)	29.1	54.5	3.7	50.8[b]		6.3	10.9	0.1	0.3	3.7		1.9	0.20
Schrenk's spruce	*Picea Schrenkiana* Fisch. & C.A. Mey	Central Asia	32.5	41.4	2.5	39.6	27.4	12.7				0.6	1.8	1	0.55
Scots pine	*Pinus sylvestris* L.	Leningrad region (RF)	28.2	53.8	2.0	51.9	20.5	11.2	-			1.6	1.0	0.6	0.23
Scots pine	*Pinus sylvestris* L.	Leningrad region (RF)	26.9			57.1[a]		10.7	7.1	1.5		4.1		2.6	0.20
Siberian fir	*Abies sibirica* Ledeb.	Siberia (RF)	29.9	51.0			23.1	5.3				0.9	2.2	1.4	0.70
Siberian larch	*Larix sibirica* Ledeb.	Siberia (RF)	29.5	45.8				9.3				1.8	1.8	5.1	1.00
Siberian pine	*Pinus sibirica* Du Tour.	Siberia (RF)	30.1	52.5			20.4	8.6				2.4	2.6	1.5	0.11
Siberian spruce	*Picea obovata* Ledeb.	Republic of Karelia (RF)	28.4			46.1[b]		9.5	8.8		3.3	1.5		1.1	0.32

a - cellulose isolated by chlorine method, free from pentosans.
b - Kürschner cellulose, free from pentosans.
c - ethanol benzol extract.
d - cellulose isolated by chlorine method.
e - cellulose isolated by Kizel' sand Semiganovskij method.
f - cellulose isolated by chlorine method without pentosans.

Table 2.16 Average ultimate strength of some wood species (Borovikov and Ugolev 1989). x/y = the value in the radial direction/ the value in the tangential direction, MC – moisture content, ⊥ – perpendicular

Wood species	Compressive strength (MPa) ⊥ to grain at proportional limit at:		Strength (MPa) in local compression ⊥ to grain at:		Ultimate tensile stress (MPa) ⊥ to grain at:	
	MC = 12%	MC ≥ 30%	MC = 12%	MC ≥ 30%	MC = 12%	MC ≥ 30%
Black alder	6.7/3.7	4.1/2.3	8.0/5.3	4.9/3.2	7.6/5.0	4.1/2.6
Ash			8.3/10.6	5.1/6.5	5.7/5.3	4.8/3.5
Aspen	5.4/3.5	3.2/2.1	6.3/4.8	3.8/2.8	4.7/3.4	3.1/1.8
Oriental Beech	6.1/7.7	3.6/3.8	11.4/10.8	6.9/6.7	5.9/3.9	3.1/2.1
Birch			–/7.7	–/4.7		
Cherry			18.8/15.1	11.4/9.2	4.7/4.6	2.6/2.5
European elm	5.7/4.6	3.4/2.8	11.0/11.0	6.7/6		
Douglas fir	3.1/5.7	1.9/3.5				
Silver fir	2.1/2.4	1.3/1.0	2.5/3.5	1.5/1.4		
Caucasian hackberry	17.2/17.0	10.6/10.4			5.4/3.7	1.0/1.9
Caucasian hornbeam	6.5/6.0	3.9/3.6	13.3/11.3	8.0/6.9	–	–
Juniper			14.3/15.7	8.6/6.5	2.5/2.0	1.5/1.2
Larch	4.3/6.1	2.6/2.4	6.1/9.3	3.6/3.8		
Lignum vitae	23.9/19.7	14.5/12.0			7.8/7.5	4.2/3.9
Lime	5.5/5.1	3.3/3.0	9.6/9.8	5.9/6.0		
Field maple	11.7/9.0	7.1/5.5			3.4/2.6	2.0/1.4
Norway maple	16.0/12.2	9.8/7.5			4.5/2.8	2.6/1.5
Wingnuts	8.8/7.4	5.3/4.5			5.1/4.1	2.8/2.2
Oak			9.1/5.3	5.5/3.2	6.5/4.2	5.5/2.8
Caucasian pear			19.4/13.8	11.8/8.3	7.1/4.5	3.8/2.4
Persimmon	23.2/17.6	14.0/10.7			11.1/11.0	6.1/5.7
Scots pine	5.1/7.5	3.0/3.0	–/13.6	–/5.6	3.2/2.6	2.0/1.5
Plane			20.5/–	12.4/–	6.8/4.2	3.6/2.2
Poplar	4.4/2.9	2.6/1.8	3.7/3.2	2.3/1.9	3.3/2.6	1.8/1.4
Brittle willow	5.7/5.5	3.4/3.3				
Yew	9.0/8.7	5.5/3.5	14.7/14.6	8.9/6.0	3.8/2.1	2.3/1.2

Table 2.17 Cleavage strength of some wood species (Borovikov and Ugolev 1989). x/y = property value at 12% MC/property value at 30% MC or more, MC—moisture content

Wood species	Cleavage strength (N/mm)	
	Radial	Tangential
Aspen	11.5/7.2	15.5/9.6
Ayan spruce	8.9/5.5	8.9/5.5
Black alder	14.1/8.7	17.0/10.6
Black locust	20.9/13.0	27.2/16.8
Black poplar	10.0/6.1	13.1/8.1
Canadian poplar	9.6/5.9	10.7/6.6
Caucasian fir	8.2/5.1	9.2/5.7
Common spruce	9.8/6.0	10.9/6.8
Common yew	12.2/7.6	14.2/8.8
Dahurian larch	12.1/7.5	10.6/6.5
Elm	16.7/10.4	18.2/11.2
English oak	16.6/10.3	22.5/14.0
European ash	22.2/13.8	23.0/14.2
European beech	22.1/13.7	–
European birch	16.5/10.3	20.6/12.8
European walnut	18.7/11.6	17.6/10.9
Field maple	22.6/14.0	27.4/17.0
Hornbeam	21.0/13.0	28.2/17.6
Japanese ash	19.9/12.4	20.0/12.5
Korean pine	8.2/5.1	10.3/6.3
Manchurian fir	7.3/4.5	9.1/5.6
Maximowicz' poplar	8.7/5.4	10.5/6.5
Mongolian birch	21.6/13.4	27.2/16.8
Norway maple		31.8/19.7
Oriental beech	16.7/10.4	25.2/15.7
Pear	22.0/13.6	28.6/17.6
Red birch	22.1/13.7	27.3/16.9
Schmidt's birch	23.5/14.5	23.5/14.5
Scots pine	11.5/7.2	11.3/7.1
Siberian fir	7.4/4.6	9.0/5.6
Siberian larch	13.4/8.3	13.3/8.2
Siberian spruce	9.7/5.7	9.3/5.8
Silver fir	7.2/4.5	8.8/5.5
Small-leaved linden	14.6/9.0	17.9/11.1
White poplar	10.3/6.3	11.9/7.4

Table 2.18 Impact hardness of some wood species (Borovikov and Ugolev 1989). x/y = property value at 12% MC/property value at 30% MC or more, MC – moisture content

Wood species	Impact hardness (J/cm^2)
Aspen	0.75/0.66
Ayan spruce	0.64/0.50
Black alder	0.77/0.67
Black locust	1.20/1.01
Black poplar	0.70/0.61
Caspian locust	1.10/0.93
Caucasian fir	0.66/0.53
Chestnut oak	1.36/1.15
Common spruce	0.75/0.59
Common yew	1.27/1.01
European beech	0.98/0.86
European birch	0.82/0.71
European walnut	1.06/0.93
Hirkan pear	0.95/0.83
Japanese ash	0.81/0.68
Khingam fir	0.58/0.46
Lankaran albitia (Persian silk tree)	0.85/0.71
Mountain pine	0.76/0.61
Norway maple	0.65/0.51
Oriental beech	0.96/0.84
Scots pine	0.73/0.58
Shir-daar maple	0.99/0.86
Siberian fir	0.53/0.42
Siberian larch	0.90/0.72
Siberian pine and Korean pine	0.63/0.50
Siberian spruce	0.65/0.52
Silver fir	0.71/0.56
Sumach ash	1.06/0.89
Sweet chestnut	0.83/0.68
Sycamore maple	0.95/0.83
Velvet maple	0.82/0.72
Zelkowa	1.52/1.28

Table 2.19 Average modulus of elasticity (E) of some wood species (Borovikov and Ugolev 1989). L – longitudinal, R – radial, T – tangential

Wood species	Compression (GPa)			Tension (GPa)			Static bending (GPa)	Shear (GPa)		
	E_L	E_R	E_T	E_L	E_R	E_T	E_b	E_{RL}	E_{TL}	E_{RT}
Aspen	12.8			15.6			11.1			
Birch	16.1	0.65	0.50	18.4	0.64	0.46	15.2	1.56	0.86	0.22
Black alder	13.0			12.1			11.0			
Black poplar	13.9			12.4			15.0			
Brittle willow	11.5			11.1			8.8			
Caucasian fir	12.7			12.7			10.5			
Caucasian hornbeam	12.4			14.8			12.2			
European ash	15.2	1.06		14.2			13.0			
Grey poplar	10.1			9.1			8.8			
Japanese ash	12.6			15.8			13.6			
Larch	14.2			14.7			15.3			
Oak	14.2	1.40	1.01	14.2	1.18	0.91	15.3	1.42	0.98	0.47
Pine	11.9	0.67	0.55	11.9	0.54	0.47	12.6	1.23	0.76	
Siberian yellow pine	13.4			10.7			10.7			
Spruce	14.4	0.64	0.40	14.5	0.66	0.46	11.0			0.05

Table 2.20 Poisson's ratio of some wood species at moisture contents between 10% and 15% (Borovikov and Ugolev 1989). L – longitudinal, R – radial, T – tangential

Wood species	μ					
	RL	TL	LR	TR	LT	RT
Ash	0.43	0.26	0.04	0.74	0.04	0.60
Beech	0.41	0.33	0.03	0.73	0.02	0.36
Birch	0.58	0.45	0.04	0.81	0.04	0.49
Oak	0.43	0.41	0.07	0.83	0.09	0.34
Pine	0.49	0.41	0.03	0.79	0.04	0.38
Spruce	0.44	0.41	0.02	0.48	0.03	0.25

Table 2.21 Average coefficients of long-term resistance (long-term strength) of some wood species (Borovikov and Ugolev 1989)

Wood species	The average ultimate strength[a] (MPa)	Time (days)				
		1	10	10^2	10^3	10^4
At the compression parallel to grain						
Aspen	43.1	0.887	0.853	0.819	0.784	0.750
Beech	53.4	0.863	0.823	0.780	0.738	0.696
Scots pine	47.7	0.832	0.780	0.728	0.678	0.626
Spruce	45.0	0.864	0.822	0.781	0.740	0.698
At the static bending						
Beech	104.0	0.814	0.766	0.712	0.662	0.611
English oak	103.0	0.806	0.752	0.699	0.646	0.592
European ash	118.0	0.860	0.821	0.782	0.743	0.704
European birch	107.0	0.781	0.721	0.660	0.600	0.540
Japanese ash	108.0	0.845	0.803	0.760	0.717	0.675
Mongolian birch	119.0	0.806	0.751	0.697	0.643	0.588
Red birch	116.0	0.800	0.744	0.689	0.633	0.577
Scots pine	84.5	0.805	0.753	0.701	0.649	0.596
Spruce	78.6	0.873	0.840	0.805	0.772	0.738
At the shear parallel to grain (average for radial and tangential)						
European ash	13.2	0.855	0.803	0.753	0.702	0.652
Japanese ash	12.6	0.850	0.794	0.740	0.687	0.634
Scots pine	7.3	0.740	0.673	0.606	0.540	0.473
Spruce	6.8	0.772	0.700	0.627	0.555	0.478
At the tension parallel to grain						
Larch	124.0	0.855	0.811	0.768	0.724	0.681

a) according to Table 2.11 The main average physical and mechanical properties of common wood species in Russia and neighboring countries (Borovikov and Ugolev 1989).

Table 2.22 Approximate values of the plastic flow limit of wood species at a moisture content of 10% to 15% (Borovikov and Ugolev 1989). MC – moisture content, // – parallel, ⊥ – perpendicular, a) the average of the ultimate tensile strength across the grain in radial and tangential directions

Wood species	Density (kg/m^3) at MC = 12%	% from ultimate strength				Compression perpendicular to grain (MPa)	
		At static bending	At compression // to grain	At tension		Radial	Tangential
				// to grain	⊥ to grain[a]		
Ash	644		67	86		6.7	
Beech	645		74	90	92		
Birch	645	65				5.9	4.6
Linden	475		82				
Oak	663	54	74			6.4	7.2
Pine	485	63	73		89	2.8	4.2
Spruce	436		88	75	94		

Table 2.23 Approximate values of swelling pressure in water of some wood species at 20°C (Borovikov and Ugolev 1989). X/Y in the table: the value in the radial direction / the value in the tangential direction, ⊥ – perpendicular

Wood species		Ovendry density (kg/m^3)	Compression ⊥ to grain (MPa)
Alder, black		490	–/0.94
Ash	sapwood	640	1.54/-
	heartwood	640	2.66/3.12
Aspen		460	0.87/1.02
Beech, Oriental		520	1.19/-
Birch, European, silver, white		560	1.02/0.84
Larch, Siberian	sapwood	530	0.72/–
	heartwood	710	0.89/–
Oak, English	sapwood	570	1.03/1.88
	heartwood	640	1.51/3.04
Pine, Scots	sapwood	520	0.80/1.19
	heartwood	520	0.91/1.77
Spruce, Norway		600	0.92/2.01

Table 2.24 Approximate moisture-related properties of some wood species (Borovikov and Ugolev 1989). X/Y in the table: the value in the radial direction, the value in the tangential direction

Wood species	Basic density (kg/m^3)	Liquid water absorption (%)	Water vapor sorption (%)	Coefficient of moisture conductivity (×10^5 cm/s)		
				20°C	60°C	80°C
Aspen	388			2.27/2.58	8.56/9.93	16.20/17.40
Beech, Oriental		126	23.2			
Birch (sapwood)	500			1.85/2.07	6.20/6.34	10.00/11.40
Black locust		80	22.8			
Larch, Dahurian			22.4			
Larch,Siberian	482	123	24.5	1.66/1.93	4.90/5.05	8.10/8.60
Spruce (heartwood)	350	167		2.65/3.26	9.05/9.20	17.7/20.0
Spruce (sapwood)	350			3.16/3.26	12.1/13.7	19.0/19.6

Table 2.25 Nail and screw withdrawal resistance for wood species at 12% moisture content (Borovikov and Ugolev 1989)

Wood species	Growing area	Nails (N/mm)			Screws (N/mm)		
		Direction:					
		Radial	Tangential	Longitudinal	Radial	Tangential	Longitudinal
Birch	Bryansk region	190.1	159.5	121.2	129	141	80
Fir	Irkutsk region	116.2	106.8	66.6	63	65	36
	Krasnoyarsk region	76.6	77.2	50.2			
Larch, Siberian	Irkutsk region	236.8	231.1	124.3	127	120	90
Pine, Scots	Bryansk region	117.4	123.7	61.5	93	88	55
	Irkutsk region	124.9	126.2	72.8	76	77	50
Pine, Siberian	Irkutsk region	81.0	104.2	65.9	88	89	58
	Krasnoyarsk region	81.0	92.9	59.6			
Spruce, Norway	Irkutsk region	126.2	120.5	50.2	89	81	43
	Krasnoyarsk region	127.5	141.3	79.1	96	90	55

Table 2.26 Abrasion (mm) of some wood species an abrasion test with polishing paper according to GOST 16,483.39–81 (Borovikov and Ugolev 1989)

Wood species	Growing area	Radial	Tangential	End
Ash	Tatarstan Republic	0.17	0.14	0.09
Beech	Tatarstan Republic	0.17	0.13	0.14
Birch	Arkhangelsk region	0.26	0.29	0.13
	Bryansk region	0.40	0.31	0.22
Fir	Irkutsk region	0.26	0.30	0.18
	Krasnoyarsk region	0.22	0.21	
Larch, Siberian	Irkutsk region	0.17	0.14	0.07
	Krasnoyarsk region	0.18	0.25	0.13
Pine, Scots	Irkutsk region	0.32	0.33	0.14
	Bryansk region	0.31	0.28	0.12
Pine, Siberian	Irkutsk region	0.28	0.18	0.28
	Krasnoyarskregion	0.28	0.18	
Spruce, Norway	Irkutsk region	0.24	0.32	0.18
	Krasnoyarsk region	0.30	0.32	0.20

Table 2.27 Dynamic modulus of elasticity and rigidity, the decrements of longitudinal and flexural vibrations of some wood species at 12% moisture content (Borovikov and Ugolev 1989). X/Y in the table: the value in the radial direction, the value in the tangential direction

Wood species	Growing area	Dynamic modulus of elasticity	Modulus of rigidity (GPa)		Logarithmic decrement of vibrations	
			G_{TL}	G_{RL}	Longitudinal	Flexural*
Ash	Tatarstan Republic	14.65	0.99	1.04	330	650/730
Aspen	Arkhangelsk region	14.35	0.55	0.81	300	280/320
Beech	Tatarstan Republic	12.65	0.91	0.99	360	
Birch	Bryansk region	18.60	0.71	0.88	560	540/610
	Arkhangelsk region	15.55	0.75	0.91	360	350/360
Fir	Arkhangelsk region	11.15	0.51	0.52	410	380/430
	Irkutsk region	11.40	0.52	0.50	290	270/270
	Krasnoyarsk region	10.90	0.43	0.51	260	260/260
Larch, Siberian	Irkutsk region	18.80	0.86	0.92	260	240/230
	Krasnoyarsk region	15.05	0.88	0.98	270	310/380
Pine, Scots	Arkhangelsk region	14.50	0.74	0.69	400	480/530
	Bryansk region	15.80	0.72	0.72	490	490/490
	Irkutsk region	12.55	0.62	0.62	200	220/210
Pine, Siberian	Irkutsk region	12.60	0.61	0.54	330	290/300
	Krasnoyarsk region	11.70	0.56	0.61	290	310/320
Spruce, Norway	Arkhangelsk region	11.55	0.56	0.59	320	280/300
	Irkutsk region	12.60	0.50	0.54	320	270/290
	Krasnoyarsk region	13.00	0.58	0.60	260	280/280

Table 2.28 Electrical conductivity (specific conductance) of oven-dry wood at 20°C, (Borovikov and Ugolev 1989)

Wood species	The flow of electric current	
	Perpendicular to grain. radial direction (S/m)	Parallel to grain (S/m)
Alder	1.0×10^{-15}	1.0×10^{-14}
Ash	3.0×10^{-15}	2.7×10^{-14}
Aspen	6.0×10^{-15}	1.2×10^{-14}
Birch	1.9×10^{-15}	4.3×10^{-15}
Hornbeam	1.2×10^{-14}	7.6×10^{-14}
Linden	1.0×10^{-14}	1.5×10^{-14}
Maple	1.5×10^{-16}	3.0×10^{-16}
Oak	7.0×10^{-15}	2.8×10^{-13}
Pine	4.3×10^{-14}	5.5×10^{-14}

Table 2.29 Specific mechanical properties of wood species (specific strength, specific hardness, specific impact bending (Borovikov and Ugolev 1989), // – parallel, a) the average for radial and tangential planes of shearing

Common name	Specific strength (MPa/(kg/m^3))				Specific impact bending	Specific end hardness
	Compression // to grain	Tension // to grain	Static bending	Shear // to grain[a]	(kJ/m^2)/(kg/m^3)	(N/mm^2)/(kg/m^3)
Alder	0.085	0.185	0.150	0.017	0.098	0.075
Ash, European	0.082	0.205	0.173	0.019	0.130	0.115
Ash, Japanese	0.078	0.224	0.165	0.019	0.095	0.100
Ash, Raywood	0.084		0.171	0.016		0.090
Ash, Red	0.077		0.151	0.018	0.220	0.101
Ash, Sumach	0.106			0.016	0.124	
Aspen	0.087	0.244	0.154	0.015	0.171	0.052
Beech	0.079	0.176	0.154	0.019	0.119	0.088
Birch, European	0.083	0.256	0.168	0.016	0.146	0.071
Birch, Mongolian	0.088	0.310	0.174	0.02	0.162	
Birch, Red	0.067	0.270	0.160	0.019	0.117	
Birch, Schmidt's	0.084		0.139	0.015	0.09	0.113
Black locust	0.091	0.213	0.184	0.174	0.236	0.117
Elm	0.07		0.142	0.014	0.142	0.084
Fir, Caucasia	0.105		0.191	0.011	0.083	
Fir, Khingam	0.103	0.259	0.185	0.020	0.095	0.083
Fir, Manchurian	0.095	0.248	0.177	0.016	0.073	
Fir, Siberian	0.106	0.176	0.180	0.017	0.077	0.073
Fir, Silver	0.103		0.175	0.016	0.082	0.065
Hornbeam	0.073	0.169	0.160	0.021	0.123	0.110
Larch	0.093	0.186	0.162	0.014	0.078	0.064
Linden	0.092	0.236	0.174	0.016	0.117	0.051
Maple	0.084		0.166	0.018	0.110	0.106
Oak, Arax	0.066		0.104	0.018		0.112
Oak, Caucasian	0.074		0.117	0.017		0.098
Oak, Chestnut	0.078		0.157	0.016	0.142	
Oak, English	0.082		0.149	0.016	0.110	0.095
Oak, Georgian	0.071		0.112	0.015		0.073
Pear	0.081		0.149	0.015	0.163	0.108
Pine, Scots	0.094	0.202	0.167	0.014	0.082	0.055
Pine, Siberian stone, Siberian yellow	0.095	0.204	0.166	0.015	0.071	0.050
Poplar	0.088	0.193	0.149	0.014	0.086	0.056
Spruce, Norway	0.101	0.226	0.176	0.015	0.088	0.057
Walnut, European	0.093		0.182	0.019	0.125	0. 104
Willow	0.084	0.218	0.155	0.019	0.122	0.060

Table 2.30 Average moisture content of green wood of different species (Borovikov and Ugolev 1989)

Species	Moisture content (%)
Alder	84
Ash, Japanese	78
Ash, European	36
Aspen	82
Beech	64
Birch, European	78
Birch, Mongolian	68
Elm	78
Fir	113
Hornbeam	60
Larch, Siberian	82
Linden	60
Linden, Amur	81
Maple	50
Oak	70
Pine, Scots	85
Pine, Siberian stone, Siberian yellow	109
Poplar	93
Spruce, Norway	91
Willow	85

Table 2.31 Mean values of density (ρ), and of elastic moduli in compression parallel to the grain (E_{c0}), and in bending (E_M), together with strength in compression parallel to the grain (f_{c0}) and in bending (f_m) for some tropical woods. Values in brackets are the coefficient of variation. MC – moisture content, green – test on green wood

Common name	Scientific name	MC (%)	ρ (kg/m³)	E_{c0} (MPa)	f_{c0} (MPa)	E_M (MPa)	f_m (MPa)	Reference
Abuí	Pouteria sp.	12	730 (7%)	13 741 (16%)	58 (2%)	13,548 (18%)	84 (19%)	Lahr et al. (2010)
Amescla	*Trattinnickia burserifolia*	12	550	9979	37			Stangerlin et al. (2017)
Andiroba	*Carapa guiamensis*	12	590 (8%)			11,380 (13%)	83,4 (12%)	Melo et al. (2000)
Angelim-amargoso	*Vatairea* spp.	15	936		78		149	IPT (2013)
Angelim-pedra	*Hymenolobium petraeum*	15	710	11 572	53		109	IPT (2013), Stangerlin et al. (2017), Melo et al. (2000)
		12	792	15 151	53			
Angelim-vermelho	*Dinizia excels*	12	1090		81	14 073	138	IPT (2013)
Cambará	*Qualea paraensis*	12	786	14 316	56			Stangerlin et al. (2017), Melo et al. (2000)
Canelão	*Ocotea velutina*	12	680	13 118	45			Stangerlin et al. (2017), Melo et al. (2000)
Cedrinho	*Erisma uncinatum*	12	622	9125	37			Stangerlin et al. (2017), Melo et al. (2000)
Cedrorana	*Cedrelinga cateniformis*	12	520		47	12 847	78	IPT (2013)
Copaíba	*Copaifera reticulata*	12	600* (9%)			11 280 (16%)	81 (18%)	Melo et al. (2000)
Cumaru	*Dipteryx odorata*	15/green	1090	19 306^{green}	94	18 547^{green}	178	IPT (2013), Stangerlin et al. (2017)
		12	1058	19 766	86			
Cupiúba	*Goupia glabra*	15/green	870	17 142^{green}		13 690^{green}	122	Gonçalves et al. (2011), IPT (2013)
		12	850 (4%)	13 583				
Fava amargosa	*Vataireopsis speciosa*	12	740			11 630	96	Gonçalez et al. (2001)
Fava-timborana	*Pseudopiptadenia psilostachya*	12	777 (9%)		72 (7%)	16 030 (14%)	104 (19%)	Balboni et al. (2018)
Garapa	*Apuleia leiocarpa*	15/green	830	14 460^{green}	54	14 107	125	IPT (2013), Gonçalves et al. (2011), Stangerlin et al. (2017), Melo et al. (2000)
		12	812 (5%)	14 333				
		12	856	13 553	66			
Itaúba	*Mezilaurus itauba*	12	995	16 606	66			Stangerlin et al. (2017), Melo et al. (2000)
Jatobá	*Hymenaea courbaril*	15/green	960	17 691^{green}	82	14 837^{green}	152	Gonçalves et al. (2011), IPT (2013), Stangerlin et al. (2017), Melo et al. (2000)
		12	954	16 792	67			
Jequitibá Rosa	*Cariniana micrantha*	12	670			13 200	110	Gonçalez et al. (2001)
Louro Preto	*Cordia glabrata*	12	940 (8%)		65,6 (10%)	8861 (28%)	111 (13%)	Moulin et al. (2016)
Mandioqueira	Ruizterania albiflora spp.	15/green	650	13 700^{green}	57	11 023^{green}	87	IPT (2013)
Matá-matá	*Eschweilera ovata*	12	982 (7%)		69 (26%)	17 870 (15%)	127 (8%)	Balboni et al. (2018)
Muiracatiara	*Astronium lecointe*	15/green	970		82	12 303^{green}	102^{green}	IPT (2013)
Murure	*Brosimum acutifolium*	12	510 (7%)			9999 (16%)	58 (17%)	Melo et al. (2000)
Oiticica-amarela	*Clarisia racemosa*	15/green	560	11 513^{green}	45	7963^{green}	80	IPT (2013)

(continued)

Table 2.31 (continued)

Common name	Scientific name	MC (%)	ρ (kg/m³)	E_{c0} (MPa)	f_{c0} (MPa)	E_M (MPa)	f_m (MPa)	Reference
Pau-jacaré	*Laetia procera*	12	660 (3%)			12,930 (9%)	74 (13%)	Melo et al. (2000)
Pau-roxo	*Peltogyne paniculata*	15/green	890	20 565^{green}	84	17 721^{green}	185	Gonçalez et al. (2001)
Peroba d'agua	*Rauwolfia paraensis*	12	710			13 420	114	Gonçalez et al. (2001)
Peroba mico	*Aspidosperma macrocarpum*	12	780			14 280	136	Gonçalez et al. (2001)
Piquiarana	*Caryocar glabrum, Caryocar villosum*	15/green	850	12 650^{green}	49	13 258^{green}	104	IPT (2013)
Quaruba	*Vochysia guianensis*	12	600		48	11 180	91	IPT (2013)
Quina Rosa	Chinchona sp.	12	840 (3%)	33 089 (21%)	59 (5%)	25 549 (7%)	112 (12%)	Lahr et al. (2010)
Sucupira preta	*Diplotropis purpurea*	12	908	15 692	65			Stangerlin et al. (2017), Melo et al. (2000)
Tachi preto	*Tachigali myrmecophilla*	12	670		58	12 945	131	IPT (2013), Melo et al. (2000)
		12	600 (14%)			11 610 (12%)	81 (13%)	
Tatajuba	*Bagassa guianensis*	12	750 (13%)	25 242 (13%)	79 (4%)	19 568 (22%)	118 (15%)	IPT (2013)
		12	820	17 304	80	15 857	138	
Tauari	*Couratari stellata*	12	520 (6%)			10 270 (13%)	65 (14%)	Melo et al. (2000)

Table 2.32 Mean values of density (ρ) and stiffness coefficient for some tropical woods obtained by ultrasound. Values in brackets are the coefficient of variation. MC – moisture content.

Common name	Scientific name	MC (%)	ρ (kg/m^3)	C_{11} (MPa)	C_{22} (MPa)	C_{33} (MPa)	C_{44} (MPa)	C_{55} (MPa)	C_{66} (MPa)	C_{12} (MPa)	C_{13} (MPa)	C_{23} (MPa)	References
Amescla	*Trattinnickia burserifolia*	12	551	16 209									Stangerlin et al. (2017)
Angelim Pedra	*Hymenolobium petraeum*	12	792	28 087									Stangerlin et al. (2017)
Cambará	*Qualea paraensis*	12	786	22 871									Stangerlin et al. (2017)
Canelão	*Ocotea velutina*	12	680	20 908									Stangerlin et al. (2017)
Cedrinho	*Erisma uncinatum*	12	622	15 470									Stangerlin et al. (2017)
Cumarú	*Dipteryx odorata*	12	1058	33 046									Stangerlin et al. (2017)
Cupiuba	*Goupia glabra*	12	850 (4.5%)	19 008 (4.7%)	3787 (6.2%)	2950 (6%)	642 (3.3%)	1089 (7.2%)	1900 (3%)	4112 (16.4%)	3150 (6.3%)	2135 (12%)	Sales et al. (2011), Gonçalves et al. (2014
		12	954 (5%)	16 410 (22%)									
Fava amargosa	*Vataireopsis speciosa*	12	740	19,77	3621	2303	111	1456	1479				Gonçalez et al. (2001)
Garapeira	*Apulleia leiocarpa*	12	812 (5.4%)	16 709 (0.6%)	3190 (2.4%)	2102 (3.4%)	568) (1.8%)	1425 (5.9%)	1897 (4.1%)	1438 (9.3%)	1275 (7.4%)	1183 (3.4%)	Gonçalves et al. (2014), Stangerlin et al. (2017)
		12	856	25 593									
Itaúba	*Mezilaurus itauba*	12	995	28 344									Stangerlin et al. (2017)
Jatobá	*Hymenaea courbaril*	12	954	30 690									Stangerlin et al. (2017)
Jequitibá Rosa	*Cariniana micrantha*	12	670	18 940	2582	1647	97	1378	1524				Gonçalez et al. (2001)
Peroba d'agua	*Rauwolfia paraensis*	12	710	20 090	3769	1933	114	1364	1436				Gonçalez et al. (2001)
Peroba mico	*Aspidosperma macrocarpum*	12	780	20 650	4422	2658	135	1392	1424				Gonçalez et al. (2001)
Sucupira-Preta	*Diplotropis purpurea*	12	908	28 932									Gonçalez et al. (2001)

Table 2.33 Average modulus of elasticity in longitudinal (E_L), radial (E_R) and tangential (E_T) axes, shear modulus in radial/tangential (G_{LT}), longitudinal/tangential (G_{LT}) and longitudinal/radial (G_{LR}) planes and Poisson's ratio (μ) in the same planes, obtained by inversion of the stiffness matrix (ultrasound test). Values in brackets are the coefficient of variation. MC – moisture content

Common name	Scientific name	MC (%)	E_L (MPa)	E_R (MPa)	E_T (MPa)	G_{RT} (MPa)	G_{LT} (MPa)	G_{LR} (MPa)	μ_{RL}	μ_{TL}	μ_{LR}	μ_{TR}	μ_{LT}	μ_{RT}	References
Angelim pedra	*Hymenolobium petraeum*	12	15 404 (4.8%))	2438 (7.9%)	1619 (0.7%)	552 (2.7%)	1117 (1.7%)	1429 (10.3%)	0.13 (25.7%)	0.08 (36.1%)	0.84 (12.3%)	0.28 (42.9%)	0.75 (31.9%)	0.42 (38.2%)	Costa (2017)
Angelim Vermelho	*Dinizia excels*	12	14 556 (1.6%)	2655 (7.0%)	1624 (3.2%)	538 (1.9%)	1120 (6.7%)	1548 (9.0%)	0.18 (17.2%)	0.11 (23.0%)	0.96 (18.9%)	0.14 (5.0%)	0.95 (22.1%)	0.23 (5.3%)	Costa (2017)
Castanheira	*Bertholletia excels*	12	11 434 (9.2%)	1839 (6.7%)	1298 (5.6%)	419 (1.1%)	955 (4.7%)	1178 (1.2%)	0.11 (5.0%)	0.06 (49.2%)	0.7 (3.9%)	0.38 (7.7%)	0.55 (60.8%)	0.54 (4.3%)	Costa (2017)
Cupiúba	*Golpia glabra*	12	14 124 (2.9%)	2019 (2.5%)	1691 (3.6%)	642 (8.1%)	1089 (8.4%)	1900 (3.7%)	0.12 (21.1%)	0.05 (39.8%)	0.80 (22.1%)	0.50 (7.4%)	0.483 (32.2%)	0.60 (4.7%)	Gonçalez et al. (2001)
Garapa	*Apuleia leiocarpa*	12	15 993 (11.9%)	2619 (7.2%)	1908 (6.5%)	552 (6.5%)	804 (1.8%)	1918 (4.8%)	0.14 (1.8%)	0.10 (15.3%)	0.87 (4.1%)	0.42 (4.6%)	0.84 (11.5%)	0.57 (5.1%)	Costa (2017)
	Apuleia leiocarpa	12	15 705 (0.8%)	2483 (3.2%)	1626 (5.3%)	568 (2.8%)	1425 (7.1%)	1897 (3.7%)	0.046 (18.5%)	0.05 (20.9%)	0.30 (21.1%)	0.35 (2.8%)	0.44 (1.6%)	0.54 (4.7%)	Gonçalez et al. (2001)
Pequiarana	*Caryocar villosum*	12	12 981 (4.8%)	2553 (5.1%)	1447 (1.8%)	506 (5.5%)	859 (3.5%)	1216 (25.1%)	0.17 (7.1%)	0.09 (40.3%)	0.88 (12.0%)	0.32 (4.1%)	0.78 (36.0%)	0.56 (4.5%)	Costa (2017)
Roxinho	*Peltogyne discolor*	12	16 260 (13.0%)	2493 (1.3%)	1874 (2.1%)	630 (1.6%)	1171 (0.7%)	1892 (3.3%)	0.11 (31.7%)	0.09 (16.6%)	0.73 (22.2%)	0.43 (6.7%)	0.77 (3.9%)	0.58 (9.4%)	Costa (2017)

Table 2.34 Physical and mechanical properties of US hardwood species from FPL's original assessment of US species (Markwardt and Wilson 1935)

Species		Place of growth	Moisture condition	No. of trees tested	No. of growth rings		Late-wood	Moisture content	Specific gravity		Specific gravity		Weight per cubic foot	Shrinkage (green to ovendry)		
Common **name**	Scientific name								At test	Oven dry	Oven dry	Green		Volumetric	Radial	Tangential
(-)	(-)	(-)	(-)	(-)	(per inch)	(per cm)	(%)	(%)	(-)	(-)	(kg/m^3)	(kg/m^3)	(lbs/ft^3)	(%)	(%)	(%)
Ash, black	*Fraxinus nigra*	Ontonagon County (MI)	Green	6	23.1	9.1	53	90.6	0.447	0.526	526	850	53	15.2	5.0	7.8
			Dry	1				9.1	0.500							
		Marathon County (WI)	Green		25.0	9.8		78.9	0.456			820	51			
			Dry					11.6	0.497							
Aspen	*Populus tremuloides*	Rusk County (WI)	Green	5	8.5	3.3		106.6	0.360	0.422	422	740	46	11.1	3.3	6.9
			Dry	1				5.2	0.421							
	Populus grandiden-tata	Sauk County (WI)	Green	5	8.2	3.2		96.4	0.354	0.412	412	690	43	11.6	3.1	7.9
			Dry	2				8.0	0.404							
Basswood	*Tilia glabra*	Marathon County (WI)	Green	3	21.8	8.6		115.1	0.316	0.374	374	670	42	14.5	6.2	8.4
			Dry	1				9.2	0.348							
Birch, paper	*Betula papyrifera*	Rusk County (WI)	Green	5	5.5	2.2		72.0	0.473	0.600	600	820	51	16.3	6.6	8.8
			Dry	2				4.2	0.582							
Birch, yellow	*Betula allegh-aniensis*	Marathon County (WI)	Green	5	19.7	7.8		71.8	0.545	0.661	661	930	58	17.0	7.9	9.0
			Dry	1				10.3	0.624							
Butternut	*Juglans cinera*	Sauk County (WI)	Green	5	9.1	3.6		102.2	0.354	0.397	397	720	45	9.4	3.6	5.7
			Dry	2				8.0	0.382							
Elm, American	*Ulmus americana*	Marathon County (WI)	Green		19.0	7.5		70.2	0.421			720	45			
			Dry					10.8	0.469							
Elm, rock	*Ulmus racemosa*	Rusk County (WI)	Green	5	27.1	10.7	50	52.7	0.569	0.658	658	860	54	14.1	4.8	8.1
			Dry	1				5.8	0.678							
Elm, slippery	*Ulmus fulva*	Sauk County (WI)	Green	5	17.2	6.8	51	90.0	0.474	0.554	554	900	56	13.4	4.9	8.7
			Dry	2				7.7	0.531							
Hackberry	*Celtis occidentalis*	Sauk County (WI)	Green	5	13.4	5.3	53	67.8	0.482	0.554	554	800	50	13.8	4.9	8.9
			Dry	2				8.7	0.540							
Haw, pear	*Crataegus tomentosa*	Sauk County (WI)	Green	2	10.6	4.2		63.4	0.623			1030	64			
			Dry	1				8.6	0.695							
Hophorn-beam	*Ostrya virginiana*	Rusk County (WI)	Green	5				52.0	0.632	0.762	762	960	60	18.6	8.2	9.6
			Dry	4	28.8	11.3		5.6	0.748							
Maple, red	*Acer rubrum*	Marathon County (WI)	Green		20.0	7.9		69.9	0.505							
			Dry					12.1	0.548							
Maple, silver	*Acer saccharinum*	Sauk County (WI)	Green	5	7.1	2.8		65.7	0.439	0.506	506	720	45	12.0	3.0	7.2
			Dry	1				8.2	0.479							
Maple, sugar	*Acer saccharum*	Marathon County (WI)	Green		22.0	8.7		62.5	0.560			910	57			
			Dry					12.5	0.621							

(continued)

Table 2.34 (continued)

Static bending												Impact Bending					
Stress at proportional limit		MOR		MOE		Work						Stress at proportional limit		Work to proportional limit		Height of 22.7 kg hammer drop causing complete failure	
						to proportional limit		to maximum load		Total							
(lb/in²)	(MPa)	(lb/in²)	(MPa)	(lb/in²) x 1000	(GPa)	((in·lb)/ in³)	(kJ/m³)	((in·lb)/ in³)	(kJ/m³)	((in·lb)/ in³)	(kJ/m³)	(lb/in²)	(MPa)	([in·lb]/ in³)	(kJ/m³)	(in)	(mm)
2,610	18.0	6,000	41.4	1,107	7.63	0.42	2.9	17.9	123.4	43.6	300.6					35	889
10,340	71.3	16,130	111.2	1,975	13.62	2.05	14.1	13.1	90.3	35.0	241.3					44	1,118
2,580	17.8	6,000	41.4	967	6.67	0.40	2.8	13.1	90.3	35.0	241.3					30	762
6,310	43.5	11,620	80.1	1,395	9.62	1.64	11.3	13.1	90.3	27.3	188.2	12,160	83.8	5.3	36	27	686
2,940	20.3	5,280	36.4	84	0.58	0.65	4.5	6.9	47.6	16.0	110.3	6,880	47.4	2.5	17	28	711
7,600	52.4	10,770	74.3	129	0.89	2.43	16.8	7.3	50.3			10,470	72.2	4.0	27	24	610
3,190	22.0	5,850	40.3	1,185	8.17	0.50	3.4	6.1	42.1	10.6	73.1	7,600	52.4	2.7	19	18	457
7,140	49.2	10,890	75.1	1,635	11.27	1.83	12.6	6.7	46.2	13.5	93.1	15,100	104.1	7.0	48	27	686
2,370	16.3	4,440	30.6	852	5.87	0.39	2.7	5.9	40.7	8.6	59.3	5,760	39.7	2.0	14	15	381
5,640	38.9	7,280	50.2	1,696	11.69	1.05	7.2	3.2	22.1	4.4	30.3	6,390	44.1	1.8	12	8	203
2,920	20.1	5,770	39.8	1,013	6.98	0.49	3.4	15.0	103.4	36.9	254.4	7,780	53.6	2.7	19	45	1,143
11,440	78.9	16,050	110.7	1,814	12.51	4.32	29.8	13.2	91.0	19.2	132.4	13,840	95.4	6.6	45	24	610
4,190	28.9	8,390	57.8	1,597	11.01	0.62	4.3	14.2	97.9	33.7	232.4	11,080	76.4	3.8	26	36	914
13,360	92.1	19,400	133.8	2,396	16.52	4.18	28.8	17.9	123.4	40.5	279.2	18,300	126.2	8.2	56	53	1,346
3,140	21.6	5,870	40.5	1,008	6.95	0.61	4.2	8.4	57.9	22.5	155.1	6,990	48.2	2.3	16	21	533
8,540	58.9	11,040	76.1	1,321	9.11	2.89	19.9	7.2	49.6	8.5	58.6	13,840	95.4	6.8	47	26	660
2,850	19.7	6,940	47.8	1,052	7.25	0.44	3.0	11.8	81.4	28.0	193.1					34	864
6,790	46.8	12,140	83.7	1,504	10.37	1.75	12.1	13.4	92.4	21.0	144.8	14,620	100.8	7.4	51	35	889
4,890	33.7	9,550	65.8	1,165	8.03	1.20	8.3	20.3	140.0	47.2	325.4	10,950	75.5	4.1	28	59	1,499
10,700	73.8	16,600	114.5	1,472	10.15	4.54	31.3	17.0	117.2	46.3	319.2	18,700	128.9	9.8	67	60	1,524
3,740	25.8	7,710	53.2	1,215	8.38	0.72	5.0	16.1	11.0	38.6	266.1	8,640	59.6	3.1	21	48	1,219
9,690	66.8	15,110	104.2	1,556	10.73	3.50	24.1	17.9	123.4	42.5	293.0	17,980	124.0	9.6	66	46	1,168
2,840	19.6	6,210	42.8	911	6.28	0.58	4.0	13.5	93.1	35.4	244.1	7,350	50.7	2.8	19	45	1,143
6,840	47.2	12,000	82.7	122	0.84	2.19	15.1	11.2	77.2	23.4	161.3	15,500	106.9	8.7	60	37	940
3,880	26.8	7,650	52.7	964	6.65	0.89	6.1	22.7	156.5	52.0	258.5						
9,020	62.2	17,250	118.9	1,368	9.43	3.30	22.8	23.8	164.1	30.3	208.9	13,500	93.1	5.5	38	20	508
4,540	31.3	8,540	58.9	1,153	7.95	1.02	7.0	13.3	91.7	39.1	269.6	10,580	72.9	3.5	24	73	1,854
13,920	96.0	18,600	128.2	2,107	14.53	5.34	36.8	14.4	99.3	27.6	190.3	16,560	114.2	7.6	52	40	1,016
4,450	30.7	8,310	57.3	1,445	9.96	0.78	5.4	9.8	67.6	20.4	140.7						
8,650	59.6	13,420	92.5	1,761	12.14	2.37	16.3	12.4	85.5	19.0	131.0	17,030	117.4	8.5	58	31	787
3,120	21.5	5,820	40.1	943	6.50	0.61	4.2	11.0	75.8	22.3	153.8	6,830	47.1	2.6	18	29	737
7,680	53.0	10,100	69.6	1,206	8.32	2.67	18.4	7.6	52.4	11.2	77.2	15,000	103.4	9.4	64	24	610
4,620	31.9	8,820	60.8	1,437	9.91	0.85	5.9	9.6	66.2	17.1	117.9						
9,110	62.8	14,830	102.2	193	1.33	2.41	16.6	13.8	95.1	21.3	146.9	17,440	120.2	8.5	58	29	737

(continued)

Table 2.34 (continued)

Species		Place of growth	Moisture condition	No. of trees tested	No. of growth rings		Late-wood	Moisture content	Specific gravity		Specific gravity		Weight per cubic foot	Shrinkage (green to ovendry)		
Common **name**	Scientific name								At test	Oven dry	Oven dry	Green		Volumetric	Radial	Tangential
(-)	(-)	(-)	(-)	(-)	(per inch)	(per cm)	(%)	(%)	(-)	(-)	(kg/m³)	(kg/m³)	(lbs/ft³)	(%)	(%)	(%)
Oak, black	*Quercus velutina*	Marathon County (WI)	Green		19.0	7.5		84.5	0.538			990	62			
			Dry					11.4	0.593							
Oak, bur	*Quercus macrocarpa*	Sauk County (WI)	Green	5	12.1	4.8	59	69.6	0.583	0.671	671	990	62	12.7	4.4	8.8
			Dry	2				10.2	0.653							
Sumac, staghorn	*Rhus hirta*	Sauk County (WI)	Green	5	8.6	3.4	61	44.7	0.449			660	41			
			Dry	3				8.0	0.481							
Willow, black	*Salix nigra*	Sauk County (WI)	Green	5	4.2	1.7		148.4	0.330	0.409	409	820	51	13.3	2.2	8.2
			Dry	2				8.2	0.380							

Species		"Place of growth"	"Moisture-condition"	Impact Bending					
"Commonname"	"Scientificname"			"Stress atpropor-tional limit"		"Work topropor-tional limit"		"Height of 22.7 kghammer drop causingcomplete failure "	
(-)	(-)	(-)	(-)	(lb/in²)	(MPa)	([in·lb]/in³)	(kJ/m³)	(in)	(mm)
Ash, black	*Fraxinus nigra*	Ontonagon County (MI)	Green					35	889
			Dry					44	1,118
		Marathon County (WI)	Green					30	762
			Dry	12,160	83.8	5.3	36	27	686
Aspen	*Populus tremuloides*	Rusk County (WI)	Green	6,880	47.4	2.5	17	28	711
			Dry	10,470	72.2	4.0	27	24	610
	Populus grandidentata	Sauk County (WI)	Green	7,600	52.4	2.7	19	18	457
			Dry	15,100	104.1	7.0	48	27	686
Basswood	*Tilia glabra*	Marathon County (WI)	Green	5,760	39.7	2.0	14	15	381
			Dry	6,390	44.1	1.8	12	8	203
Birch, paper	*Betula papyrifera*	Rusk County (WI)	Green	7,780	53.6	2.7	19	45	1,143
			Dry	13,840	95.4	6.6	45	24	610
Birch, yellow	*Betula alleghaniensis*	Marathon County (WI)	Green	11,080	76.4	3.8	26	36	914
			Dry	18,300	126.2	8.2	56	53	1,346
Butternut	*Juglans cinera*	Sauk County (WI)	Green	6,990	48.2	2.3	16	21	533
			Dry	13,840	95.4	6.8	47	26	660
Elm, American	*Ulmus americana*	Marathon County (WI)	Green					34	864
			Dry	14,620	100.8	7.4	51	35	889
Elm, rock	*Ulmus racemosa*	Rusk County (WI)	Green	10,950	75.5	4.1	28	59	1,499
			Dry	18,700	128.9	9.8	67	60	1,524
Elm, slippery	*Ulmus fulva*	Sauk County (WI)	Green	8,640	59.6	3.1	21	48	1,219
			Dry	17,980	124.0	9.6	66	46	1,168
Hackberry	*Celtis occidentalis*	Sauk County (WI)	Green	7,350	50.7	2.8	19	45	1,143
			Dry	15,500	106.9	8.7	60	37	940
Haw, pear	*Crataegus tomentosa*	Sauk County (WI)	Green						
			Dry	13,500	93.1	5.5	38	20	508
Hophornbeam	*Ostrya virginiana*	Rusk County (WI)	Green	10,580	72.9	3.5	24	73	1,854
			Dry	16,560	114.2	7.6	52	40	1,016
Maple, red	*Acer rubrum*	Marathon County (WI)	Green						
			Dry	17,030	117.4	8.5	58	31	787

(continued)

Table 2.34 (continued)

Static bending												Impact Bending					
Stress at proportional limit		MOR		MOE		Work						Stress at proportional limit		Work to proportional limit		Height of 22.7 kg hammer drop causing complete failure	
						to proportional limit		to maximum load		Total							
(lb/in²)	(MPa)	(lb/in²)	(MPa)	(lb/in²) x 1000	(GPa)	((in·lb)/in³)	(kJ/m³)	((in·lb)/in³)	(kJ/m³)	((in·lb)/in³)	(kJ/m³)	(lb/in²)	(MPa)	([in·lb]/in³)	(kJ/m³)	(in)	(mm)
3,720	25.6	7,650	52.7	1,121	7.73	0.71	4.9	13.2	91.0	33.4	230.3	12,250	84.5	5.6	38	35	889
8,220	56.7	14,670	101.1	1,641	11.31	2.31	15.9	14.2	97.9	24.3	167.5	15,470	106.7	7.3	50	30	762
3,640	25.1	7,180	49.5	877	6.05	0.89	6.1	10.7	73.8	26.1	180.0	10,020	69.1	4.7	32	44	1,118
7,000	48.3	10,900	75.2	106	0.73	2.79	19.2	9.6	66.2	16.2	111.7	15,340	105.8	8.6	59	27	686
3,030	20.9	5,840	40.3	809	5.58	0.67	4.6	10.8	74.5	42.4	292.3						
10,450	72.1	12,110	83.5	135	0.93	4.50	31.0	7.8	53.8	15.6	107.6						
1,360	9.4	4,240	29.2	637	4.39	0.46	3.2	8.7	60.0	14.4	99.3	6,170	42.5	2.2	15	27	686
5,020	34.6	7,550	52.1	1,006	6.94	2.26	15.6	3.7	25.5	5.7	39.3	9,940	68.5	4.7	32	12	305

"Compressionparallel to grain"				"Compression-perpendicular to grain"		"Jankahardness"				"Shear parallelto grain, i.e. maximumshearing strength"		"Cleavage, i.e.load to causesplitting"		"Tension perpendicularto grain, i.e. maximumtensile strength"	
"Stress atproportional limit"		"Maximumcrushing strength"		"Stress atproportional limit"		End		Side							
(lb/in²)	(MPa)	(lb/in²)	(MPa)	(lb/in²)	(MPa)	(lb)	(N)	(lb)	(N)	(lb/in²)	(MPa)	lb/(in of width)	(N/mm)	(lb/in²)	(MPa)
1,720	11.9	2,340	16.1	409	2.82	610	2,713	552	2,455	866	5.97	266	46.6	490	3.38
6,620	45.6	8,190	56.5	1,270	8.76	1,338	5,952	994	4,422	1794	12.37	397	69.5	760	5.24
		2,260	15.6	452	3.12	565	2,513	490	2,180	854	5.89	292	51.1	490	3.38
3,950	27.2	5,590	38.5	893	6.16	1,101	4,897	792	3,523	1660	11.45	402	70.4	696	4.80
1,600	11.0	2,160	14.9	203	1.40	266	1,183	318	1,415	625	4.31	116	20.3	182	1.25
4,320	29.8	6,440	44.4	552	3.81	848	3,772	420	1,868	890	6.14	224	39.2	380	2.62
2,210	15.2	2,720	18.8	269	1.85	443	1,971	366	1,628	813	5.61	216	37.8	394	2.72
5,100	35.2	5,100	35.2	646	4.45	710	3,158	462	2,055	1305	9.00	236	41.3	380	2.62
1,180	8.1	1,830	12.6	191	1.32	283	1,259	220	979	554	3.82	146	25.6	286	1.97
3,420	23.6	4,800	33.1	540	3.72	461	2,051	364	1,619	1432	9.87	278	48.7	406	2.80
1,640	11.3	2,210	15.2	304	2.10	400	1,779	486	2,162	786	5.42	239	41.9	382	2.63
6,140	42.3	9,470	65.3	912	6.29	1,487	6,615	1,278	5,685	1627	11.22				
2,540	17.5	3,400	23.4	439	3.03	827	3,679	754	3,354	1146	7.90	290	50.8	486	3.35
7,160	49.4	9,560	65.9	1,340	9.24	1,542	6,859	1,280	5,694	1428	9.85	490	85.8	1,112	7.67
2,130	14.7	2,580	17.8	258	1.78	414	1,842	394	1,753	762	5.25	225	39.4	419	2.89
5,680	39.2	7,180	49.5	764	5.27	621	2,762	514	2,286	1417	9.77	227	39.8	401	2.76
		2,700	18.6	292	2.01	536	2,384	486	2,162	825	5.69			578	3.99
4,040	27.9	5,840	40.3	727	5.01	892	3,968	679	3,020	1447	9.98	321	56.2	644	4.44
3,000	20.7	3,820	26.3	813	5.61	1,013	4,506	988	4,395	1276	8.80	406	71.1	662	4.56
5,700	39.3	9,280	64.0	2,109	14.54	1,861	8,278	1,686	7,500	2218	15.29	332	58.1	410	2.83
2,660	18.3	3,180	21.9	468	3.23	715	3,180	653	2,905	1090	7.52	373	65.3	614	4.23
5,740	39.6	7,950	54.8	1,254	8.65	1,144	5,089	836	3,719	1754	12.09	294	51.5		
1,930	13.3	2,520	17.4	475	3.28	740	3,292	677	3,011	1058	7.29	331	58.0	607	4.19
4,250	29.3	6,400	44.1	1,330	9.17	1,154	5,133	887	3,946	1786	12.31	316	55.3	542	3.74
		3,110	21.4	980	6.76	1,218	5,418	1,204	5,356	1356	9.35				
4,750	32.8	8,420	58.1	1,880	12.96	2,270	10,097	1,675	7,451						
2,570	17.7	3,750	25.9	733	5.05	1,157	5,147	1,170	5,204	1374	9.47	334	58.5	450	3.10
8,890	61.3	11,750	81.0	2,303	15.88	3,148	14,003	2,394	10,649	2114	14.58				
		3,680	25.4	605	4.17	766	3,407	748	3,327	1232	8.49	267	46.8		
5,110	35.2	6,610	45.6	1,291	8.90	1,531	6,810	1,024	4,555	1789	12.33	476	83.4	809	5.58

(continued)

Table 2.34 (continued)

Species		"Place of growth"	"Moisture-condition"	Impact Bending					
"Commonname"	"Scientificname"			"Stress atproportional limit"		"Work toproportional limit"		"Height of 22.7 kghammer drop causingcomplete failure "	
(-)	(-)	(-)	(-)	(lb/in^2)	(MPa)	([in·lb]/in^3)	(kJ/m^3)	(in)	(mm)
Maple, silver	*Acer saccharinum*	Sauk County (WI)	Green	6,830	47.1	2.6	18	29	737
			Dry	15,000	103.4	9.4	64	24	610
Maple, sugar	*Acer saccharum*	Marathon County (WI)	Green						
			Dry	17,440	120.2	8.5	58	29	737
Oak, black	*Quercus velutina*	Marathon County (WI)	Green	12,250	84.5	5.6	38	35	889
			Dry	15,470	106.7	7.3	50	30	762
Oak, bur	*Quercus macrocarpa*	Sauk County (WI)	Green	10,020	69.1	4.7	32	44	1,118
			Dry	15,340	105.8	8.6	59	27	686
Sumac, staghorn	*Rhus hirta*	Sauk County (WI)	Green						
			Dry						
Willow, black	*Salix nigra*	Sauk County (WI)	Green	6,170	42.5	2.2	15	27	686
			Dry	9,940	68.5	4.7	32	12	305

"Compressionparallel to grain"				"Compression-perpendicular to grain"		"Jankahardness"				"Shear parallelto grain, i.e. maximumshearing strength"		"Cleavage, i.e.load to causesplitting"		"Tension perpendicularto grain, i.e. maximumtensile strength"	
"Stress atproportional limit"		"Maximumcrushing strength"		"Stress atproportional limit"		End		Side							
(lb/in^2)	(MPa)	(lb/in^2)	(MPa)	(lb/in^2)	(MPa)	(lb)	(N)	(lb)	(N)	(lb/in^2)	(MPa)	lb/(in of width)	(N/mm)	(lb/in^2)	(MPa)
1,930	13.3	2,490	17.2	456	3.14	671	2,985	592	2,633	1053	7.26	202	35.4	564	3.89
5,640	38.9	6,600	45.5	1,181	8.14	1,376	6,121	746	3,318	1714	11.82	353	61.8	489	3.37
		4,020	27.7	870	6.00	965	4,293			1434	9.89				
5,360	37.0	7,370	50.8	1,755	12.10	1,909	8,492	1,342	5,970	2112	14.56	603	105.6		
		3,080	21.2	802	5.53	847	3,768	1,060	4,715	1292	8.91				
4,650	32.1	7,120	49.1	1,246	8.59	1,598	7,108	1,208	5,373	2118	14.60	488	85.5	894	6.16
2,380	16.4	3,290	22.7	836	5.76	1,158	5,151	1,108	4,929	1354	9.34	428	75.0	804	5.54
3,810	26.3	6,640	45.8	1,672	11.53	1,460	6,494	1,419	6,312	1918	13.22	312	54.6	653	4.50
		2,680	18.5	477	3.03	671	2,985	590	2,624						
5,660	39.0	7,750	53.4	1,368	9.43	965	4,293	718	3,194						
1,060	7.3	1,710	11.8	237	1.63	391	1,739	384	1,708	686	4.73	248	43.4	431	2.97
3,220	22.2	5,780	39.9	784	5.41	794	3,532	548	2,438	1342	9.25	371	65.0		

Table 2.35 Shrinkage values of wood from the USA (Forest Products Laboratory 2021)

Common name	Shrinkage[a] (%) from green to absolutely dry		
	Radial	Tangential	Volumetric
Hardwoods			
Alder, red	4.4	7.3	12.6
Ash			
Black	5.0	7.8	15.2
Blue	3.9	6.5	11.7
Green	4.6	7.1	12.5
Oregon	4.1	8.1	13.2
Pumpkin	3.7	6.3	12
White	4.9	7.8	13.3
Aspen			
Bigtooth	3.3	7.9	11.8
Quaking	3.5	6.7	11.5
Basswood, American	6.6	9.3	15.8
Beech, American	5.5	11.9	17.2
Birch			
Alaska paper	6.5	9.9	16.7
Gray	5.2		14.7
Paper	6.3	8.6	16.2
River	4.7	9.2	13.5
Sweet	6.5	9.0	15.6
Yellow	7.3	9.5	16.8
Buckeye, yellow	3.6	8.1	12.5
Butternut	3.4	6.4	10.6
Cherry, black	3.7	7.1	11.5
Chestnut, American	3.4	6.7	11.6
Cottonwood			
Balsam poplar	3.0	7.1	10.5
Black	3.6	8.6	12.4
Eastern	3.9	9.2	13.9
Elm			
American	4.2	9.5	14.6
Cedar	4.7	10.2	15.4
Rock	4.8	8.1	14.9
Slippery	4.9	8.9	13.8
Winged	5.3	11.6	17.7
Hackberry	4.8	8.9	13.8
Hickory, pecan	4.9	8.9	13.6
Hickory, true			
Mockernut	7.7	11	17.8
Pignut	7.2	11.5	17.9
Shagbark	7.0	10.5	16.7
Shellbark	7.6	12.6	19.2
Holly, American	4.8	9.9	16.9
Honeylocust	4.2	6.6	10.8
Locust, black	4.6	7.2	10.2
Madrone, Pacific	5.6	12.4	18.1
Magnolia			
Cucumbertree	5.2	8.8	13.6
Southern	5.4	6.6	12.3
Sweetbay	4.7	8.3	12.9
Maple			

Common name	Shrinkage[a] (%) from green to absolutely dry		
	Radial	Tangential	Volumetric
Softwoods			
Cedar			
Yellow	2.8	6.0	9.2
Atlantic white	2.9	5.4	8.8
Eastern redcedar	3.1	4.7	7.8
Incense	3.3	5.2	7.7
Northern white	2.2	4.9	7.2
Port-Orford	4.6	6.9	10.1
Western redcedar	2.4	5.0	6.8
Douglas-fir,			
Coast[b]	4.8	7.6	12.4
Interior north[b]	3.8	6.9	10.7
Interior west[b]	4.8	7.5	11.8
Fir			
Balsam	2.9	6.9	11.2
California red	4.5	7.9	11.4
Grand	3.4	7.5	11.0
Noble	4.3	8.3	12.4
Pacific silver	4.4	9.2	13.0
Subalpine	2.6	7.4	9.4
White	3.3	7.0	9.8
Hemlock			
Eastern	3.0	6.8	9.7
Mountain	4.4	7.1	11.1
Western	4.2	7.8	12.4
Larch, western	4.5	9.1	14
Pine			
Eastern white	2.1	6.1	8.2
Jack	3.7	6.6	10.3
Loblolly	4.8	7.4	12.3
Lodgepole	4.3	6.7	11.1
Longleaf	5.1	7.5	12.2
Pitch	4.0	7.1	10.9
Pond	5.1	7.1	11.2
Ponderosa	3.9	6.2	9.7
Red	3.8	7.2	11.3
Shortleaf	4.6	7.7	12.3
Slash	5.4	7.6	12.1
Sugar	2.9	5.6	7.9
Virginia	4.2	7.2	11.9
Western white	4.1	7.4	11.8
Redwood			
Old growth	2.6	4.4	6.8
Young growth	2.2	4.9	7.0
Spruce			
Black	4.1	6.8	11.3
Engelmann	3.8	7.1	11.0
Red	3.8	7.8	11.8
Sitka	4.3	7.5	11.5
Tamarack	3.7	7.4	13.6

[a]Expressed as a percentage of the green dimension

(continued)

Table 2.35 (continued)

Common name	Shrinkage[a] (%) from green to absolutely dry		
	Radial	Tangential	Volumetric
Hardwoods			
Bigleaf	3.7	7.1	11.6
Black	4.8	9.3	14.0
Red	4.0	8.2	12.6
Silver	3.0	7.2	12.0
Striped	3.2	8.6	12.3
Sugar	4.8	9.9	14.7
Oak, red			
Black	4.4	11.1	15.1
Laurel	4.0	9.9	19
Northern red	4.0	8.6	13.7
Pin	4.3	9.5	14.5
Scarlet	4.4	10.8	14.7
Southern red	4.7	11.3	16.1
Water	4.4	9.8	16.1
Willow	5.0	9.6	18.9
Oak, white			
Bur	4.4	8.8	12.7
Chestnut	5.3	10.8	16.4
Live	6.6	9.5	14.7
Overcup	5.3	12.7	16
Post	5.4	9.8	16.2
Swamp, chestnut	5.2	10.8	16.4
White	5.6	10.5	16.3
Persimmon, common	7.9	11.2	19.1
Sassafras	4.0	6.2	10.3
Sweetgum	5.3	10.2	15.8
Sycamore, American	5.0	8.4	14.1
Tanoak	4.9	11.7	17.3
Tupelo			
Black	5.1	8.7	14.4
Water	4.2	7.6	12.5
Walnut, black	5.5	7.8	12.8
Willow, black	3.3	8.7	13.9
Yellow-poplar	4.6	8.2	12.7

Common name	Shrinkage[a] (%) from green to absolutely dry		
	Radial	Tangential	Volumetric
Softwoods			

[b]Coast type Douglas-fir is defined as Douglas-fir growing in the States of Oregon and Washington West of the summit of the Cascade Mountains. Interior West includes the State of California and all counties in Oregon and Washington East of but adjacent to the Cascade summit. Interior North includes the remainder of Oregon and Washington and the States of Idaho, Montana, and Wyoming

Table 2.36 Shrinkage values of some woods imported into the USA[a] (Forest Products Laboratory 2021)

Species	Scientific name	Shrinkage[b] from green to ovendry moisture content (%)			Location[c]
Common name		Radial	Tangential	Volumetric	
Afrormosia	*Pericopsis elata*	3.0	6.4	10.7	AF
Albarco	*Cariniana* spp.	2.8	5.4	9.0	AM
Andiroba	*Carapa guianensis*	3.1	7.6	10.4	AM
Angelin	*Andira inermis*	4.6	9.8	12.5	AM
Angelique	*Dicorynia guianensis*	5.2	8.8	14.0	AM
Apitong	*Dipterocarpus* spp.	5.2	10.9	16.1	AS
Avodire	*Turreanthus africanus*	4.6	6.7	12.0	AF
Azobe	*Lophira alata*	8.4	11.0	17	AM
Balata	*Manilkara bidentata*	6.3	9.4	16.9	AM
Balsa	*Ochroma pyramidale*	3.0	7.6	10.8	AM
Banak	*Virola* spp.	4.6	8.8	13.7	AM
Benge	*Guibourtia arnoldiana*	5.2	8.6	13.8	AF
Bubinga	*Guibourtia spp.*	5.8	8.4	14.2	AF
Bulletwood	*Manilkara bidentata*	6.3	9.4	16.9	AM
Caribbean pine	*Pinus caribaea*	6.3	7.8	12.9	AM
Cativo	*Prioria copaifera*	2.4	5.3	8.9	AM
Ceiba	*Ceiba pentandra*	2.1	4.1	10.4	AM
Cocobolo	*Dalbergia retusa*	2.7	4.3	7.0	AM
Courbaril	*Hymenaea courbaril*	4.5	8.5	12.7	AM
Cuangare	*Dialyanthera* spp.	4.2	9.4	12.0	AM
Degame	*Calycophyllum candidissimum*	4.8	8.6	13.2	AM
Determa	Ocotea rubra	3.7	7.6	10.4	AM
Ebony	*Diospyros* spp.	5.4	8.8	14.2	AS
Ebony	*Diospyros* spp.	9.2	10.8	20.0	AF
Ekop	*Tetraberlinia tubmaniana*	5.6	10.2	15.8	AF
Gmelina	*Gmelina arborea*	2.4	4.9	8.8	AS
Goncalo alves	*Astronium graveolens*	4.0	7.6	10.0	AM
Greenheart	*Ocotea rodiaei*	8.8	9.6	17.1	AM
Hura	*Hura crepitans*	2.7	4.5	7.3	AM
Ilomba	*Pycnanthus angolensis*	4.6	8.4	12.8	AF
Imbuia	*Phoebe porosa*	2.7	6.0	9.0	AM
Ipe	*Tabebuia* spp.	6.6	8.0	13.2	AM
Iroko	*Chlorophora excelsa, Chlorophora regia*	2.8	3.8	8.8	AF
Jarrah	*Eucalyptus marginata*	7.7	11.0	18.7	AS
Jelutong	*Dyera costulata*	2.3	5.5	7.8	AS
Kaneelhart	*Licaria* spp.	5.4	7.9	12.5	AM
Kapur	*Dryobalanops* spp.	4.6	10.2	14.8	AS
Karri	*Eucalyptus diversicolor*	7.8	12.4	20.2	AS
Kempas	*Koompassia malaccensis*	6.0	7.4	14.5	AS
Keruing	*Dipterocarpus* spp.	5.2	10.9	16.1	AS
Lauan, light red and red	*Shorea* spp.	4.6	8.5	14.3	AS
Lauan, dark red	*Shorea* spp.	3.8	7.9	13.1	AS
Lauan, white	*Pentacme contorta*	4.0	7.7	11.7	AS
Limba	*Terminalia superba*	4.5	6.2	10.8	AF
Macawood	*Platymiscium* spp.	2.7	3.5	6.5	AM
Mahogany, African	*Khaya* spp.	2.5	4.5	8.8	AF
Mahogany, true	*Swietenia macrophylla*	3.0	4.1	7.8	AM
Manbarklak	*Eschweilera* spp.	5.8	10.3	15.9	AM
Manni	*Symphonia globulifera*	5.7	9.7	15.6	AM
Marishballi	*Licania* spp.	7.5	11.7	17.2	AM
Meranti, white	*Shorea* spp.	3.0	6.6	7.7	AS

(continued)

Table 2.36 (continued)

Species	Scientific name	Shrinkage[b] from green to ovendry moisture content (%)			Location[c]
Common name		Radial	Tangential	Volumetric	
Meranti, yellow	*Shorea* spp.	3.4	8.0	10.4	AS
Merbau	*Intsia bijuga, Intsia palembanica*	2.7	4.6	7.8	AS
Mersawa	*Anisoptera* spp.	4.0	9.0	14.6	AS
Mora	*Mora* spp.	6.9	9.8	18.8	AM
Obeche	*Triplochiton scleroxylon*	3.0	5.4	9.2	AF
Ocota pine	*Pinus oocarpa*	4.6	7.5	12.3	AM
Okoume	*Aucoumea klaineana*	4.1	6.1	11.3	AF
Opepe	*Nauclea* spp.	4.5	8.4	12.6	AF
Ovangkol	*Guibourta ehie*	4.5	8.2	12.0	AF
Para-angelium	*Hymenolobium excelsum*	4.4	7.1	10.2	AM
Parana pine	*Araucaria angustifolia*	4.0	7.9	11.6	AS
Pau Marfim	*Balfourodendron riedelianum*	4.6	8.8	13.4	AM
Peroba de campos	*Paratecoma peroba*	3.8	6.6	10.5	AM
Peroba Rosa	*Aspidosperma* spp.	3.8	6.4	11.6	AM
Piquia	*Caryocar* spp.	5.0	8.0	13.0	AM
Pilon	*Hyeronima* spp.	5.4	11.7	17.0	AM
Primavera	*Cybistax donnell-smithii*	3.1	5.1	9.1	AM
Purpleheart	*Peltogyne* spp.	3.2	6.1	9.9	AM
Ramin	*Gonystylus* spp.	4.3	8.7	13.4	AS
Roble	*Quercus* spp.	6.4	11.7	18.5	AM
Roble	*Tabebuia* spp. (Roble group)	3.6	6.1	9.5	AM
Rosewood, Brazilian	*Dalbergia nigra*	2.9	4.6	8.5	AM
Rosewood, Indian	*Dalbergia latifolia*	2.7	5.8	8.5	AS
Rubberwood	*Hevea brasiliensis*	2.3	5.1	7.4	AM
Sande	*Brosimum spp. Utile group*	4.6	8.0	13.6	AM
Sapele	*Entandrophragma cylindricum*	4.6	7.4	14.0	AF
Sepetir	*Pseudosindora* spp., *Sindora* spp.	3.7	7.0	10.5	AS
Spanish cedar	*Cedrela* spp.	4.2	6.3	10.3	AM
Sucupira	*Diplotropis purpurea*	4.6	7.0	11.8	AM
Teak	*Tectona grandis*	2.5	5.8	7.0	AS
Wallaba	*Eperua* spp.	3.6	6.9	10.0	AM

[a]Shrinkage values taken from world literature and may not represent a true species average.
[b]Expressed as a percentage of the green dimension.
[c]AF is Africa; AM is Tropical America; AS is Asia and Oceania.

Table 2.37 Thermal conductivity of selected hardwoods and softwoods from USA[a] (Forest Products Laboratory 2021). Standard deviation in brackets, MC – moisture content

Common name	Specific Gravity	Conductivity W m^{-1} K^{-1} (BTU in. h^{-1} ft^{-2} °F$-^{1}$)		Resistivity K m W^{-1} (h ft^{2} °F BTU^{-1} in.$^{-1}$)	
		Ovendry	12% MC	Ovendry	12% MC
Hardwoods					
Ash					
Black	0.53	0.12 (0.84)	0.15 (1.00)	8.2 (1.2)	6.8 (1.0)
White	0.63	0.14 (0.98)	0.17 (1.20)	7.1 (1.0)	5.8 (0.8)
Aspen					
Big tooth	0.41	0.10 (0.68)	0.12 (0.82)	10.0 (1.5)	8.5 (1.2)
Quaking	0.40	0.10 (0.67)	0.12 (0.80)	10.0 (1.5)	8.6 (1.2)
Basswood, American	0.38	0.09 (0.64)	0.11 (0.77)	11.0 (1.6)	9.0 (1.3)
Beech, American	0.68	0.15 (1.00)	0.18 (1.30)	6.6 (1.0)	5.4 (0.8)
Birch					
Sweet	0.71	0.16 (1.10)	0.19 (1.30)	6.4 (0.9)	5.2 (0.8)
Yellow	0.66	0.15 (1.00)	0.18 (1.20)	6.8 (1.0)	5.6 (0.8)
Cherry, black	0.53	0.12 (0.84)	0.15 (1.00)	8.2 (1.2)	6.8 (1.0)
Chestnut, American	0.45	0.11 (0.73)	0.13 (0.89)	9.4 (1.4)	7.8 (1.1)
Cottonwood					
Black	0.35	0.09 (0.60)	0.10 (0.72)	12.0 (1.7)	9.6 (1.4)
Eastern	0.43	0.10 (0.71)	0.12 (0.85)	9.8 (1.4)	8.1 (1.2)
Elm					
American	0.54	0.12 (0.86)	0.15 (1.00)	8.1 (1.2)	6.7 (1.0)
Rock	0.67	0.15 (1.00)	0.18 (1.30)	6.7 (1.0)	5.5 (0.8)
Slippery	0.56	0.13 (0.88)	0.15 (1.10)	7.9 (1.1)	6.5 (0.9)
Hackberry	0.57	0.13 (0.90)	0.16 (1.10)	7.7 (1.1)	6.4 (0.9)
Hickory, pecan	0.69	0.15 (1.10)	0.19 (1.30)	6.6 (1.0)	5.4 (0.8)
Hickory, true					
Mockernut	0.78	0.17 (1.20)	0.21 (1.40)	5.9 (0.9)	4.8 (0.7)
Shagbark	0.77	0.17 (1.20)	0.21 (1.40)	5.9 (0.9)	4.9 (0.7)
Magnolia, southern	0.52	0.12 (0.83)	0.14 (1.00)	8.4 (1.2)	6.9 (1.0)
Maple					
Black	0.60	0.14 (0.94)	0.16 (1.10)	7.4 (1.1)	6.1 (0.9)
Red	0.56	0.13 (0.88)	0.15 (1.10)	7.9 (1.1)	6.5 (0.9)
Silver	0.50	0.12 (0.80)	0.14 (0.97)	8.6 (1.2)	7.1 (1.0)
Sugar	0.66	0.15 (1.00)	0.18 (1.20)	6.8 (1.0)	5.6 (0.8)
Oak, red					
Black	0.66	0.15 (1.00)	0.18 (1.20)	6.8 (1.0)	5.6 (0.8)
Northern red	0.65	0.14 (1.00)	0.18 (1.20)	6.9 (1.0)	5.7 (0.8)
Southern red	0.62	0.14 (0.96)	0.17 (1.20)	7.2 (1.0)	5.9 (0.9)
Oak, white					
Bur	0.66	0.15 (1.00)	0.18 (1.20)	6.8 (1.0)	5.6 (0.8)
White	0.72	0.16 (1.10)	0.19 (1.30)	6.3 (0.9)	5.2 (0.8)
Sweetgum	0.55	0.13 (0.87)	0.15 (1.10)	8.0 (1.2)	6.6 (1.0)
Sycamore, American	0.54	0.12 (0.86)	0.15 (1.00)	8.1 (1.2)	6.7 (1.0)
Tupelo					
Black	0.54	0.12 (0.86)	0.15 (1.00)	8.1 (1.2)	6.7 (1.0)
Water	0.53	0.12 (0.84)	0.15 (1.00)	8.2 (1.2)	6.8 (1.0)
Yellow-poplar	0.46	0.11 (0.75)	0.13 (0.90)	9.3 (1.3)	7.7 (1.1)
Softwoods					
Bald cypress	0.47	0.11 (0.76)	0.13 (0.92)	9.1 (1.3)	7.5 (1.1)
Cedar					
Atlantic white	0.34	0.09 (0.59)	0.10 (0.70)	12.0 (1.7)	9.9 (1.4)

(continued)

Table 2.37 (continued)

Common name	Specific Gravity	Conductivity W m^{-1} K^{-1} (BTU in. h^{-1} ft^{-2} °F^{-1})		Resistivity K m W^{-1} (h ft^2 °F BTU^{-1} in.$^{-1}$)	
		Ovendry	12% MC	Ovendry	12% MC
Eastern red	0.48	0.11 (0.77)	0.14 (0.94)	8.9 (1.3)	7.4 (1.1)
Northern white	0.31	0.08 (0.55)	0.09 (0.65)	13.0 (1.8)	11.0 (1.5)
Port-Orford	0.43	0.10 (0.71)	0.12 (0.85)	9.8 (1.4)	8.1 (1.2)
Western red	0.33	0.08 (0.57)	0.10 (0.68)	12.0 (1.7)	10.0 (1.5)
Yellow	0.46	0.11 (0.75)	0.13 (0.90)	9.3 (1.3)	7.7 (1.1)
Douglas-fir					
Coast	0.51	0.12 (0.82)	0.14 (0.99)	8.5 (1.2)	7.0 (1.0)
Interior north	0.50	0.12 (0.80)	0.14 (0.97)	8.6 (1.2)	7.1 (1.0)
Interior west	0.52	0.12 (0.83)	0.14 (1.00)	8.4 (1.2)	6.9 (1.0)
Fir					
Balsam	0.37	0.09 (0.63)	0.11 (0.75)	11.0 (1.6)	9.2 (1.3)
White	0.41	0.10 (0.68)	0.12 (0.82)	10.0 (1.5)	8.5 (1.2)
Hemlock					
Eastern	0.42	0.10 (0.69)	0.12 (0.84)	10.0 (1.4)	8.3 (1.2)
Western	0.48	0.11 (0.77)	0.14 (0.94)	8.9 (1.3)	7.4 (1.1)
Larch, western	0.56	0.13 (0.88)	0.15 (1.10)	7.9 (1.1)	6.5 (0.9)
Pine					
Eastern white	0.37	0.09 (0.63)	0.11 (0.75)	11.0 (1.6)	9.2 (1.3)
Jack	0.45	0.11 (0.73)	0.13 (0.89)	9.4 (1.4)	7.8 (1.1)
Loblolly	0.54	0.12 (0.86)	0.15 (1.00)	8.1 (1.2)	6.7 (1.0)
Lodgepole	0.43	0.10 (0.71)	0.12 (0.85)	9.8 (1.4)	8.1 (1.2)
Longleaf	0.62	0.14 (0.96)	0.17 (1.20)	7.2 (1.0)	5.9 (0.9)
Pitch	0.53	0.12 (0.84)	0.15 (1.00)	8.2 (1.2)	6.8 (1.0)
Ponderosa	0.42	0.10 (0.69)	0.12 (0.84)	10.0 (1.4)	8.3 (1.2)
Red	0.46	0.11 (0.75)	0.13 (0.90)	9.3 (1.3)	7.7 (1.1)
Shortleaf	0.54	0.12 (0.86)	0.15 (1.00)	8.1 (1.2)	6.7 (1.0)
Slash	0.61	0.14 (0.95)	0.17 (1.20)	7.3 (1.1)	6.0 (0.9)
Sugar	0.37	0.09 (0.63)	0.11 (0.75)	11.0 (1.6)	9.2 (1.3)
Western white	0.40	0.10 (0.67)	0.12 (0.80)	10.0 (1.5)	8.6 (1.2)
Redwood					
Old growth	0.41	0.10 (0.68)	0.12 (0.82)	10.0 (1.5)	8.5 (1.2)
Young growth	0.37	0.09 (0.63)	0.11 (0.75)	11.0 (1.6)	9.2 (1.3)
Spruce					
Black	0.43	0.10 (0.71)	0.12 (0.85)	9.8 (1.4)	8.1 (1.2)
Engelmann	0.37	0.09 (0.63)	0.11 (0.75)	11.0 (1.6)	9.2 (1.3)
Red	0.42	0.10 (0.69)	0.12 (0.84)	10.0 (1.4)	8.3 (1.2)
Sitka	0.42	0.10 (0.69)	0.12 (0.84)	10.0 (1.4)	8.3 (1.2)
White	0.37	0.09 (0.63)	0.11 (0.75)	11.0 (1.6)	9.2 (1.3)

[a]Values in this table are approximate and should be used with caution; actual conductivities may vary by as much as 20%. The specific gravities do not also represent species averages.

Table 2.38 Poisson's ratios for various species at approximately 12% moisture content (Forest Products Laboratory 2021)

Common name	μ_{LR}	μ_{LT}	μ_{RT}	μ_{TR}	μ_{RL}	μ_{TL}
Hardwoods						
Ash, white	0.371	0.440	0.684	0.360	0.059	0.051
Aspen, quaking	0.489	0.374		0.496	0.054	0.022
Balsa	0.229	0.488	0.665	0.231	0.018	0.009
Basswood	0.364	0.406	0.912	0.346	0.034	0.022
Birch, yellow	0.426	0.451	0.697	0.426	0.043	0.024
Cherry, black	0.392	0.428	0.695	0.282	0.086	0.048
Cottonwood, eastern	0.344	0.420	0.875	0.292	0.043	0.018
Mahogany, African	0.297	0.641	0.604	0.264	0.033	0.032
Mahogany, Honduras	0.314	0.533	0.600	0.326	0.033	0.034
Maple, sugar	0.424	0.476	0.774	0.349	0.065	0.037
Maple, red	0.434	0.509	0.762	0.354	0.063	0.044
Oak, red	0.350	0.448	0.560	0.292	0.064	0.033
Oak, white	0.369	0.428	0.618	0.300	0.074	0.036
Sweetgum	0.325	0.403	0.682	0.309	0.044	0.023
Walnut, black	0.495	0.632	0.718	0.367	0.052	0.036
Yellow-poplar	0.318	0.392	0.703	0.329	0.030	0.019
Softwoods						
Baldcypress	0.338	0.326	0.411	0.356		
Cedar, northern white	0.337	0.340	0.458	0.345		
Cedar, western red	0.378	0.296	0.484	0.403		
Douglas-fir	0.292	0.449	0.390	0.374	0.036	0.029
Fir, subalpine	0.341	0.332	0.437	0.336		
Hemlock, western	0.485	0.423	0.442	0.382		
Larch, western	0.355	0.276	0.389	0.352		
Pine						
Loblolly	0.328	0.292	0.382	0.362		
Lodgepole	0.316	0.347	0.469	0.381		
Longleaf	0.332	0.365	0.384	0.342		
Pond	0.280	0.364	0.389	0.320		
Ponderosa	0.337	0.400	0.426	0.359		
Red	0.347	0.315	0.408	0.308		
Slash	0.392	0.444	0.447	0.387		
Sugar	0.356	0.349	0.428	0.358		
Western white	0.329	0.344	0.410	0.334		
Redwood	0.360	0.346	0.373	0.400		
Spruce, sitka	0.372	0.467	0.435	0.245	0.040	0.025
Spruce, Engelmann	0.422	0.462	0.530	0.255	0.083	0.058

Table 2.39 Elastic ratios for various species at approximately 12% moisture content[a] (Forest Products Laboratory 2021)

Common name	E_T/E_L	E_R/E_L	G_{LR}/E_L	G_{LT}/E_L	G_{RT}/E_L
Hardwoods					
Ash, white	0.080	0.125	0.109	0.077	
Balsa	0.015	0.046	0.054	0.037	0.005
Basswood	0.027	0.066	0.056	0.046	
Birch, yellow	0.050	0.078	0.074	0.068	0.017
Cherry, black	0.086	0.197	0.147	0.097	
Cottonwood, eastern	0.047	0.083	0.076	0.052	
Mahogany, African	0.050	0.111	0.088	0.059	0.021
Mahogany, Honduras	0.064	0.107	0.066	0.086	0.028
Maple, sugar	0.065	0.132	0.111	0.063	
Maple, red	0.067	0.140	0.133	0.074	
Oak, red	0.082	0.154	0.089	0.081	
Oak, white	0.072	0.163	0.086		
Sweetgum	0.050	0.115	0.089	0.061	0.021
Walnut, black	0.056	0.106	0.085	0.062	0.021
Yellow-poplar	0.043	0.092	0.075	0.069	0.011
Softwoods					
Baldcypress	0.039	0.084	0.063	0.054	0.007
Cedar, northern white	0.081	0.183	0.210	0.187	0.015
Cedar, western red	0.055	0.081	0.087	0.086	0.005
Douglas-fir	0.050	0.068	0.064	0.078	0.007
Fir, subalpine	0.039	0.102	0.070	0.058	0.006
Hemlock, western	0.031	0.058	0.038	0.032	0.003
Larch, western	0.065	0.079	0.063	0.069	0.007
Pine					
Loblolly	0.078	0.113	0.082	0.081	0.013
Lodgepole	0.068	0.102	0.049	0.046	0.005
Longleaf	0.055	0.102	0.071	0.060	0.012
Pond	0.041	0.071	0.050	0.045	0.009
Ponderosa	0.083	0.122	0.138	0.115	0.017
Red	0.044	0.088	0.096	0.081	0.011
Slash	0.045	0.074	0.055	0.053	0.010
Sugar	0.087	0.131	0.124	0.113	0.019
Western white	0.038	0.078	0.052	0.048	0.005
Redwood	0.089	0.087	0.066	0.077	0.011
Spruce, Sitka	0.043	0.078	0.064	0.061	0.003
Spruce, Engelmann	0.059	0.128	0.124	0.120	0.010

[a]E_L may be approximated by increasing the modulus of elasticity values in Table 2.34 by 10%.

Table 2.40 Strength properties of some commercially important woods grown in the USA[a] (Forest Products Laboratory 2021)

Common names	Moisture content	Specific gravity[(b)]	Static bending			Impact bending	Compression parallel to grain	Compression perpendicular to grain	Shear parallel to grain	Tension perpen-dicular to grain	Side hardness[(e)]
			Modulus of rupture	Modulus of elasticity[(c)]	Work to max. load						
Hardwoods	(%)		(kPa)	(MPa)	(kJ/m³)	(mm)	(kPa)	(kPa)	(kPa)	(kPa)	(N)
Alder, red	Green	0.37	45 000	8 100	55	560	20 400	1 700	5 300	2 700	2 000
	12	0.41	68 000	9 500	58	510	40 100	3 000	7 400	2 900	2 600
Ash											
Black	Green	0.45	41 000	7 200	83	840	15 900	2 400	5 900	3 400	2 300
	12	0.49	87 000	11 000	103	8 90	41 200	5 200	10 800	4 800	3 800
Blue	Green	0.53	66 000	8 500	101		28 800	5 600	10 600		
	12	0.58	95 000	9 700	99		48 100	9 800	14 000		
Green	Green	0.53	66 000	9 700	81	890	29 000	5 000	8 700	4 100	3 900
	12	0.56	97 000	11 400	92	810	48 800	9 000	13 200	4 800	5 300
Oregon	Green	0.50	52 000	7 800	84	990	24 200	3 700	8 200	4 100	3 500
	12	0.55	88 000	9 400	99	840	41 600	8 600	12 300	5 000	5 200
White	Green	0.55	66 000	9 900	108	970	27 500	4 600	9 300	4 100	4 300
	12	0.60	103 000	12 000	115	1 090	51 100	8 000	13 200	6 500	5 900
Aspen											
Bigtooth	Green	0.36	37 000	7 700	39		17 200	1 400	5 000		
	12	0.39	63 000	9 900	53		36 500	3 100	7 400		
Quaking	Green	0.35	35 000	5 900	44	560	14 800	1 200	4 600	1 600	1 300
	12	0.38	58 000	8 100	52	530	29 300	2 600	5 900	1 800	1 600
Basswood, American	Green	0.32	34 000	7 200	37	410	15 300	1 200	4 100	1 900	1 100
	12	0.37	60 000	10 100	50	410	32 600	2 600	6 800	2 400	1 800
Beech, American	Green	0.56	59 000	9 500	82	1 090	24 500	3 700	8 900	5 000	3 800
	12	0.64	103 000	11 900	104	1 040	50 300	7 000	13 900	7 000	5 800
Birch											
Paper	Green	0.48	44 000	8 100	112	1 240	16 300	1 900	5 800	2 600	2 500
	12	0.55	85 000	11 000	110	860	39 200	4 100	8 300		4 000
Sweet	Green	0.60	65 000	11 400	108	1 220	25 800	3 200	8 500	3 000	4 300
	12	0.65	117 000	15 000	124	1 190	58 900	7 400	15 400	6 600	6 500
Yellow	Green	0.55	57 000	10 300	111	1 220	23 300	3 000	7 700	3 000	3 600
	12	0.62	114 000	13 900	143	1 400	56 300	6 700	13 000	6 300	5 600
Butternut	Green	0.36	37 000	6 700	57	610	16 700	1 500	5 200	3 000	1 700
	12	0.38	56 000	8 100	57	610	36 200	3 200	8 100	3 000	2 200
Cherry. black	Green	0.47	55 000	9 000	88	840	24 400	2 500	7 800	3 900	2 900
	12	0.50	85 000	10 300	79	740	49 000	4 800	11 700	3 900	4 200
Chestnut, American	Green	0.40	39 000	6 400	48	610	17 000	2 100	5 500	3 000	1 900
	12	0.43	59 000	8 500	45	480	36 700	4 300	7 400	3 200	2 400
Cottonwood											
Balsam poplar	Green	0.31	27 000	5 200	29		11 700	1 000	3 400		
	12	0.34	47 000	7 600	34		27 700	2 100	5 400		
Black	Green	0.31	34 000	7 400	34	510	15 200	1 100	4 200	1 900	1 100
	12	0.35	59 000	8 800	46	560	31 000	2 100	7 200	2 300	1 600
Eastern	Green	0.37	37 000	7 000	50	530	15 700	1 400	4 700	2 800	1 500
	12	0.40	59 000	9 400	51	510	33 900	2 600	6 400	4 000	1 900
Elm											
American	Green	0.46	50 000	7 700	81	970	20 100	2 500	6 900	4 100	2 800
	12	0.50	81 000	9 200	90	990	38 100	4 800	10 400	4 600	3 700
Rock	Green	0.57	66 000	8 200	137	1 370	26 100	4 200	8 800		
	12	0.63	102 000	10 600	132	1 420	48 600	8 500	13 200		
Slippery	Green	0.48	55 000	8 500	106	1 190	22 900	2 900	7 700	4 400	2 900
	12	0.53	90 000	10 300	117	1 140	43 900	5 700	11 200	3 700	3 800

(continued)

Table 2.40 (continued)

Common names	Moisture content	Specific gravity[(b)]	Static bending			Impact bending	Compression parallel to grain	Compression perpendicular to grain	Shear parallel to grain	Tension perpen-dicular to grain	Side hardness[(e)]
			Modulus of rupture	Modulus of elasticity[(c)]	Work to max. load						
Hardwoods	(%)		(kPa)	(MPa)	(kJ/m^3)	(mm)	(kPa)	(kPa)	(kPa)	(kPa)	(N)
Hackberry	Green	0.49	45 000	6 600	100	1 220	18 300	2 800	7 400	4 300	3 100
	12	0.53	76 000	8 200	88	1 090	37 500	6 100	11 000	4 000	3 900
Hickory, pecan											
Bitternut	Green	0.6	71 000	9 700	138	1 680	31 500	5 500	8 500		
	12	0.66	118 000	12 300	125	1 680	62 300	11 600			
Nutmeg	Green	0.56	63 000	8 900	157	1 370	27 400	5 200	7 100		
	12	0.60	114 000	11 700	173	—	47 600	10 800			
Pecan	Green	0.60	68 000	9 400	101	1 350	27 500	5 400	10 200	4 700	5 800
	12	0.66	94 000	11 900	95	1 120	54 100	11 900	14 300		8 100
Water	Green	0.61	74 000	10 800	130	1 420	32 100	6 100	9 900		
	12	0.62	123 000	13 900	133	1 350	59 300	10 700			
Hickory, true[(d)]											
Mockernut	Green	0.64	77 000	10 800	180	2 240	30 900	5 600	8 800		6 400
	12	0.72	132 000	15 300	156	1 960	61 600	11 900	12 000		8 800
Pignut	Green	0.66	81 000	11 400	219	2 260	33 200	6 300	9 400		6 800
	12	0.75	139 000	15 600	210	1 880	63 400	13 700	14 800		9 500
Shagbark	Green	0.64	76 000	10 800	163	1 880	31 600	5 800	10 500		6 500
	12	0.72	139 000	14 900	178	1 700	63 500	12 100	16 800		8 400
Shellbark	Green	0.62	72 000	9 200	206	2 640	27 000	5 600	8 200		7 400
	12	0.69	125 000	13 000	163	2 240	55 200	12 400	14 500		8 100
Honeylocust	Green	0.60	70 000	8 900	87	1 190	30 500	7 900	11 400	6 400	6 200
	12		101 000	11 200	92	1 190	51 700	12 700	15 500	6 200	7 000
Locust, black	Green	0.66	95 000	12 800	106	1 120	46 900	8 000	12 100	5 300	7 000
	12	0.69	134 000	14 100	127	1 450	70 200	12 600	17 100	4 400	7 600
Magnolia											
Cucumbertree	Green	0.44	51 000	10 800	69	760	21 600	2 300	6 800	3 000	2 300
	12	0.48	85 000	12 500	84	890	43 500	3 900	9 200	4 600	3 100
Southern	Green	0.46	47 000	7 700	106	1 370	18 600	3 200	7 200	4 200	3 300
	12	0.50	77 000	9 700	88	740	37 600	5 900	10 500	5 100	4 500
Maple											
Bigleaf	Green	0.44	51 000	7 600	60	580	22 300	3 100	7 700	4 100	2 800
	12	0.48	74 000	10 000	54	710	41 000	5 200	11 900	3 700	3 800
Black	Green	0.52	54 000	9 200	88	1 220	22 500	4 100	7 800	5 000	3 700
	12	0.57	92 000	11 200	86	1 020	46 100	7 000	12 500	4 600	5 200
Red	Green	0.49	53 000	9 600	79	810	22 600	2 800	7 900		3 100
	12	0.54	92 000	11 300	86	810	45 100	6 900	12 800		4 200
Silver	Green	0.44	40 000	6 500	76	740	17 200	2 600	7 200	3 900	2 600
	12	0.47	61 000	7 900	57	640	36 000	5 100	10 200	3 400	3 100
Sugar	Green	0.56	65 000	10 700	92	1 020	27 700	4 400	10 100		4 300
	12	0.63	109 000	12 600	114	990	54 000	10 100	16 100		6 400
Oak, red											
Black	Green	0.56	57 000	8 100	84	1 020	23 900	4 900	8 400		4 700
	12	0.61	96 000	11 300	94	1 040	45 000	6 400	13 200		5 400
Cherrybark	Green	0.61	74 000	12 300	101	1 370	31 900	5 200	9 100	5 500	5 500
	12	0.68	125 000	15 700	126	1 240	60 300	8 600	13 800	5 800	6 600
Laurel	Green	0.56	54 000	9 600	77	990	21 900	3 900	8 100	5 300	4 400
	12	0.63	87 000	11 700	81	990	48 100	7 300	12 600	5 400	5 400

(continued)

Table 2.40 (continued)

Common names	Moisture content	Specific gravity[b]	Static bending			Impact bending	Compression parallel to grain	Compression perpendicular to grain	Shear parallel to grain	Tension perpen-dicular to grain	Side hardness[e]
			Modulus of rupture	Modulus of elasticity[c]	Work to max. load						
Hardwoods	(%)		(kPa)	(MPa)	(kJ/m³)	(mm)	(kPa)	(kPa)	(kPa)	(kPa)	(N)
Northern red	Green	0.56	57 000	9 300	91	1 120	23 700	4 200	8 300	5 200	4 400
	12	0.63	99 000	12 500	100	1 090	46 600	7 000	12 300	5 500	5 700
Pin	Green	0.58	57 000	9 100	97	1 220	25 400	5 000	8 900	5 500	4 800
	12	0.63	97 000	11 900	102	1 140	47 000	7 000	14 300	7 200	6 700
Scarlet	Green	0.60	72 000	10 200	103	1 370	28 200	5 700	9 700	4 800	5 300
	12	0.67	120 000	13 200	141	1 350	57 400	7 700	13 000	6 000	6 200
Southern red	Green	0.52	48 000	7 900	55	740	20 900	3 800	6 400	3 300	3 800
	12	0.59	75 000	10 300	65	660	42 000	6 000	9 600	3 500	4 700
Water	Green	0.56	61 000	10 700	77	990	25 800	4 300	8 500	5 700	4 500
	12	0.63	106 000	13 900	148	1 120	46 700	7 000	13 900	6 300	5 300
Willow	Green	0.56	51 000	8 900	61	890	20 700	4 200	8 100	5 200	4 400
	12	0.69	100 000	13 100	101	1 070	48 500	7 800	11 400		6 500
Oak, white											
Bur	Green	0.58	50 000	6 100	74	1 120	22 700	4 700	9 300	5 500	4 900
	12	0.64	71 000	7 100	68	740	41 800	8 300	12 500	4 700	6 100
Chestnut	Green	0.57	55 000	9 400	65	890	24 300	3 700	8 300	4 800	4 000
	12	0.66	92 000	11 000	76	1 020	47 100	5 800	10 300		5 000
Live	Green	0.80	82 000	10 900	85		37 400	14 100	15 200		
	12	0.88	127 000	13 700	130		61 400	19 600	18 300		
Overcup	Green	0.57	55 000	7 900	87	1 120	23 200	3 700	9 100	5 000	4 300
	12	0.63	87 000	9 800	108	970	42 700	5 600	13 800	6 500	5 300
Post	Green	0.60	56 000	7 500	76	1 120	24 000	5 900	8 800	5 400	5 000
	12	0.67	91 000	10 400	91	1 170	45 300	9 900	12 700	5 400	6 000
Swamp chestnut	Green	0.60	59 000	9 300	88	1 140	24 400	3 900	8 700	4 600	4 900
	12	0.67	96 000	12 200	83	1 040	50 100	7 700	13 700	4 800	5 500
Swamp white	Green	0.64	68 000	11 000	100	1 270	30 100	5 200	9 000	5 900	5 200
	12	0.72	122 000	14 100	132	1 240	59 300	8 200	13 800	5 700	7 200
White	Green	0.60	57 000	8 600	80	1 070	24 500	4 600	8 600	5 300	4 700
	12	0.68	105 000	12 300	102	940	51 300	7 400	13 800	5 500	6 000
Sassafras	Green	0.42	41 000	6 300	49		18 800	2 600	6 600		
	12	0.46	62 000	7 700	60		32 800	5 900	8 500		
Sweetgum	Green	0.46	49 000	8 300	70	910	21 000	2 600	6 800	3 700	2 700
	12	0.52	86 000	11 300	82	810	43 600	4 300	11 000	5 200	3 800
Sycamore, American	Green	0.46	45 000	7 300	52	660	20 100	2 500	6 900	4 300	2 700
	12	0.49	69 000	9 800	59	660	37 100	4 800	10 100	5 000	3 400
Tanoak	Green	0.58	72 000	10 700	92		32 100				
	12										
Tupelo											
Black	Green	0.46	48 000	7 100	55	760	21 000	3 300	7 600	3 900	2 800
	12	0.50	66 000	8 300	43	560	38 100	6 400	9 200	3 400	3 600
Water	Green	0.46	50 000	7 200	57	760	23 200	3 300	8 200	4 100	3 200
	12	0.50	66 000	8 700	48	580	40 800	6 000	11 000	4 800	3 900
Walnut, black	Green	0.51	66 000	9 800	101	940	29 600	3 400	8 400	3 900	4 000
	12	0.55	101 000	11 600	74	860	52 300	7 000	9 400	4 800	4 500
Willow, black	Green	0.36	33 000	5 400	76		14 100	1 200	4 700		
	12	0.39	54 000	7 000	61		28 300	3 000	8 600		
Yellow-poplar	Green	0.40	41 000	8 400	52	660	18 300	1 900	5 400	3 500	2 000
	12	0.42	70 000	10 900	61	610	38 200	3 400	8 200	3 700	2 400

(continued)

Table 2.40 (continued)

Common names	Moisture content	Specific gravity[(b)]	Static bending			Impact bending	Compression parallel to grain	Compression perpendicular to grain	Shear parallel to grain	Tension perpen-dicular to grain	Side hardness[(e)]
			Modulus of rupture	Modulus of elasticity[(c)]	Work to max. load						
Hardwoods	(%)		(kPa)	(MPa)	(kJ/m³)	(mm)	(kPa)	(kPa)	(kPa)	(kPa)	(N)
Softwoods											
Baldcypress	Green	0.42	46 000	8 100	46	640	24 700	2 800	5 600	2 100	1 700
	12	0.46	73 000	9 900	57	610	43 900	5 000	6 900	1 900	2 300
Cedar											
Atlantic white	Green	0.31	32 000	5 200	41	460	16 500	1 700	4 800	1 200	1 300
	12	0.32	47 000	6 400	28	330	32 400	2 800	5 500	1 500	1 600
Eastern redcedar	Green	0.44	48 000	4 500	103	890	24 600	4 800	7 000	2 300	2 900
	12	0.47	61 000	6 100	57	560	41 500	6 300	—	—	4 000
Incense	Green	0.35	43 000	5 800	44	430	21 700	2 600	5 700	1 900	1 700
	12	0.37	55 000	7 200	37	430	35 900	4 100	6 100	1 900	2 100
Northern white	Green	0.29	29 000	4 400	39	380	13 700	1 600	4 300	1 700	1 000
	12	0.31	45 000	5 500	33	300	27 300	2 100	5 900	1 700	1 400
Port-Orford	Green	0.39	45 000	9 000	51	530	21 600	2 100	5 800	1 200	1 700
	12	0.43	88 000	11 700	63	710	43 100	5 000	9 400	2 800	2 800
Western redcedar	Green	0.31	35 900	6 500	34	430	19 100	1 700	5 300	1 600	1 200
	12	0.32	51 700	7 700	40	430	31 400	3 200	6 800	1 500	1 600
Yellow	Green	0.42	44 000	7 900	63	690	21 000	2 400	5 800	2 300	2 000
	12	0.44	77 000	9 800	72	740	43 500	4 300	7 800	2 500	2 600
Douglas-fir[e]											
Coast	Green	0.45	53 000	10 800	52	660	26 100	2 600	6 200	2 100	2 200
	12	0.48	85 000	13 400	68	790	49 900	5 500	7 800	2 300	3 200
Interior West	Green	0.46	53 000	10 400	50	660	26 700	2 900	6 500	2 000	2 300
	12	0.50	87 000	12 600	73	810	51 200	5 200	8 900	2 400	2 900
Interior North	Green	0.45	51 000	9 700	56	560	23 900	2 500	6 600	2 300	1 900
	12	0.48	90 000	12 300	72	660	47 600	5 300	9 700	2 700	2 700
Interior South	Green	0.43	47 000	8 000	55	380	21 400	2 300	6 600	1 700	1 600
	12	0.46	82 000	10 300	62	510	43 000	5 100	10 400	2 300	2 300
Fir											
Balsam	Green	0.33	38 000	8 600	32	410	18 100	1 300	4 600	1 200	1 300
	12	0.35	63 000	10 000	35	510	36 400	2 800	6 500	1 200	1 700
California red	Green	0.36	40 000	8 100	44	530	19 000	2 300	5 300	2 600	1 600
	12	0.38	72 400	10 300	61	610	37 600	4 200	7 200	2 700	2 200
Grand	Green	0.35	40 000	8 600	39	560	20 300	1 900	5 100	1 700	1 600
	12	0.37	61 400	10 800	52	710	36 500	3 400	6 200	1 700	2 200
Noble	Green	0.37	43 000	9 500	41	480	20 800	1 900	5 500	1 600	1 300
	12	0.39	74 000	11 900	61	580	42 100	3 600	7 200	1 500	1 800
Pacific silver	Green	0.40	44 000	9 800	41	530	21 600	1 500	5 200	1 700	1 400
	12	0.43	75 800	12 100	64	610	44 200	3 100	8 400		1 900
Subalpine	Green	0.31	34 000	7 200			15 900	1 300	4 800		1 200
	12	0.32	59 000	8 900			33 500	2 700	7 400		1 600
White	Green	0.37	41 000	8 000	39	560	20 000	1 900	5 200	2 100	1 500
	12	0.39	68 000	10 300	50	510	40 000	3 700	7 600	2 100	2 100
Hemlock											
Eastern	Green	0.38	44 000	7 400	46	530	21 200	2 500	5 900	1 600	1 800
	12	0.40	61 000	8 300	47	530	37 300	4 500	7 300		2 200
Mountain	Green	0.42	43 000	7 200	76	810	19 900	2 600	6 400	2 300	2 100
	12	0.45	79 000	9 200	72	810	44 400	5 900	10 600		3 000

(continued)

Table 2.40 (continued)

Common names	Moisture content	Specific gravity[(b)]	Static bending			Impact bending	Compression parallel to grain	Compression perpendicular to grain	Shear parallel to grain	Tension perpen-dicular to grain	Side hardness[(e)]
			Modulus of rupture	Modulus of elasticity[(c)]	Work to max. load						
Hardwoods	(%)		(kPa)	(MPa)	(kJ/m³)	(mm)	(kPa)	(kPa)	(kPa)	(kPa)	(N)
Western	Green	0.42	46 000	9 000	48	560	23 200	1 900	5 900	2 000	1 800
	12	0.45	78 000	11 300	57	580	49 000	3 800	8 600	2 300	2 400
Larch, western	Green	0.48	53 000	10 100	71	740	25 900	2 800	6 000	2 300	2 300
	12	0.52	90 000	12 900	87	890	52 500	6 400	9 400	3 000	3 700
Pine											
Eastern white	Green	0.34	34 000	6 800	36	430	16 800	1 500	4 700	1 700	1 300
	12	0.35	59 000	8 500	47	460	33 100	3 000	6 200	2 100	1 700
Jack	Green	0.40	41 000	7 400	50	660	20 300	2 100	5 200	2 500	1 800
	12	0.43	68 000	9 300	57	690	39 000	4 000	8 100	2 900	2 500
Loblolly	Green	0.47	50 000	9 700	57	760	24 200	2 700	5 900	1 800	2 000
	12	0.51	88 000	12 300	72	760	49 200	5 400	9 600	3 200	3 100
Lodgepole	Green	0.38	38 000	7 400	39	510	18 000	1 700	4 700	1 500	1 500
	12	0.41	65 000	9 200	47	510	37 000	4 200	6 100	2 000	2 100
Longleaf	Green	0.54	59 000	11 000	61	890	29 800	3 300	7 200	2 300	2 600
	12	0.59	100 000	13 700	81	860	58 400	6 600	10 400	3 200	3 900
Pitch	Green	0.47	47 000	8 300	63		20 300	2 500	5 900		
	12	0.52	74 000	9 900	63		41 000	5 600	9 400		
Pond	Green	0.51	51 000	8 800	52		25 200	3 000	6 500		
	12	0.56	80 000	12 100	59		52 000	6 300	9 500		
Ponderosa	Green	0.38	35 000	6 900	36	530	16 900	1 900	4 800	2 100	1 400
	12	0.40	65 000	8 900	49	480	36 700	4 000	7 800	2 900	2 000
Red	Green	0.41	40 000	8 800	42	660	18 800	1 800	4 800	2 100	1 500
	12	0.46	76 000	11 200	68	660	41 900	4 100	8 400	3 200	2 500
Sand	Green	0.46	52 000	7 000	66		23 700	3 100	7 900		
	12	0.48	80 000	9 700	66		47 700	5 800			
Shortleaf	Green	0.47	51 000	9 600	57	760	24 300	2 400	6 300	2 200	2 000
	12	0.51	90 000	12 100	76	840	50 100	5 700	9 600	3 200	3 100
Slash	Green	0.54	60 000	10 500	66		26 300	3 700	6 600		
	12	0.59	112 000	13 700	91		56 100	7 000	11 600		
Spruce	Green	0.41	41 000	6 900			19 600	1 900	6 200		2 000
	12	0.44	72 000	8 500			39 000	5 000	10 300		2 900
Sugar	Green	0.34	34 000	7 100	37	430	17 000	1 400	5 000	1 900	1 200
	12	0.36	57 000	8 200	38	460	30 800	3 400	7 800	2 400	1 700
Virginia	Green	0.45	50 000	8 400	75	860	23 600	2 700	6 100	2 800	2 400
	12	0.48	90 000	10 500	94	810	46 300	6 300	9 300	2 600	3 300
Western white	Green	0.36	32 000	8 200	34	480	16 800	1 300	4 700	1 800	1 200
	12	0.35	67 000	10 100	61	580	34 700	3 200	7 200		1 900
Redwood											
Old-growth	Green	0.38	52 000	8 100	51	530	29 000	2 900	5 500	1 800	1 800
	12	0.40	69 000	9 200	48	480	42 400	4 800	6 500	1 700	2 100
Young-growth	Green	0.34	41 000	6 600	39	410	21 400	1 900	6 100	2 100	1 600
	12	0.35	54 000	7 600	36	380	36 000	3 600	7 600	1 700	1 900
Spruce											
Black	Green	0.38	42 000	9 500	51	610	19 600	1 700	5 100	700	1 500
	12	0.42	74 000	11 100	72	580	41 100	3 800	8 500		2 400
Engelmann	Green	0.33	32 000	7 100	35	410	15 000	1 400	4 400	1 700	1 150
	12	0.35	64 000	8 900	44	460	30 900	2 800	8 300	2 400	1 750

(continued)

Table 2.40 (continued)

Common names	Moisture content	Specific gravity[b]	Static bending			Impact bending	Compression parallel to grain	Compression perpendicular to grain	Shear parallel to grain	Tension perpen-dicular to grain	Side hardness[e]
			Modulus of rupture	Modulus of elasticity[c]	Work to max. load						
Hardwoods	(%)		(kPa)	(MPa)	(kJ/m³)	(mm)	(kPa)	(kPa)	(kPa)	(kPa)	(N)
Red	Green	0.37	41 000	9 200	48	460	18 800	1 800	5 200	1 500	1 600
	12	0.40	74 000	11 400	58	640	38 200	3 800	8 900	2 400	2 200
Sitka	Green	0.37	39 000	8 500	43	610	18 400	1 900	5 200	1 700	1 600
	12	0.40	70 000	10 800	65	640	38 700	4 000	7 900	2 600	2 300
White	Green	0.33	34 000	7 900	41	560	16 200	1 400	4 400	1 500	1 200
	12	0.36	65 000	9 600	53	510	35 700	3 000	6 700	2 500	1 800
Tamarack	Green	0.49	50 000	8 500	50	710	24 000	2 700	5 900	1 800	1 700
	12	0.53	80 000	11 300	49	580	49 400	5 500	8 800	2 800	2 600

[a]Results of tests on clear specimens in the green and air-dried conditions, converted to metric units directly from Table 5–3b in Forest Products Laboratory (2010). Definition of properties: impact bending is height of drop that causes complete failure, using a 22.7 kg (50 lb) hammer; compression parallel to grain is also called the maximum crushing strength; compression perpendicular to grain is fibre stress at proportional limit; shear is maximum shearing strength; tension is maximum tensile strength; and side hardness is hardness measured when the load is perpendicular to the grain.

[b]Specific gravity is based on oven-dry weight and volume when green or at 12% moisture content.

[c]Modulus of elasticity (MOE) measured from a simply supported, centre-loaded beam, on a span depth ratio of 14/1. To correct for shear deflection, the modulus should be increased by 10%.

[d]Values for side hardness of the true hickories are from Bendtsen and Ethington (1975).

[e]Coast Douglas-fir is defined as Douglas-fir growing in Oregon and Washington State west of the Cascade Mountains summit. Interior West includes California and all counties in Oregon and Washington east of, but adjacent to, the Cascade summit; Interior North, the remainder of Oregon and Washington plus Idaho, Montana, and Wyoming; and Interior South, Utah, Colorado, Arizona, and New Mexico.

References

Anon (1962) Wood. Indicators of physical and mechanical properties. Technical guidance materials. Committee of standards at Council of Ministers of USSR, Moscow (in Russian)

Babiak M, Cunderlik I, Kurjatko S (2000) Physical parameters related to moisture movements and dimensional stability. Final report on the project activities. Technology for high quality products from black locust. Inco-Copernicus Project No. PL 96–4144; contract No. ERB IC15-CT 960713. Inco-Copernicus, Brussels

Balboni BM, Silva T, Andrade FWC et al (2018) Physical-mechanical characterization of two amazon woods coming from the second cutting cycle. Anais da Academia Brasileira de Ciência 90(4):3656–3572

Beech E, Rivers M, Oldfield S et al (2017) GlobalTreeSearch: the first complete global database of tree species and country distributions. J Sustain For 36(5):454–489

Bendtsen BA, Ethington RL (1975) Mechanical properties of 23 species of eastern hardwoods. USDA Forest Service. Forest Products Laboratory. Research note FPL-0230. Madison WI

Borovikov AM, Ugolev BN (1989) Wood Handbook. Lesnaya Promyshlennost, Moscow (in Russian)

Bunster J (2006) Commercial timbers of Mozambique - Technological Catalogue. Traforest Lda, Maputo

Carrasco J, Oporto G, Zondlo J et al (2013) Torrefaction kinetics of red oak (*Quercus rubra*) in a fluidized reactor. BioResources 8(4):5067–5082

CEN (2019) EN 14081-1:2016+A1:2019, Timber structures–Strength graded structural timber with rectangular cross section–Part 1: General requirements. CEN–European Committee for Standardization, Brussels

Cheng J (1985) Wood science. China Forestry Publishing House, Beijing

Cheng J, Yang J, Liu P (1985) Chinese wood. China Forestry Publishing House Beijing

Chudnoff M (1979) Tropical timbers of the world. Forest Products Laboratory, USDA Forest Service, Madison WI

Costa RA (2017) Determinação das constantes elásticas da madeira considerando sua ortotropia (Determination of the elastic constants of wood considering its orthotropy). Dissertation, Universidade Federal de Minas Gerais (UFMG), Belo Horizonte

de Oliveira Araújo S, Rocha Vital B, Oliveira B et al (2016) Physical and mechanical properties of heat treated wood from *Aspidosperma populifolium*, *Dipteryx odorata* and *Mimosa scabrella*. Maderas-Cienc Technol 18(1):143-156

FAO (2015) Global forest resources assessment. Food and Agriculture Organization of the United Nations (FAO), Rome

Fodor F, Lankveld C. Németh R (2017) Testing common hornbeam (*Carpinus betulus* L.) acetylated with the Accoya method under industrial conditions. iFo 10: 948–954

Forest Products Laboratory (2010) Wood handbook. Wood as an engineering material (Centential Edition). General Technical Report FPL-GTR-190. Department of Agriculture, Forest Service, Madison WI

Forest Products Laboratory (2021) Wood handbook. Wood as an engineering material. General Technical Report FPL-GTR-282. Department of Agriculture, Forest Service, Madison WI

García R, Pizarro C, Lavín AG et al (2014) Spanish biofuels heating value estimation. Part I: Ultimate analysis data. Fuel 117:1130–1138

Gonçalez JC, Valle AT, Costa AF (2001) Estimativas das constantes elásticas da madeira por meio de ondas de ultrassom (Estimates of the elastic constants of the wood by means of ultrasonic waves). Cerne 7(2):81–92

Gonçalves R, Trinca AJ, Cerri DGP (2011). Comparison of elastic constants of wood determined by ultrasonic wave propagation and static compression test. Wood Fib Sci 43(1):64–75

Gonçalves R, Trinca AJ, Pelli BP (2014). Elastic constants of wood determined by ultrasound using three geometries of specimens. Wood Sci Techn 48:269–287

Grabner M (2017). WerkHolz. Eigenschaften und historische Nutzung 60 mitteleuropäischer Baum- und Straucharten (Workwood. Properties and historical use of 60 Central European tree and shrub species). Norbert, Kessel

Günther B, Gebauer K, Barkowski R et al (2012) Calorific value of selected wood species and wood products. Eur. J. Wood Wood Prod 70(5):755–757

IPT (2013) Catálogo de madeiras brasileiras. São Paulo: Instituto de Pesquisas Tecnológicas (Catalog of Brazilian woods). Report No 4371, Technological Research Institute (ITP), Sao Paulo

ITTO (2019) International Tropical Timber Organization: Lesser Used species Database. http://www.tropicaltimber.info. (Accessed on April 18, 2019)

ITTO (2023) International Tropical Timber Organization: http://www.tropicaltimber.info/specie/cumaru-dipteryx-spp/#lower-content (Accessed on 2 January, 2023)

Jankowska A, Kozakiewicz P (2014) Comparison of thermal properties of selected wood species intended to woodwork windows production. Annals of Warsaw University of Life Sciences-SGGW. For Wood Technol 85:101–105

Jenkins BM, Ebeling JM (1985) Thermochemical properties of biomass fuels. California agriculture-California Agricultural Experiment Station (USA)

Jiang X, Chen Y, Yin Y (2010) Atlas of gymnosperms woods of China. Science Press, Beijing

Jiang J, Bachtiar EV, Lu J et al (2017) Moisture-dependent orthotropic elasticity and strength properties of Chinese fir wood. Eur J Wood Wood Prod 75(6):927–938

Jänicke S, Beech E, Rivers MC (2019) Exploring the diversity and conservation status of tree species with TreeeX. Environ 78(21):Article ID 627

Kamperidou V, Barboutis I (2017) Mechanical performance of thermally modified black pine (*Pinus nigra* L.) wood. EJPAU 20(1):Article ID 2

Keenan RJ, Reams GA, Achard F et al (2015) Dynamics of global forest area: Results from the FAO Global Forest Resources Assessment 2015. For Ecol Manag 352:9-20

Knigge K, Schulz H (1966) Grundriss der Forstbenutzung (Ground plan of forest use). Parey, Hamburg

Kollmann F (1941) Schriftenreihe Eigenschaften und Verwertung der deutschen Nutzhölser, Erster Band: Die Esche und ihre Holz (Series of published properties and utilization of the German Nutzhölser, first volume: the ash and its wood). Verlag von Julius Springer, Berlin

Kollmann FFP, Côté Jr WA (1968) Principles of wood science and technology (Vol. 1). Springer, Berlin

Komán Sz (2018) Energy-related characteristics of poplars and black locust. BioResources 13(2):4323–4331

Kučera LJ, Gfeller B (1994) Einheimische und Fremdlandische Nutzhölzer - Merkmale und Eigenschaften der in der Schweiz gebrauchlichsten Holzarten (Native and foreign timber - characteristics and properties of the types of wood most commonly used in Switzerland). Berner Fachhochschule Hochschule für Architektur, Bau und Holz, Biel

Köhl M, Lasco R, Cifuentes M et al (2015) Changes in forest production, biomass and carbon: Results from the 2015 UN FAO Global Forest Resource Assessment. For Ecol Manag 352:21–34

Lahr FAR, Chahud E, Mantese AC et al (2010) Propriedades Físicas e Mecânicas de três espécies de madeira provenientes de áreas certificadas da Floresta Amazônica (Physical and Mechanical Properties of three wood species from certified areas of the Amazon Forest). Construindo 2(2):34–37

Le CIRAD (2012). Centre de coopération internationale en recherche agronomique pour le développement (Center for International Cooperation in Agricultural Research for Development). https://tropix.cirad.fr/FichiersComplementaires/EN/America/CUMARU.pdf

Lemmens RHMJ, Louppe D, Oteng-Amoako AA (eds) (2012) Plant resources of tropical Africa. 7(2) - Timbers 2, PROTA Foundation (CTA), Wageningen

Li J (2014) Wood science. Science Press, Beijing

Li L, Yu Y, Guo J (2006) Bamboo chemistry and utilization. Zhejiang University Press, Hangzhou

Limama A, Zerizera A, Quenardb D et al (2016). Experimental thermal characterization of bio-based materials (Aleppo Pine wood, cork and their composites) for building insulation. Energy Build 116:89–95

Lin JG, He SD, Lin SD et al (1999). Study on the variation pattern of basic density and mechanical properties of *Dendrocalamus latiflorus* timber. J Bamboo Res 18:58–62

Markwardt LJ, Wilson TRC (1935) Strength and related properties of woods grown in the United States. USDA Technical Bulletin No. 479, U.S. Department of Agriculture, Washington, D.C.

Matsunaga M, Minato K (1998) Physical and mechanical properties required for violin bow materials II: Comparison of the processing properties and durability between pernambuco and substitutable wood species. J Wood Sci 44(2):142–146

Mazlan MA, Uemura Y, Osman N et al (2015). Characterizations of bio-char from fast pyrolysis of meranti wood sawdust. J Phys, Conference Series 622(1):Article ID 012054

Melo JE, Pellicane PJ, Souza MR (2000) Goodness-of-fit analysis on wood properties data from six Brazilian tropical hardwoods. Wood Sci Technol 34:83–97

Molnár S (2004) Faanyagismeret (Wood Science). Szaktudás Kiadó Ház Rt, Budapest

Molnár S, Bariska M (2002) Magyarország ipari fái (Wood species of Hungary). Szaktudás Kiadó Ház, Budapest

Molnár S, Komán Sz (2006) A nyárfajták faminőségi és fatechnológiai tulajdonságai és felhasználásuk (Wood quality and technological properties of Poplars). In: Tóth B (ed) Nemesnyárfajták ismertetője: Útmutató a nemesnyár fajták kiválasztásához (Brochure to hybrid poplars: A guide to selection of hybrid Poplars). Magyarország ERTI, Budapest

Molnár S, Farkas P, Börcsok et al (eds) (2016) Földünk ipari fái (Industrial woods of the world). ERFARET Nonprofit Kft, Sopron

Moreira AP, Silveira E, Almeida DH et al (2017) Toughness and impact strength in dynamic bending of wood as a function of the modulus of elasticity and the strength in compression to the grain. Int J Mater Eng 7(4):61–67

Moulin JC, Rodrigues BP, Oliveira JTS et al (2016) Propriedades tecnológicas do lenho de louro-preto (Technological properties of Louro Preto wood). Pesquisa Florestal Brasileira 36(88):415–421

Németh R (2002) A hidrotermikus kezelés hatása az akác faanyagának szorpciós tulajdonságaira. Doktori értekezés. Soproni Egyetem (The effect of hydrothermal treatment on the sorption properties of black locust wood). Dissertation, University of Sopron

Németh R, Fehér S, Peszlen et al (2000) Static and dynamic strength properties. Final report on the project activities. Technology for high quality products from black locust. Inco-Copernicus Project No. PL 96-4144; contract No. ERB IC15-CT 960713. Inco-Copernicus, Brussels

Niemz P (1993) Physik des Holzes und der Holzwerkstoffe (Physics of wood and wood materials). DRW-Verlag, Leinfelden-Echterdingen

Niemz P, Sonderegger W (2017) Holzphysik - Physik des Holzes und der Holzwerkstoffe (Physics of wood and wood materials). Fachbuchverlag Leipzig im Carl Hanser Verlag, Leipzig

Niemz P, Teischinger A, Sandberg D (eds) (2023) Springer Handbook of Wood Science and Technology. Springer Nature, Cham

Nikitin NI (1962) Chemistry of wood and cellulose. USSR Academy of Sciences, Institute of Macromolecular Compounds, Moscow-Leningrad (in Russian)

Peng H, Jiang J, Lu J et al (2016) Influence of density and equilibrium moisture content on the hardness anisotropy of wood. For Prod J 66(7–8):443–452

Prütz G (1941) Häufigkeitsverteilung und Raumgewichtsabhängigkeit wichtiger Eigenschaften bei vielen Holzarten (Frequency distribution and density dependence of important properties in various wood species). Kolonialforstl Mitteilungen 298–313

Qiu J (2016) Comprehensive evaluation of properties for important wood and bamboo in Yunnan province. China Zhijian Publishing House, Beijing

Ravenshorst GJP, Gard WF, Van De Kuilen JWG (2013) The importance of characterisation and sampling of tropical wood species with regard to strength and durability classification. Heron 53(2–3):201–228

Rong XW (2008) Wood elasticity and viscoelasticity. Anhui Agricultural University, Hefei

Rousodimos G (1997) Physical and mechanical properties of the wood of Greek forest species. Report of Forest Research Institute of Athens, Data from: http://www.fria.gr/EngPage/files/physical_mechanical_properties.pdf

Sales A, Candian M, Cardin VS (2011) Evaluation of the mechanical properties of Brazilian lumber (*Goupia glabra*) by nondestructive techniques. Constr Build Mater 25:1450–1454

Sandberg D, Kitek Kuzman M, Gaff M (2018) Kompozitní výrobky na bázi dřeva - Dřevo jako kompozitní a konstrukční materiál (Engineered Wood Products - Wood as an Engineering and Architectural Material). Czech University of Life Sciences Prague (CULS), Prague

Stangerlin DM, da Costa AF, Garlet A et al (2013) Resistência natural da madeira de três espécies amazônicas submetidas ao ataque de fungos apodrecedores (Natural resistance of the wood of three Amazonian species subjected to the attack of rotting fungi). Revista Ciência da Madeira, Brazilian J Wood Sci 4(1):15–32

Stangerlin DM, Cavalcante CFP, Costa CA et al (2017) Propriedades mecânicas de madeiras amazônicas estimadas por meio de métodos de propagação de ondas ultrassonoras (Mechanical properties of Amazonian woods estimated by means of ultrasonic wave propagation methods). Nativa 5:628–633

Thunell B (1945) Trä – Dess byggnad och felaktigheter (Wood – Its construction and imperfections). Byggnadsstandardiseringen, Stockholm

Trendelenburg R (1935) Schwankungen des Raumgewichts wichtiger Nadelhölzer nach Wuchsort, Standort und Einzelstamm (Variations in the density of important softwoods depending on where they grew, location and individual trunk). Z. VDI 85–89

Vasubsbu M, Nagaraju B, Kumar JV et al (2015) Experimental measurement of thermal conductivity of wood species in India: effect of density and porosity. Int J Sci Environ Technol 4(5):1360–1364

Voulgaridis EV, Passialis CN (1995) Characteristics and technological properties of the wood of mediterranean evergreen hardwoods. Forêt méditerranéenne 1:3–12

Wagenführ R (2007) Holzatlas (Wood atlas). Fachbuchverlag Leipzig im Carl Hanser Verlag, München

Walker JC (2006) Primary wood processing: principles and practice. 2nd edn. Springer, Dordrecht

Wegst UGK (2006) Wood for sound. Am J Bot 93 (10):1439–1448

Wegst UGK, Ashby M (2007) The mechanical efficiency of natural materials. Philos Mag 84(21):2167–2186

Xu W (1988) Poplar. Heilongjiang People's Publishing House, Heilongjiang

Yao X, Liang J, Ma N et al (1992) The microstructure of main Chinese bamboos. Dalian Publishing House, Dalian

Yao X, Yi T, Ma N et al (2002) Bambooculm anatomy of China. Science Press, Beijing

Yu YM, Fang W, Lin XC et al (2005). The physico-mechnical properties of *Pleioblastus amarus* bamboo wood. J Southwest For College 25:64–67

Prof.Dr. Galina Gorbacheva (Sect. 2.3.4) is an Associate Professor of the Department of Wood Science and Technology, Mytishchi Branch, Bauman Moscow State Technical University. She received her PhD in wood physics from the Moscow State Forest University (MSFU) in 2004. Dr. Gorbacheva is Scientific Secretary of Regional Coordinating Council on Wood Science, and a leading researcher in projects funded by the Ministry of Education and Science of the Russian Federation, a Member of TC 144 "Building materials and constructions", TC 078 "Timber" at the Federal Agency on Technical Regulating and Metrology of the Russian Federation, and a consultant for Federal Customs Service of Russia, the federal and regional bodies of state of the Russian Federation, the objects of cultural and historical heritage.

Prof.Dr. Andrade Egas (Sect. 2.3.1) is an Associate Professor at the Division of Wood Science and Technology of the Department of Forestry Engineering at Eduardo Mondlane University, Mozambique. He received his PhD in wood mechanical processing at the University of Pinar del Río, Cuba 1998. Dr. Egas has coordinated several joint research projects about wood science and technology with international institutions, including Swedish, Finnish and Brazilian universities and research institutions. He is the author and co-author of over 40 technical reports and articles. His current research is focused on wood properties and wood product design.

Prof.Dr. Raquel Gonçalves (Sect. 2.3.5) is currently a Professor at the University of Campinas (UNICAMP), Brazil, and leads the research group Nondestructive Testing of Wood of the School of Agricultural Engineering at the UNICAMP. She is a contributing member of various technical committees, works on nondestructive testing standards for timber grading and on transferring nondestructive testing technologies to Brazil's timber industry, and also works closely with a private technology deployment firm in developing ultrasound and ultrasonic tomography equipment for use in evaluating wooden poles, wood product grading systems, and tree inspection. Dr. Gonçalves coordinates the Nondestructive Testing Laboratory (LabEND) at UNICAMP's School of Agricultural Engineering and has guided numerous graduate and post-doctoral candidates.

Prof.Dr. Jiali Jiang (Sect. 2.3.2) has been a researcher for wood physics research work units at the Chinese Academy of Forestry since 2009. She currently serves as an associate professor at RIWI. Her current research focuses are the characterisation and evaluation of wood viscoelastic properties and the wood–water relationship. Dr. Jiang received the IUFRO Outstanding Doctoral Research Award (ODRA) in 2010. She holds a BSc (Wood Science and Engineering) from Beijing Forestry University and an MSc and PhD in Wood Science and Technology from the Chinese Academy of Forestry.

Prof.Dr.-Ing.habil. Dr.h.c. Peter Niemz (Sects. 2.1–2.2) studied at the Faculty of Mechanical Engineering of TU Dresden. From 1972 onwards, he worked at the Institute of Wood Technology Dresden (IHD) and at TU Dresden. From 1993 to 1996, he served as a Full Professor at the Universidad Austral de Chile in Valdivia before joining the Swiss Federal Institute of Technology (ETH Zurich) in 1997. He was Professor of Wood Physics from 2002 until his retirement in 2015.

Prof.Dr. Róbert Németh (Sect. 2.3.3) is a professor at the Institute of Wood Science/University of Sopron. He has been teaching and researching since 1994 and earned his PhD in 2003. He teaches wood modification, international timber utilisation, wood utilisation, wood physics, and wood–water relationships. Since 2011, he has been Director of the Institute of Wood Sciences, and since 2014, he has served as a university professor. He has been the principal investigator of several Hungarian and EU-founded projects. He has been actively involved in management committees of different COST actions. His main research areas are the material characteristics of broad-leaved tree species, the hydrothermal treatment of wood materials, the modification of wood (thermal, chemical), wood quality, wood sorption properties, wood permeability, and wood drying.

Prof.Dr. Robert J. Ross (Sect. 2.3.6) has served as Project Leader for several research work units at the USDA Forest Service, Forest Products Laboratory since 1988. He currently serves as Assistant Director (Acting) at FPL. His current research focuses on developing and using nondestructive evaluation technologies for various wood products and structures. He has written or co-authored more than 300 technical reports and articles about nondestructive evaluation and jointly holds two dozen U.S. and foreign patents. He served as team leader and editor of the Centennial Edition of Wood Handbook – Wood as an Engineering Material, FPL's flagship publication. He holds adjunct faculty appointments at Beijing Forestry University and Mississippi State University and is a research professor at Michigan Technological University.

Prof.Dr. Dick Sandberg (Sects. 2.1–2.2) received his Doctorate in Mechanical Engineering, Wood Technology and Processing from the Royal Institute of Technology (KTH), Stockholm, Sweden, in 1998. He is a Chaired Professor of Design and Fabrication of Timber Construction at the Norwegian University of Science and Technology (NTNU). He specialises in using new measurement technology, advanced information processing, and wood communication from felling to finished product. He is an Elected Fellow of the International Academy of Wood Science.

Res.Eng. Christopher Adam Senalik (Sect. 2.3.6) is a Research General Engineer with the USDA Forest Service, Forest Products Laboratory (FPL). In 2005, he left a career in heavy truck collision forensic engineering to pursue his doctorate in nondestructive evaluation of wood products and utility poles. While at FPL, his focus has been on the non-destructive testing of wood, the analysis of timber bridge structures and abutments, and moisture cycling effects on the strength of wood composites. He is currently involved in evaluating the use of ground-penetrating radar for inspecting wood structures. Adam is a technical advisor to the American Lumber Standards Committee (ALSC) Board of Review.

3 Properties of Modified Wood

Dick Sandberg and Andreja Kutnar

3.1 Wood Modification

Wood modification is an all-encompassing term that refers to the application of a chemical, mechanical, physical or biological method to alter the properties of the material, thereby enhancing the physical, mechanical, or aesthetic properties of sawn timber, veneer or wood particles used in the production of wood composites. A wood-modification process produces a material that can be disposed of at the end of the product´s life cycle without presenting any environmental hazards greater than those associated with the disposal of unmodified wood. Modification is thus applied to overcome weaknesses in the wood material that are primarily related to moisture sensitivity, low dimensional stability, low hardness and wear resistance, low resistance to biodeterioration caused by fungi, termites or marine borers, and low resistance to UV irradiation.

Figure 3.1 presents various modification methods, their mechanisms of wood protection, and the extent of commercialisation of the modification processes is highlighted (cf. Jones et al. 2019). Bent, densified, and moulded wood are examples of modified wood that have been available on the market for at least 150 years. In contrast, thermally modified timber, as well as acetylated and furfurylated timber, are examples of modified wood that have become more recently available on the market. As environmental concerns have grown considerably in recent decades, interest in bio-based modifications in different areas of society has increased and new processes and materials are constantly being developed worldwide.

D. Sandberg (✉)
Department of Manufacturing and Civil Engineering, Norwegian University of Science and Technology (NTNU), Gjøvik, Norway
e-mail: dick.sandberg@ntnu.no

A. Kutnar
Slovenia and InnoRenew CoE, University of Primorska, Koper and Izola, Slovenia
e-mail: andreja.kutnar@upr.si andreja.kutnar@innorenew.eu

3.2 Wood Modified by a Chemical Process

Chemical treatments can be divided into active and passive modification processes. In active modification, wood constituents are chemically altered, whereas passive modification involves filling the lumen and/or cell wall without reacting with the wood constituents within the cell wall (Fig. 3.2). It may also involve condensation of the added chemical itself, increasing its molecular weight.

The incorporation of chemicals in the lumens generally does not affect moisture sorption by the cell wall. Still, it increases the weight of the wood sample. Therefore, a reduced equilibrium moisture content (EMCR) was introduced, which relates the weight of moisture adsorbed to the weight of wood before modification:

$$EMC_R = \frac{m_{wet,treated} - m_{dry,treated}}{m_{dry,untreated}} \quad (3.1)$$

where $m_{wet,treated}$ and $m_{dry,treated}$ are the weights of conditioned and dry modified wood, and $m_{dry,untreated}$ is the dry weight of the wood sample before modification.

3.2.1 Biological Properties

A comprehensive review has assessed the decay risk according to levels of moisture exclusion efficiency (MEE), anti-swelling efficiency (ASE) and ASE* (cf. Sandberg et al. 2021). An analysis of the modification methods undertaken (Table 3.1) enables the estimation of threshold levels for MEE, ASE, and ASE*, as well as the respective weight gain (WPG) required for each treatment (Engelund Thybring 2013).

Several studies have highlighted that acetylated wood exhibit high biological resistance, with the biological durability of acetylated wood reaching the highest Durability Class (DC 1), comparable to that of highly durable tropical species. This means that it can even be used in ground

P. Niemz et al. (eds.), *Wood Material and Processing Data*, https://doi.org/10.1007/978-3-031-38329-8_3

Fig. 3.1 Chemical and thermal modification methods, principles and degree of commercialisation (Sandberg et al. 2021). X – available product, (X) – introduced or being used in large-scale experiments

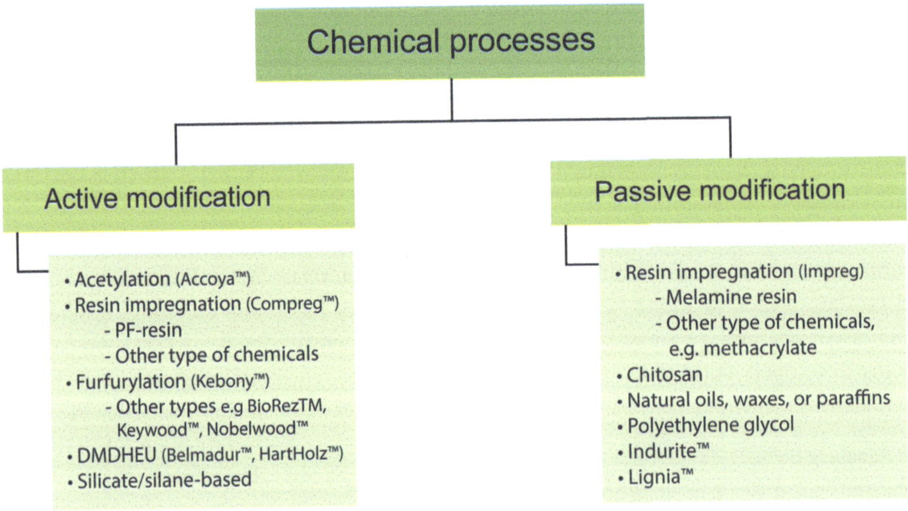

Fig. 3.2 Passive and active chemical modification processes (Sandberg et al. 2021)

contact conditions and competes with traditional biohazard treatments, as shown in Figs. 3.3 and 3.4.

3.2.2 Mechanical Properties

See Tables 3.2, 3.3, and 3.4, Fig. 3.5.

Table 3.1 Estimated threshold conditions for decay in various wood modifications (Engelund Thybring 2013). WPG – weight percentage gain, MEE – moisture exclusion efficiency, and ASE – anti-swelling efficiency, ASE* – an alternative measure of ASE, where the volume increase resulting from various wood modification methods has been deducted from the dry volume of the unreacted wood

Modification	Threshold limit (%)			
	WPG	MEE	ASE	ASE*
Acetylation	20	42	63	60
DMDHEU	25	43	45	43
Furfurylation	35	40	74	
Glutaralde-hyde	10	24	50	48
Glyoxal	>50			
Thermal modification	–15	42	46	

3.2.3 Physical and Chemical Properties

See Figs. 3.6, 3.7, and 3.8, Tables 3.5 and 3.6.

3.3 Wood Modified by a Thermal Process

The properties of wood can be enhanced by eco-friendly methods such as thermo-hydro-mechanical (THM) treatments, which combine temperature, moisture, and mechanical action. The use of a THM treatment on an industrial scale dates back to 1850, when Michael Thonet presented an invention for bending solid wood. Other THM treatments have now reached an industrial scale, with new markets and thermal processing techniques are continually being developed.

Thermal modification processes can be divided into four major categories: thermo (T) treatment, thermo-hydro (TH) treatment, thermo-mechanical (TM) treatment, and thermo-hydro-mechanical (THM) treatment (Fig. 3.9).

The term set-recovery (SR) is used to describe the reswelling of densified wood that occurs after the spring-back, when the material is re-moistened:

$$SR = \frac{R'_c - R_c}{R_o - R_c} \tag{3.2}$$

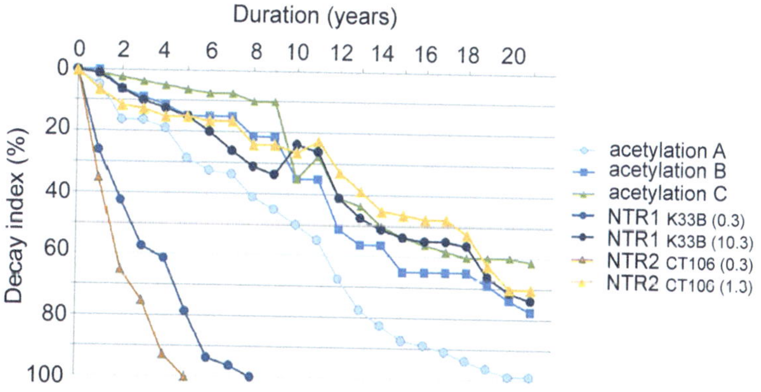

Fig. 3.3 Rot index after soil-buried tests of acetylated wood: A (high WPG), B (medium WPG), C (low WPG) and CCA-types of preservative treated ones: K33B (low charge), K33B (high charge), CT 106 (low charge), CT 106 (high charge) (from Sandberg et al. 2021). WPG – weight percentage gain

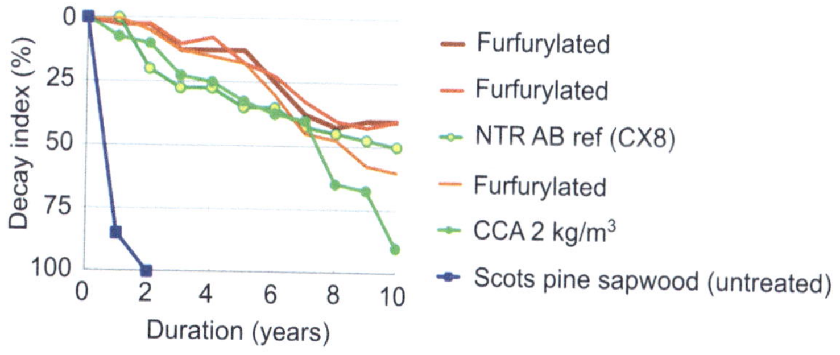

Fig. 3.4 Decay resistance of furfurylated wood in ground contact tested according to the EN 252 standard (CEN 2014) and of preservative-impregnated timber (Kebony 2020)

Table 3.2 Properties of chemically modified timber from some selected processes. ASE – anti-swelling efficiency MOE – modulus of elasticity, MOR –modulus of rupture

Treatment	Wood species	Moisture content at 20 °C and 65% RH	Owen-dry density	MOE	MOR	Brinell hardness	ASE	Reference
		(%)	(kg/m³)	(GPa)	(MPa)	(N/mm²)	(%)	
Acetylation	European beech, poplar spp., Scots pine, radiata pine	4.6–5.2	469–541	7.5–9.4	73–87			Bongers and Beckers (2003)
Furfurylation (furfuryl alcohol mixture (FA 70 mix) from Kebony™)	Maritime pine	7.3	850	10.8	176		1.2–3.4	Esteves et al. (2011)
Treatment with thermosetting resins: melamine formaldehyde (MF), phenol formaldehyde (PF) or urea formaldehyde (UF)	Burflower-tree (*Anthocephalus cadamba* Miq.)			Increase over untreated: MF: 10% PF: 12% UF: 6%	MF: 32.8 PF: 33.4 UF: 30.9		PF: 70.6 MF: 68.2 UF: 48.5	Deka and Saikia (2000)
Styrene impregnating with glycidyl methacrylate (GMA) as a crosslinker	Rubber wood			7.3	120		53	Devi et al. (2003)
Treatment with phenol-formaldehyde (PF) resin (solid content of 20%) at T = 80–180 °C and under compression at 2 MPa	Japanese cedar	Oven-dry	950–1070	17.9–22.6	148–255			Shams and Yano (2004)
Compreg™ (PF-resin impregnated and densified veneer)	Different species	Low	1200–1400	17	Up to 200	10–20 times increase compared to that of untreated wood	~70	see e.g. Sandberg et al. (2021)

(continued)

Table 3.2 (continued)

Treatment	Wood species	Moisture content at 20 °C and 65% RH	Owen-dry density	MOE	MOR	Brinell hardness	ASE	Reference
		(%)	(kg/m^3)	(GPa)	(MPa)	(N/mm^2)	(%)	
Impreg™ (uncompressed phenol-, melamine- or urea-based resin- impregnated wood, cured under mild acidic or alkaline conditions)	Different species	Low	15% to 20% greater than untreated wood	Small changes compared to untreated wood		Increase considerably more than proportional to density increase	~70	see e.g. Sandberg et al. (2021)
DMDHEU (cyclic N-methylol compound 1,3-dimethylol-4,5-dihydroxy-ethyleneurea)	Mainly Scots pine and European beech		The increase in density depends on treatment conditions		Slightly lower than untreated wood	Up to doubled	up to 70	see e.g. Sandberg et al. (2021)

Silicates, silanes, and chitosan are mainly used to treat decking, water bridges, poles, cladding and similar construction components to protect the surface and reduce mould growth. The treatment does not influence mechanical properties.

where R_0 is the oven-dry dimension of the specimen in the densification direction before densification, R_c is the dimension of the specimen after densification, and R'_c is the dimension of the densified specimen after the wet-dry cycling. The set-recovery can vary between 0 and 100%, where zero indicates that there is no set-recovery, i.e., that the compression is completely fixed, and 100% indicates that the recovery is total, i.e. the material has the same dimension as before densification.

3.3.1 Biological Properties

Charred Wood

Charring the wood surface with heat is said to make the wood more durable and to protect it against insect attack; however, the effect depends on the species and the degree of charring. Generalising the properties is difficult as the number of studies is limited. Table 3.7 gives an example of the mass loss due to attack by brown-rot and white-rot fungi.

Thermally Modified Timber (TMT)

Compared to unmodified timber, TMT is distinguished by an increased resistance to wood-destructive fungi, but the properties of TMT are challenging to generalise as they are highly dependent on the species, on the type of modification process, and on the process conditions. TMT is in general not recommended for use in structures that come into direct contact with soil or water. Table 3.8 shows the application of various classes of ThermoWood®, the most common thermal-modification process.

Table 3.3 Dry and wet strength and stiffness of untreated and acetylated radiata pine, from Dreher et al. (1964) and Minato et al. (2003) in Sandberg et al. (2021). MOE – modulus of elasticity, MOR – modulus of rupture

Treatment	Dry MOE (MPa)	Wet MOE (MPa)	Difference (%)	Dry MOR (MPa)	Wet MOR (MPa)	Difference (%)
Untreated	10 540	6760	−36	63.6	39.4	−62
Acetylated	10 602	9690	−8.6	64.4	58.0	−10

Table 3.4 Mechanical properties of Compreg™ products (Ryu et al. 1991). Average of 10 replicates, standard deviations in parentheses. PF – phenol-formaldehyde resin, WPG – weight-percentage gain, WPG_E – estimated weight-percentage gain, TS – thickness swelling, ASE – anti-swelling efficiency, ASE_N – normalised anti-swelling efficiency, IS – irreversible swelling, MOE – modulus of elasticity, MW – molecular weight

Treatment	Conc	WPG (%)	WPG_E (%)	TS (%)	ASE (%)	ASE_N (%)	IS (%)	MOE (%)
PF high MW	5	3.6 (0.7)	9.1 (0.9)	21.8 (4.3)	70.6 (5.4)	20.3 (3.6)	2.1 (4.4)	159 (24)
	10	11.3 (2.2)	17.2 (2.8)	19.0 (4.1)	74.7 (5.0)	6.8 (1.2)	1.5 (3.0)	150 (17)
	20	23.0 (2.6)	32.6 (4.6)	9.2 (4.8)	86.1 (6.5)	3.8 (0.6)	2.4 (7.7)	
PF low MW	5	4.1 (1.4)	17.1 (2.8)	39.7 (5.6)	50.3 (6.6)	13.8 (6.0)	20.6 (10.2)	131 (31)
	10	5.8 (1.2)	16.2 (1.8)	33.1 (5.0)	56.8 (5.4)	10.1 (2.0)	9.3 (6.0)	154 (23)
	20	16.5 (2.1)	32.9 (2.8)	25.8 (3.5)	66.5 (4.3)	4.1 (0.5)	7.5 (6.7)	138 (21)
Control				76.2 (11.5)			20.3 (10.8)	208 (19)

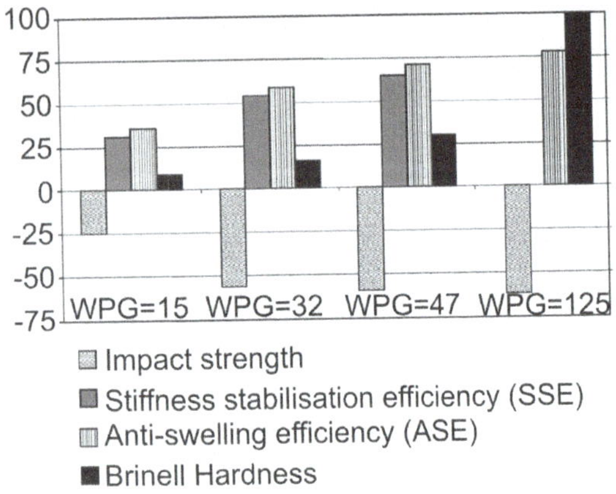

Fig. 3.5 Strength properties of furfurylated wood (Lande et al. 2004)

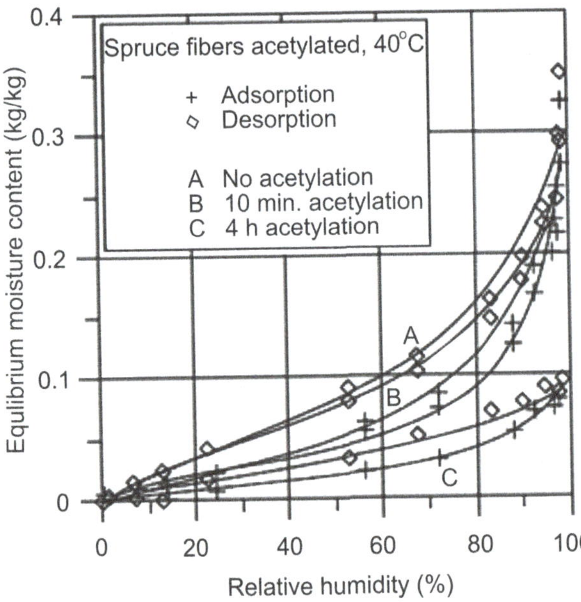

Fig. 3.6 Sorption isotherms for acetylated Norway spruce fibres (Strømdahl 2000)

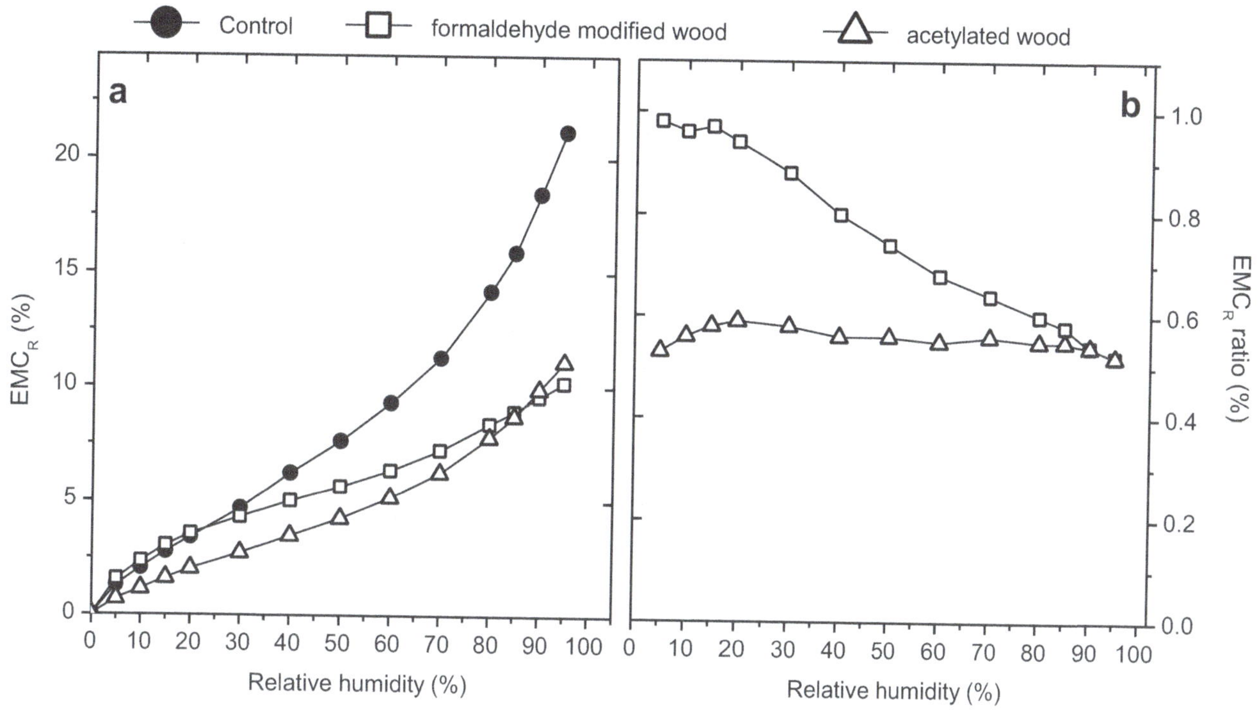

Fig. 3.7 Moisture content as a function of relative humidity (RH) for acetylated wood, formaldehyde-modified wood, and untreated Scots pine sapwood flour: (a) moisture adsorption isotherms at 25 °C, and (b) EMC_R ratios as a function of RH of the modified wood with respect to their untreated controls calculated from the EMC_R values at each RH (Mai and Militz 2023)

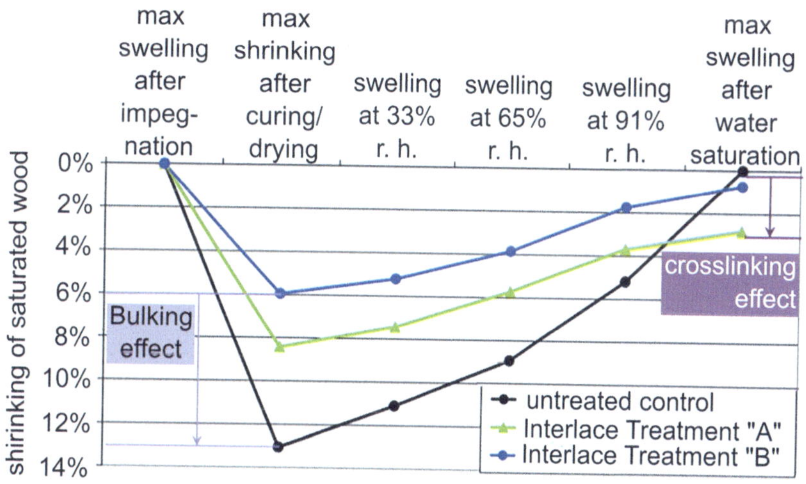

Fig. 3.8 Reduction of swollen structure by a cross-linking reaction after treatment with DMDHEU (Krause 2006). r.h. – relative humidity

Table 3.5 Changes in dimensions of Scots pine during acetylation. WPG – weight percentage gain (Sandberg et al. 2021)

WPG (%)	Increase in wood volume[1)]	Calculated volume of added chemical[2)]
17.5	3.0	2.9
22.8	3.9	4.0

[1)]Difference in oven-dry volume between reacted and non-reacted wood.

[2)]Density used in volume calculations: acetic anhydride 1049 kg/m^3.

Table 3.6 Stability of acetylated solid radiata pine wood during four wetting cycles in water (Rowell 2012). WPG – weight percentage gain, ASE – anti-shrinkage efficiency

	WPG (%)	ASE (%)	ASE2 (%)	ASE3 (%)	ASE4 (%)	Weight loss after test (%)
Acetic anhydride	22.5	70.3	71.4	70.6	69.2	<0.2

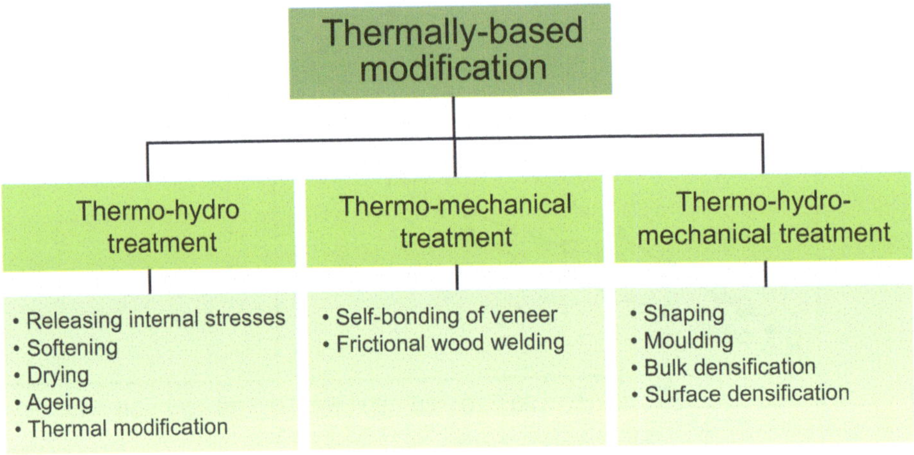

Fig. 3.9 Thermal modification processes (Sandberg et al. 2021)

Table 3.7 Mass loss of charred surfaces of Norway spruce and silver birch surfac-exposed to decay brown-rot fungus *Coniophora puteana* and white-rot fungus *Trametes versicolor* (Kymäläinen et al. 2022)

	Mass loss (%)	
	Coniophora puteana	*Trametes versicolor*
Norway spruce		
Untreated	56	19
Charred by contact heating	32	9
Charred by flame heating	40	10
Silver birch		
Untreated	50	60
Charred by contact heating	14	23
Charred by flame heating	46	42

Table 3.8 Use Classes (UC) according to the EN 335 (CEN 2013) standard of TMT according to the ThermoWood® process (International Thermowood Association 2021). UC 3 = exterior use, exposed to weather, and UC 2 = outdoor use under the roof

ThermoWood product	Species	Use Class
Thermo-D	Norway spruce, Scots pine	UC 3
Thermo-D	European ash, ayous, frake	UC 3
Thermo-D	Hardwoods	UC 2
Thermo-S	Norway spruce, Scots pine	UC 2
Thermo-S	Hardwoods	UC 2

Densified Wood

See Fig. 3.10 and Table 3.9.

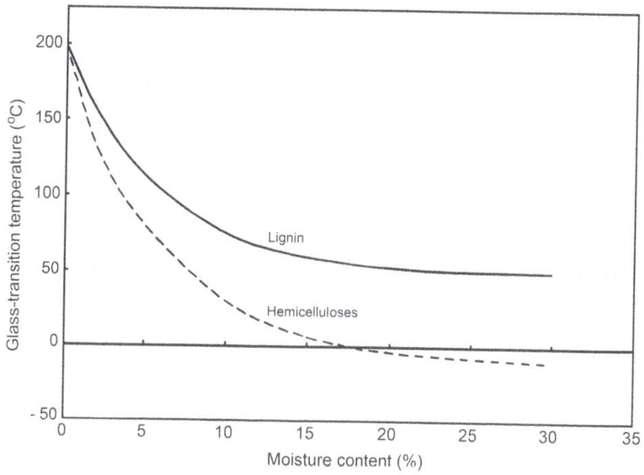

Fig. 3.10 Apparent glass transition temperature of in situ lignin and hemicelluloses as a function of moisture content (Kelley et al. 1987)

3.3.2 Mechanical Properties

Thermally Modified Timber (TMT)

See Table 3.10.

Self-Bonded Wood

See Table 3.11.

Friction-Welded Wood

See Table 3.12.

Densified Wood

See Tables 3.13, 3.14, 3.15, and 3.16, Figs. 3.11, 3.12, 3.13, 3.14, and 3.15.

3.3.3 Physical and Chemical Properties

Thermally Modified Timber (TMT)

See Table 3.17, Figs. 3.16 and 3.17.

Self-Bonding of Wood

See Table 3.18.

Friction Welding of Wood

See Figs. 3.18, 3.19, and 3.20.

Densified Wood

See Figs. 3.21, 3.22, 3.23, 3.24, 3.25, 3.26, 3.27, and 3.28, Tables 3.19, 3.20, and 3.21.

Table 3.9 Suitable steam plasticising conditions for wood (Stevens and Turner 1970)

Process parameter	Unit	Values (bold type indicates preferred values)
Temperature	(°C)	80 … **100 … 110** … 160
Pressure	(MPa)	0.10 … **0.17 … 0.22** … 0.50
Duration	(min./mm wood thickness)	**1** … 2
Moisture content of the wood	(%)	10 … **15 … 25** … 27

Table 3.10 Properties of thermally modified timber from some selected processes. ASE – antiswelling efficiency, MOE – modulus of elasticity, MOR – modulus of rupture, OHT – oil heat treatment, RH – relative humidity (Sandberg et al. 2021)

Process	Wood species	Treatment	MC at 20 °C and RH 65%	Own-dry density	MOE	MOR	Brinell Hardness	ASE radial/tangential	Reference
			(%)	(kg/m³)	(GPa)	(MPa)	(N/mm²)	(%)	
Mühlböck™	European beech, Norway spruce	Gas atmosphere: 180 °C for 4 h and 220 °C for 4 h	Beech: 4.9–7.1 Spruce: 5.3–6.8	Beech: 600–652 Spruce: 350–400	Beech: 12.9–14.9 Spruce: 11.0–13.2	Beech: 94–99 Spruce: 60–85			Arnold (2010)
PLATO™	European beech, Norway spruce, Douglas fir, Scotch pine, radiata pine	PLATO-process, 160–190 °C				Beech: 94–91 Spruce: 87–78 D. fir: 83–73 S. pine: 90–74 r. pine: 72–60		Beech: 10 (r), 13 (t) Spruce: 11 (r), 40 (t) D. fir: 13 (r), 23 (t) S. pine: 33 (r), 41 (t) r. pine: 35 (r), 40 (t)	Militz and Tjeerdsma (2000)
	Korean paulownia, Japanese pine, tulip tree, creamy bark birch	Heat treatment at 175 °C and 200 °C; vacuum		Paulownia: 190–220 Pine: 450–500 Tulip: 370–440 Birch: 570–650	Paulownia: 5.0–5.4 Pine: 10.3–13.1 Tulip: 10.7–11.3 Birch: 12.3–13.7	Paulownia: 14.0–25.9 Pine: 64.7–85.8 Tulip: 32.8–57.2 Birch: 65.2–98.3	Paulownia: 22.1–30.6 Pine: 84.1–84.8 Tulip: 62.1–81.9 Birch: 95.7–118.8		Won et al. (2012)
	Norway spruce, Scots pine	Heat treatment at 165–185 °C; super atmospheric pressure (8–10 bar)	Spruce: 6.0 Pine: 4.1	Spruce: 400 Pine: 500	Spruce: 11.2 Pine: 10.7	Spruce: 39.3 Pine: 85.9	Pine: 18.4		Boonstra et al. (2007)
ThermoWood™	Norway spruce	Thermal treatment in steam at 213 °C for 3 h	4.6	400	11.2	26.0			van Blokland et al. (2019)
OHT	Norway spruce, Scots pine	Oil heat treatment at 180–200 °C			10.0–11.0	50.0–81.0			Rapp and Sailer (2001)

Table 3.11 Mechanical properties of self-bonded veneer panels manufactured according to the Cristescu process Cristescu (2015) at different temperatures. Pressing time 300 s and 5 MPa compressive pressure (Sandberg et al. 2021)

Property	Unit	Standard	200 °C	225 °C	250 °C	Reference
			(mild)	(medium)	(harsh)	
Density	(kg/cm^3)	EN 323 (CEN 1993a)	670 ± 20	744 ± 39	901 ± 19	Cristescu et al. (2015a,b)
Mass loss due to temperature	(%)		0.5	1.2	4.3	Cristescu et al. (2015a)
Three-point bending test						Cristescu et al. (2015b)
Maximum load	(N)	EN 310 (CEN 1993b)	912 ± 52	1897 ± 74	2148 ± 39	
Ultimate strength	(MPa)	EN 310	1.9 ± 0.0	4.9 ± 1.9	19.8 ± 3.8	
Type of failure			Inter-laminar shear	Inter-laminar shear	Tension	
Shear strength	(MPa)	EN 314 (CEN 2004)	2.0 ± 0.2	5.1 ± 0.4	4.6 ± 0.0	Cristescu et al. (2015b)
Brinell hardness	(kg/mm^2)	EN 1534 (CEN 2010)	4.3 ± 0.0	6.4 ± 2.7	10.8 ± 0.7	Cristescu et al. (2015a)

Table 3.12 Tensile-shear strength (TS) according to the EN 205 standard (CEN 2016), with the maximum temperature (MT) and maximum average temperature (MAT) measured after 4 seconds of welding at different average temperature increase rates (TIR). Each value in the table is the average of 10 specimens ± standard deviation (Frihart et al. 2023)

Species	Fibre orientation	TS	MT	MAT	TIR
		(MPa)	(°C)	(°C)	(°C/second)
Norway maple	Tangential (0°)	6.5 ± 1.6	265 ± 3	216 ± 6.4	45.7
Norway maple	Radial (90°)	8.1 ± 0.8	284 ± 8	223 ± 10.5	47.5
Norway maple	Radial (45°)	9.0 ± 0.9	282 ± 13	223 ± 6.2	47.5

Table 3.13 Mechanical properties of European beech densified in an open system (Huguenin and Navi 1995). MOE – modulus of elasticity

Type of wood	Density (kg/m^3)	Impact resistance[1)]		Bending properties		MOE
		Tangential	Radial	σ_{Rup} [2)]	σ_{max} [3)]	E_L [4)]
		(J/cm^2)	(J/cm^2)	(MPa)	(MPa)	(MPa)
Beech for joinery purposes	630	3.8	8.5	68	29	6.9
Beech to be densified	820	8.2	10	117	53	11.0
Densified beech	1 130	14	15	200	95	22.9

[1)] Energy per unit area necessary to cause rupture under radial or tangential loading.
[2)] Crack opening under stress in mode I (radial loading).
[3)] Linear elastic stress limit (radial loading).
[4)] Modulus of elasticity in longitudinal tension.

Table 3.14 Properties of Sonowood™ (densified solid wood) from the Swiss Wood Solutions company (Sandberg et al. 2021). MC – moisture content

Property	Unit	Norway maple	Norway spruce
Density	(kg/m^3)	1200–1400	1300–1400
Brinell hardness	(N/mm^2)	>80	>100
Color		Mocha	Caramel
Dimensional stability (difference in swelling)	(% per % MC change)	Height ~ 0.70 Width ~ 0.30	Height ~ 0.75 Width ~ 0.33
Sound velocity	(m/s)	>4400	>5500
Damping (Log. decrement)		~0.053	~0.040
Elastic modulus	(GPa)	>23	>39

Table 3.15 Properties of densified wood from some selected processes (Sandberg et al. 2021). MOE – modulus of elasticity, MOR – modulus of rupture, RH – relative humidity, VTC – viscoelastic thermal compression process

Process	Wood species	Treatment	MC at 20 °C and RH 65%	Absolutely dry density	MOE	MOR	Brinell Hardness	Set recovery	Reference
			(%)	(kg/m³)	(GPa)	(MPa)	(N/mm²)	(%)	
VTC	Hybrid poplar (*Populus deltoides × Populus trichocarpa*)	175–200 °C; steam pressure 860 kPa; compression 1.38 MPa	7.0	550–800	12–20	101–154			Kutnar and Kamke (2008)
VTC	Hybrid poplar (*Populus deltoides × Populus trichocarpa*)	150, 160, 170 °C ; steam: saturated, superheated, transient conditions; compression 5.5 MPa	6.8–8.7	730–1200	12–32	139–276			Kutnar and Kamke (2012)
VTC	Hybrid poplar (*Populus deltoides × Populus trichocarpa*)	175–200 °C ; steam pressure 860 kPa; compression 1.3 MPa		1150–1250			19–32		Rautkari et al. (2011)
Surface densification	Douglas fir Scots pine	150 °C		650–700			25–26		Rautkari et al. (2013)
Veneer densification	American aspen Hybrid poplar (*Populus maximowiczii x Populus balsamifera*)	140–220 °C ; steam pressure 550 kPa; compression 4.5–9.0 MPa	5.0–7.0	700–800	10–20	100–140	25–35		Fang et al. (2012)
Densification	Scots pine	150 °C ; steam post-treatment at 200 °C	oven-dry	750–900			19–24		Laine et al. (2016a)
Suraface densification	Radiata pine (acetylated)	surface densification at T 150 °C	3.3				27–29	22–26	Laine et al. (2016b)
Densification combined with thermal modification	Black Sea fir Large-leafed linden Black poplar	Densification at 140 °C ; ThermoWood post-treatment at 185 °C and 212 °C		Fir: 500–700 Linden: 580–660 Poplar: 670–870	Fir: 8.5–10.6 Linden: 8.0–10.0 Poplar: 9.6–12.1	Fir: 51–91 Linden: 47–86 Poplar: 65–107			Pelit et al. (2018)
Densification	Norway spruce (Nano-silver impregnated)	THM treatment at 150–175 °C		470–530	7.8–8.0	85–100		6–7	Rassam et al. 2012)

(continued)

Table 3.15 (continued)

Process	Wood species	Treatment	MC at 20 °C and RH 65%	Absolutely dry density	MOE	MOR	Brinell Hardness	Set recovery	Reference
			(%)	(kg/m^3)	(GPa)	(MPa)	(N/mm^2)	(%)	
Densification combined with thermal modification	Black poplar European beech Norway spruce English oak European ash	THM treatment at 130 °C + post thermal modification at 240 °C , 20% vacuum	3.3	Poplar: 610 Spruce: 550 Oak: 670 Ash: 810 Beech: 970	Poplar: 17.8 Spruce: 3.8 Oak: 6.9 Ash: 11.3 Beech: 13.4	Poplar: 101 Spruce: 25 Oak: 39; Ash: 104 Beech: 112		0	Wehsener et al. (2018)
VTC on veneer	Western hemlock	150, 160, 170, 175 °C + annealing at T = 200 °C	9.9–10.6	540–1030	14–24	107–229		7–20	Liu et al (2013)
Densification combined ammonia softening	European beech	Lignamon process		885–1185 (at 20 °C /65% RH)	22.5	175	35		Paril et al. (2014)

Table 3.16 Mechanical properties of resin-impregnated and compressed-laminated wood and insulating wood[1]; Lignostone™ from the Röchling company (Sandberg et al. 2021)

Property	Unit	DIN 7707[1)]	Parallel veneer		Cross-wise veneer		Tangential packed veneer	
			KP 20,212	KP 20,214	KP 20,222	KP 20,224	KP 20,242	KP 20,244
Density	(kg/m³)		850	1250	950	1250	950	1250
Flexural strength[2), 3)]	(MPa)	⊥	140	200	110	130	130	180
Modulus of elasticity[2), 3)]	(GPa)	⊥	11	16	9	11	11	13
Compressive strength	(MPa)	⊥	100	120	200	230	120	140
	(MPa)	//	55	90	70	90	80	100

[1)]Name according to the DIN 7707-1 standard (DIN 1979).
[2)]The fibres of the outside veneers must run in the longitudinal direction of the test specimen.
[3)]Parallel laminated types must be present in the tension zones of at least four longitudinal layers.
⊥ = perpendicular to the lamination.
// = parallel to the lamination.

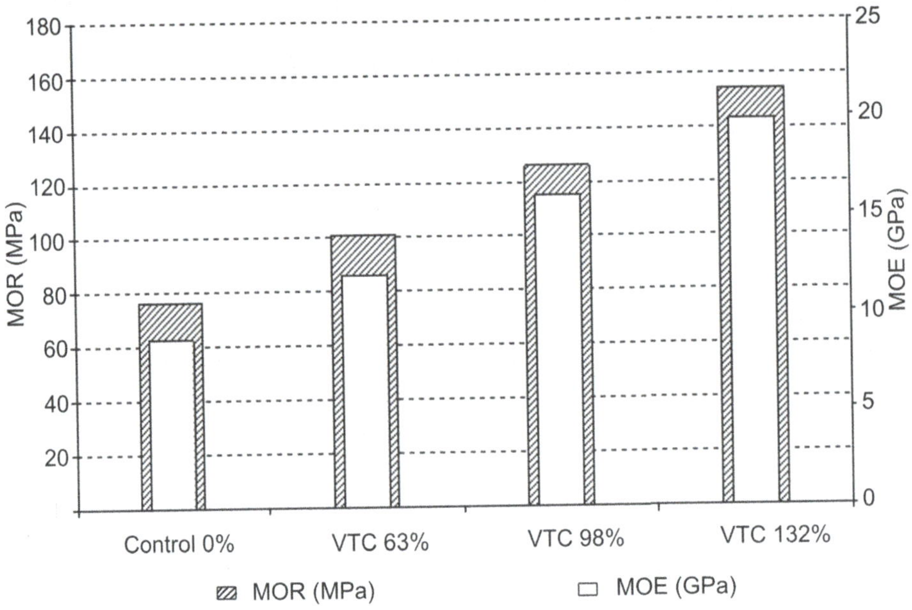

Fig. 3.11 Bending strength of densified wood. Modulus of elasticity (MOE) and modulus of rupture (MOR), for non-densified (control) and viscoelastic thermal compressed (VTC) wood with different degrees of densification, 63%, 98%, and 132% (Kutnar and Kamke (2008)

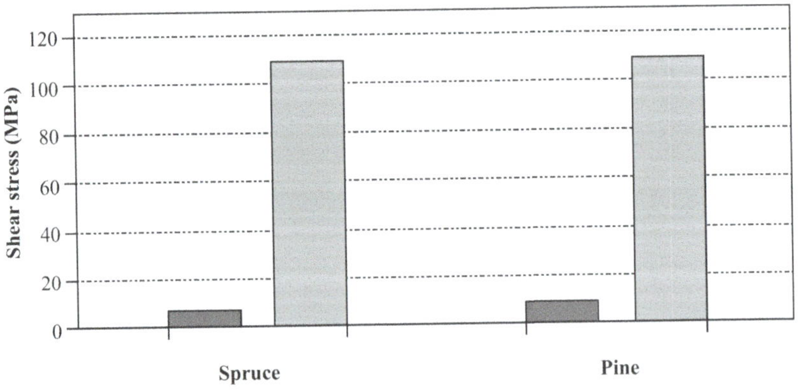

Fig. 3.12 Average shear strength in the LR-plane of non-densified (black) and densified (grey) maritime pine and Norway spruce in closed systems (Navi et al. 2000)

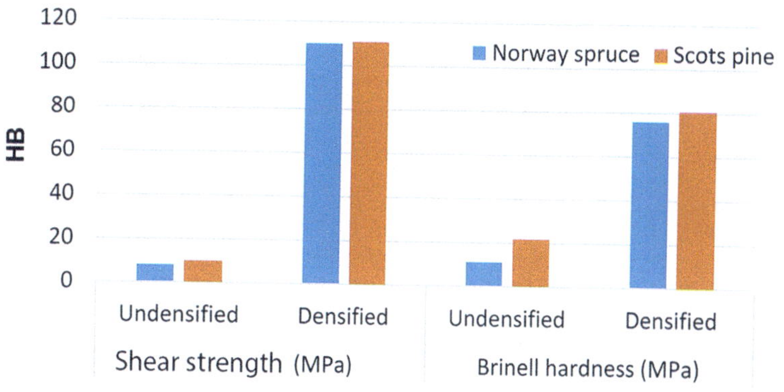

Fig. 3.13 Shear strength and Brinell hardness (HB) of un-densified and densified wood in a thermo-hydro-mechanical (THM) closed system (Navi et al. 2000)

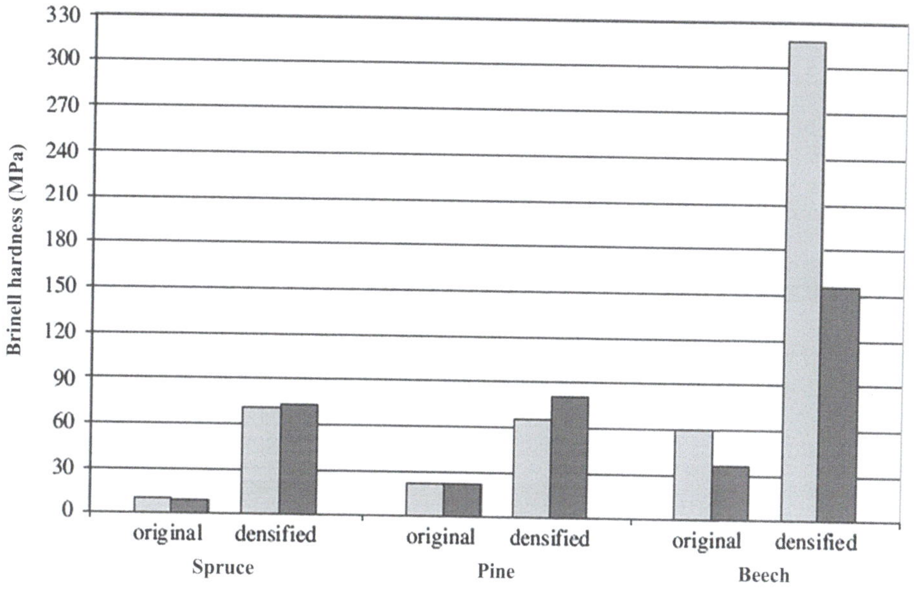

Fig. 3.14 Average Brinell hardness in a closed system of undensified (original) and densified wood for European beech, maritime pine and Norway spruce, (Navi et al. 2000). Light grey is radial section, and dark grey is tangential section

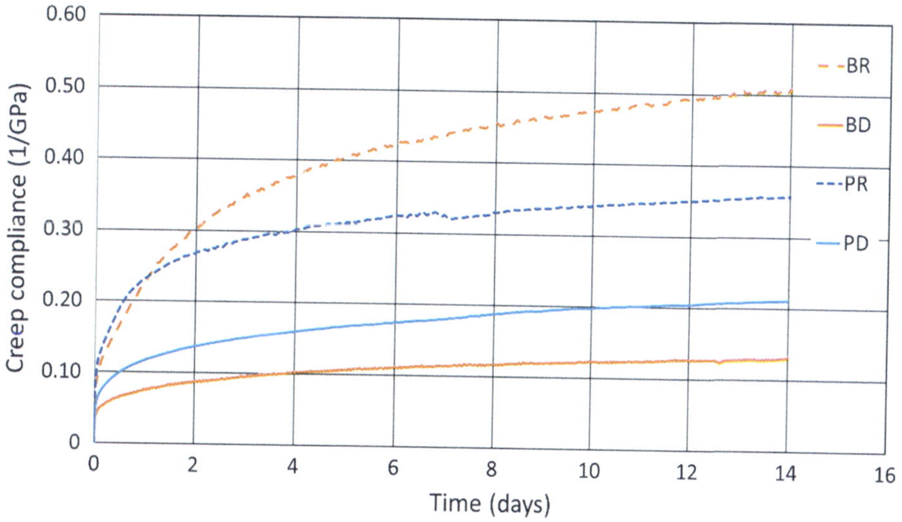

Fig. 3.15 Creep compliance as a function of time for undensified references (R) and densified (D) European beech (B) and Scots pine (P) (Han et al. 2022a)

Table 3.17 Sugar analysis of American aspen before and after thermal modification with saturated steam at 220 °C in a closed system for 8 min (Rowell et al. 2009)

	Unit	Before modification	After after modification
Weight loss	(%)	0	2.20
Arabinan	(%)	0.56	0.20
Galactan	(%)	0.68	0.58
Rhamnan	(%)	0.30	0.14
Glucan	(%)	42.40	42.20
Xylan	(%)	16.30	14.60
Mannan	(%)	1.60	1.50
Total carbohydrate	(%)	61.80	59.20

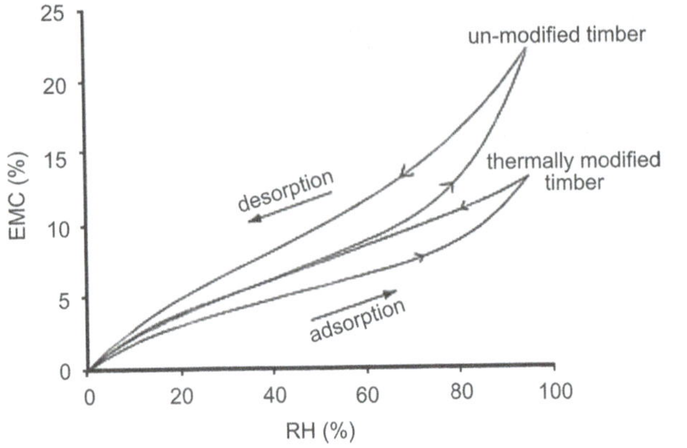

Fig. 3.16 Sorption–desorption isotherms of non-modified wood and thermally modified timber (Sandberg et al. 2021)

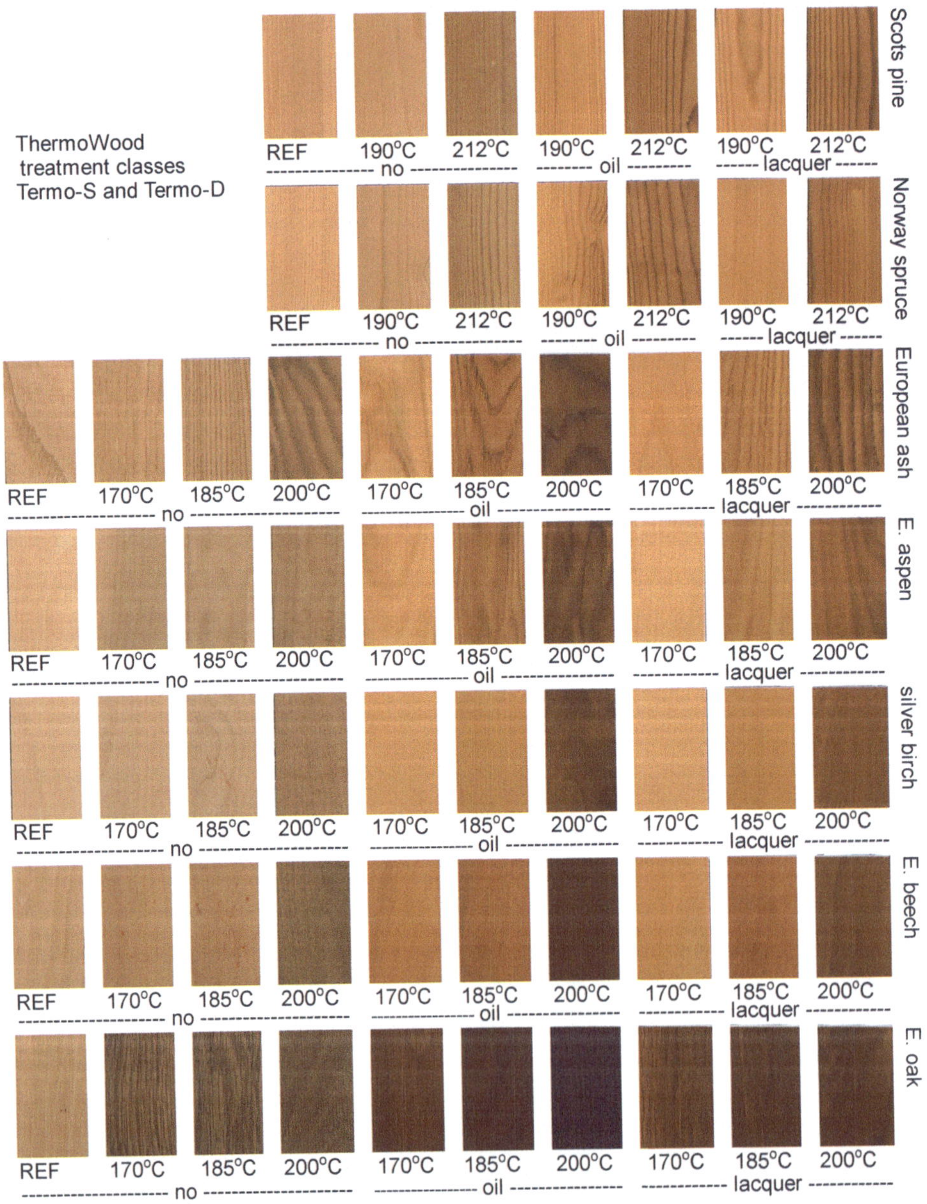

Fig. 3.17 Colours of different species treated by the ThermoWood® processes, and thereafter treated with a transparent oil or lacquer (Sandberg et al. 2021)

Table 3.18 Physical properties of self-bonded veneer panels manufactured according to the Cristescu process at different temperatures. Pressing time 300 s under 5 MPa compressive pressure. EMC – equilibrium moisture content, RH – relative humidity (Sandberg et al. 2021)

Property	Unit	Standard	200 °C (mild)	225 °C (medium)	250 °C (harsh)	Reference
Density	(kg/cm³)	EN 323 (CEN 1993a)	670±20	744±39	901±19	Cristescu et al. (2015a,b)
Mass loss due to temperature	(%)		0.5	1.2	4.3	Cristescu et al. (2015a)
Thickness swelling	(%)	EN 317 (CEN 1993c)		22.7±3.2	16.0±2.8	Cristescu et al. (2015b)
Water absorption[1)]	(%)		72.2±0.0	61.5±1.5	40.9±0.8	Cristescu et al. (2015b)
pH	(%)		4.75	4.62	4.54	Cristescu et al. (2015a)
EMC at 20°C, 50% RH	(%)		5.1	3.7	2.4	Cristescu et al. (2015a)
Water-resistant bond line		EN 317 (CEN (1993c)	No	No	Yes	Cristescu et al. (2015b)

[1)]Water uptake after soaking in water at a temperature of 20 °C for 48 h.

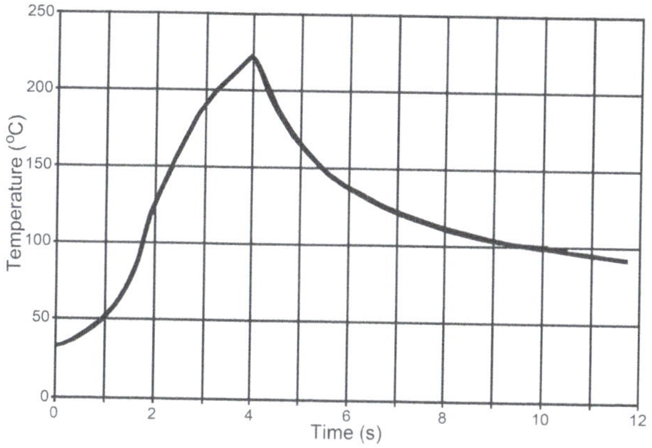

Fig. 3.18 Profile of temperature increase during welding for a joint manufactured with wood cut in the longitudinal radial (45°) grain orientation with a welding time (WT) of 4 s (Ganne-Chedeville et al. 2006)

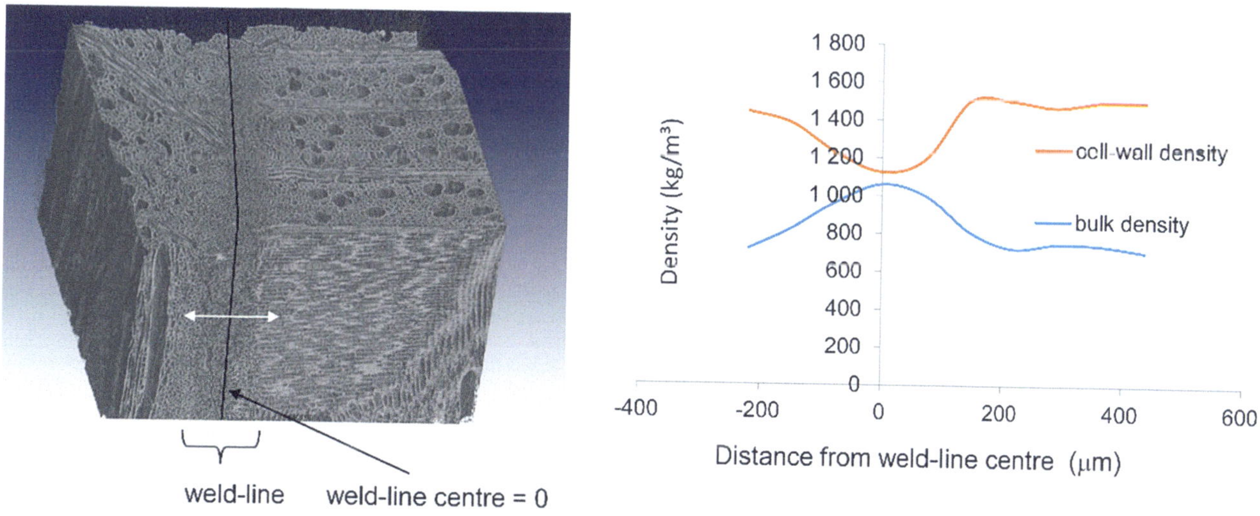

Fig. 3.19 Density profile around the welding zone (European beech) based on the grey values obtained by -CT scanning. μ-CT image of a weld-line and from welding of unaffected wood (left), and density measurement where zero indicates the centre of the weld-line (right), adapted from Vaziri et al. (2015)

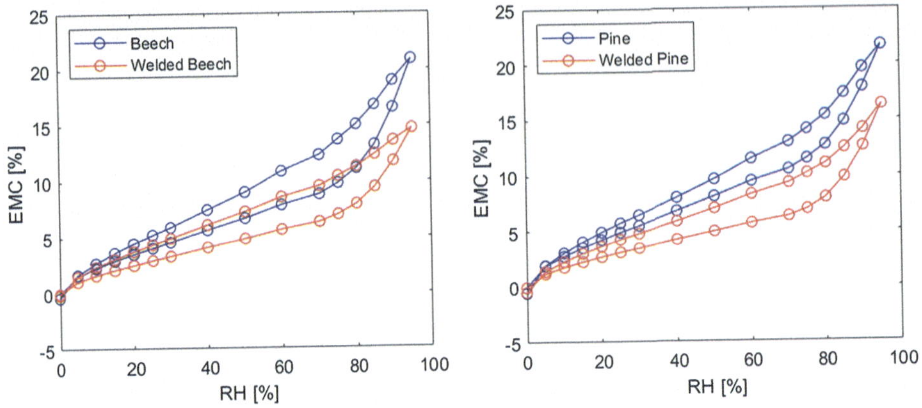

Fig. 3.20 Sorption isotherms of the welded bond-line and normal wood of European beech and Scots pine measured in the hygroscopic range of 0–95% relative humidity (RH) (Vaziri et al. 2023)

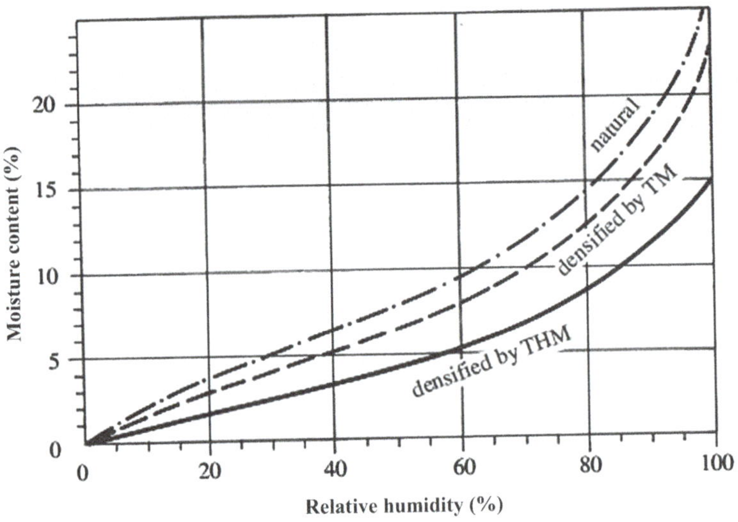

Fig. 3.21 Adsorption isotherms for initial, thermo-mechanical (TM) and thermo-hydro-mechanical (THM) treated European beech (Navi et al. 2000)

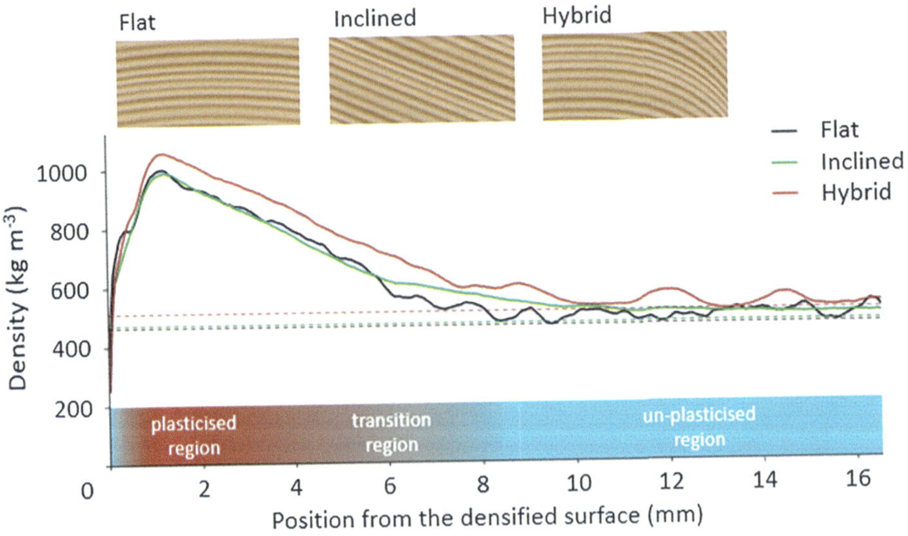

Fig. 3.22 Typical density profiles (continuous lines) of surface-densified Scots pine. The dashed lines show the density profile before densification. The regions of the specimens were plasticised to different degrees during densification, which is schematically presented by the colour-gradient (Scharf et al. 2023)

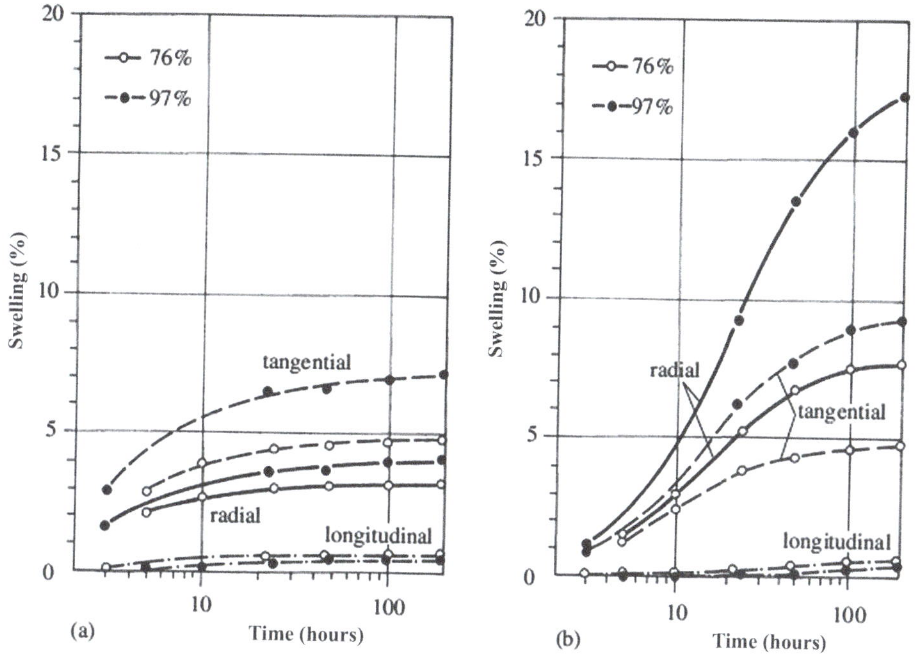

Fig. 3.23 The swelling (for densified wood including set-recovery) of European beech as a function of time (logarithmic scale) during humidification (76% or 97% RH) in an open system: (a) beech before densification, and (b) beech after densification (Huguenin and Navi 1995)

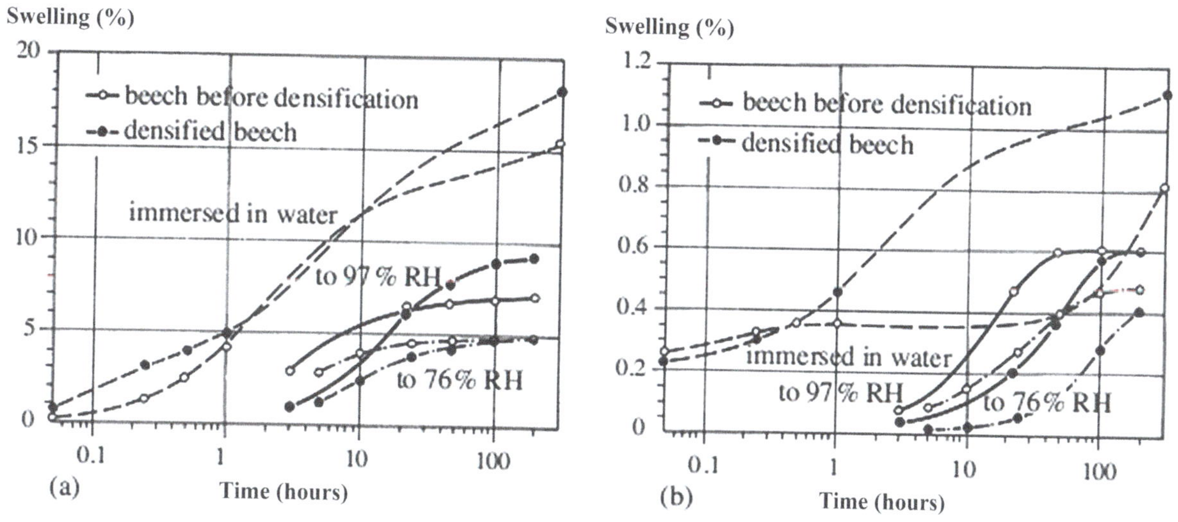

Fig. 3.24 The swelling (for densified wood including set-recovery) of European beech as a function of time (logarithmic scale) during humidification (76% or 97% RH) or immersed in water: (a) tangential swelling, and (b) longitudinal swelling (Huguenin and Navi 1995)

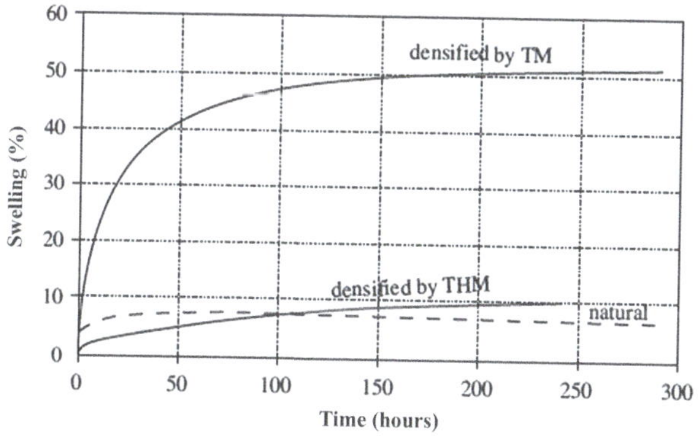

Fig. 3.25 Swelling/set-recovery in the radial direction of initial, TM- and THM-post-treated wood immersed in water at 18 °C (Navi et al. 2000)

Fig. 3.26 Combinations of the THM post-treatment parameters that reduce the set-recovery to less than 10% Navi and Sandberg (2012) in Neyses (2019)

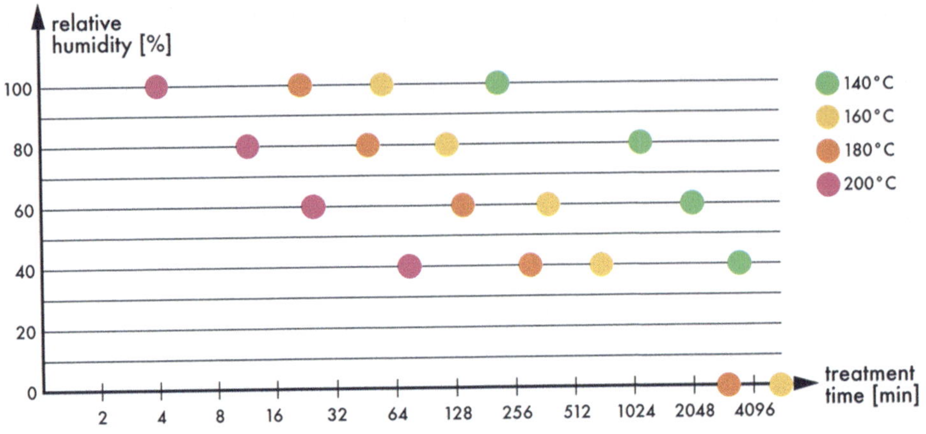

Fig. 3.27 Set-recovery of surface-densified specimens pre-treated with ionic liquids or superbases, after two wet-dry cycles (Neyses 2019; Neyses et al. 2019). The densification temperature is shown by the colour code. BMIMCL is 1-butyl-3-methylimidazolium chloride, DBN is 1,5-diazabicyclo[4.3.0] non-5-ene, DBU is 1,8-diazabicyclo[5.4.0] undec-7-ene, and TMG is 1,1,3,3-tetramethylguanidine

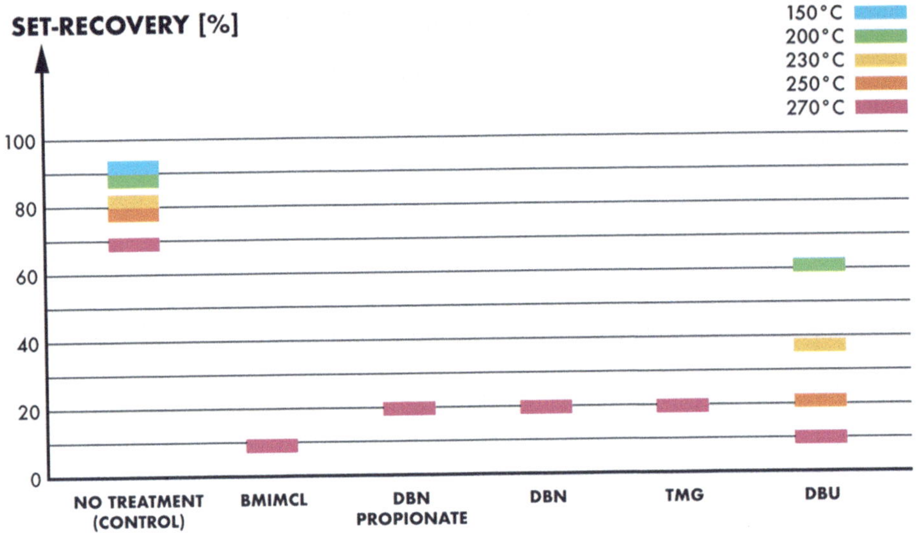

Fig. 3.28 Brinell hardness of surface-densified specimens pre-treated with ionic liquids or superbases (Neyses 2019; Neyses et al. 2019). BMIMCL is 1-butyl-3-methylimidazolium chloride, DBN is 1,5-diazabicyclo[4.3.0] non-5-ene, DBU is 1,8-diazabicyclo[5.4.0] undec-7-ene, and TMG is 1,1,3,3-tetramethylguanidine

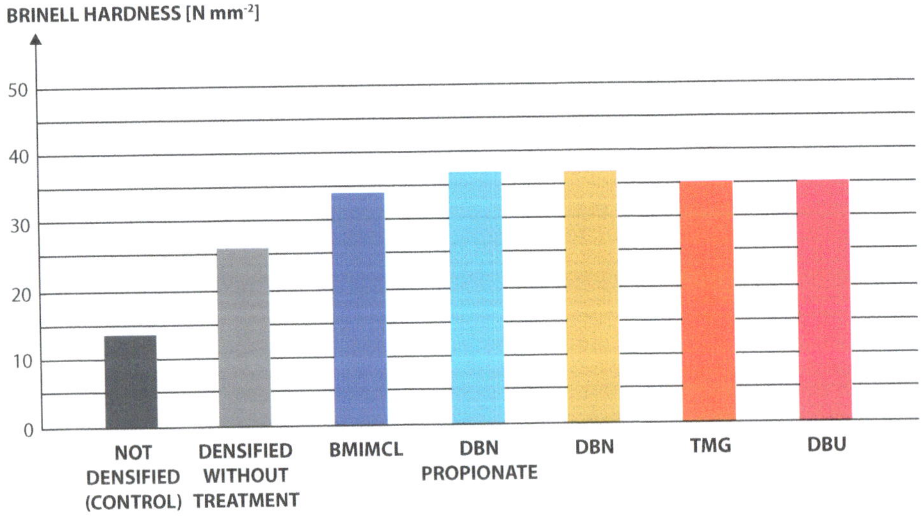

Table 3.19 Fixation of set-recovery (SR) after thermo-hydro-mechanical densification in a closed system. Time in red represents the minimum post-treatment time necessary to eliminate the set-recovery (set-recovery < 2%). The uncertainty of evaluating set-recovery values is about ± 1% (Navi and Sandberg 2012). RH – relative humidity

RH/T	140 °C	SR (%)	160 °C	SR (%)	180 °C	SR (%)	200 °C	SR (%)
100%	30 min	32.8	15 min	16.3	5 min	14.4	1 min	9.6
	60 min	19.7	30 min	6.1	10 min	2.7	2 min	12.0
	90 min	12.7	45 min	4.4	15 min	2.8	3 min	3.5
	120 min	5.6	**60 min**	**1.3**	**20 min**	**1.3**	**4 min**	**1.5**
	150 min	3.5	75 min	0.3	25 min	1.3	5 min	1.6
	210 min	**4.0**			30 min	0.9	6 min	1.3
80%	600 min	4.6	80 min	8.1	25 min	18.1	5 min	8.2
	1200 min	**2.0**	100 min	2.3	35 min	9.1	9 min	1.8
			120 min	**2.6**	40 min	**1.9**	**13 min**	**1.1**
			140 min	2.5	45 min	2.7	17 min	0.5
					50 min	1.9	20 min	0.6
60%	1200 min	5.7	330 min	2.3	80 min	5.7	10 min	2.1
	2000 min	6.6	**420 min**	**2.7**	90 min	4.0	**25 min**	**0.1**
			600 min	2.3	100 min	3.7	40 min	0.5
					120 min	2.7	55 min	1.1
					140 min	**1.2**	65 min	0.1
40%	1200 min	16.9	600 min	3.1	150 min	8.2	65 min	1.8
	2400 min	7.1	720 min	3.6	240 min	2.1	**80 min**	**1.6**
	3600 min	6.4			270 min	3.0	95 min	1.4
					300 min	1.5	110 min	1.4
					330 min	**2.0**	120 min	1.8
0	192 h	21.4	96 h	10.7	24 h	17.5	Charring of the specimens	
			144 h	7.0	48 h	5.5		
			192 h	5.7	72 h	5.0		

Table 3.20 Physical properties of resin-impregnated and compressed-laminated wood and insulating wood[1]; Lignostone™ from the Röchling company (Sandberg et al. 2021)

Property	Unit	DIN 7707[1)]	Parallel veneer		Cross-wise veneer		Tangential packed veneer	
			KP 20212	KP 20214	KP 20222	KP 20224	KP 20242	KP 20244
Density	(kg/m^3)		850	1250	950	1250	950	1250
Oil absorption	(%)		30	7	25	7	25	7
Moisture content	(%)		5	5	5	5	5	5
Operating temperature limit	(°C)		105	100	105	100	105	100
Volume resistivity	(| x cm)		10^{12}	10^{12}	10^{12}	10^{12}	10^{12}	10^{12}
Electric strength	(kV/3 mm)	90 °C ⊥	45–55	45–55	45–55	45–55	50–55	50–55
	(kV/25 mm)	90 °C //	70–90	70–90	70–90	70–90	80–90	80–90
Dissipation factor	(tan δ)	50 Hz, 25 °C	0.01	0.01	0.01	0.01	0.01	0.01
Relative permittivity	(e_r)		3.7	4.1	3.7	4.1	3.7	4.1

[1)]Name according to the DIN 7707-1 standard (DIN 1979).
⊥ = perpendicular to the lamination.
// = parallel to the lamination.

Table 3.21 Physical and flexural properties of untreated and treated densified Scots pine. Densified in an open system at 170–200 °C, 5 specimens in each group, and three-point bending based on EN 408 (CEN 2012) (Han et al. 2022b). EMC – equilibrium moisture content, MOE – modulus of elasticity, MOR – modulus of rupture

Group	After press opening	After 2 weeks conditioning at 20 ± 2°C and 65 ± 5% relative humidity						
	Thickness	Thickness	Density	EMC	MOE	MOR	Water uptake	Set-recovery
	(mm)	(mm)	(kg m^{-3})	(%)	(GPa)	(MPa)	(%)	(%)
Untrteated reference			480 (±0.04)	9.4 (±1.6)	12.4 (±1.9)	101.5 (±10.4)	47.5 (±11.7)	
Densified	9.85 ± 0.05)	10.67 (±0.18)	940 (±0.04)	7.0 (±0.7)	14.0 (±3.4)	166.8 (±25.9)	81.1 (±21.3)	72.3 (±4.1)
Resin-impregnated* and thereafter densified	9.83 (±0.05)	10.09 (±0.06)	980 (±0.04)	5.2 (±0.7)	19.4 (±4.1)	212.7 (±38.5)	23.8 (±14.5)	24.7 (±11.3)
Densified and thereafter thermally modified	9.83 (±0.03)	10.12 (±0.12)	910 (±0.03)	6.2 (±0.6)	17.0 (±4.3)	157.4 (±40.4)	35.3 (±14.0)	6.5 (±1.2)

*Low-molecular-weight phenol–formaldehyde resin-impregnated wood (Metadynea Austria GmbH, Krems an der Donau, Austria).

References

Arnold M (2010) Effect of moisture on the bending properties of thermally modified beech and spruce. J Mater Sci 45:669–680

Bongers HPM, Beckers EPJ (2003) Mechanical properties of acetylated solid wood treated on pilot plant scale. In: Van Acker J, Hill C (eds) Proceedings of the first European conference on wood modification: ECWM, Ghent

Boonstra MJ, Van Acker J, Tjeerdsma BF et al (2007) Strength properties of thermally modified softwoods and its relation to polymeric structural wood constituents. Anal For Sci 64(7):679–690

CEN (1993a) EN 323 Determination of density. CEN–European Committee for Standardization, Brussels

CEN (1993b) EN 310 Determination of modulus of elasticity in bending and of bending strength. CEN–European Committee for Standardization, Brussels

CEN (1993c) EN 317 Determination of swelling in thickness after immersion in water. CEN–European Committee for Standardization, Brussels

CEN (2004) EN 622-2, Fibreboards - Specifications - Part 2: Requirements for hardboards. CEN–European Committee for Standardization, Brussels

CEN (2010) EN 1534, Wood flooring - Determination of resistance to indentation: Test method. CEN–European Committee for Standardization, Brussels

CEN (2012) EN 408, Timber structures - Structural timber and glued laminated timber - Determination of some physical and mechanical properties. CEN–European Committee for Standardization, Brussels

CEN (2013) EN 335, Durability of wood and wood-based products – Use Classes: Definitions, Application to solid wood and wood-based products. CEN–European Committee for Standardization, Brussels

CEN (2014) EN 252, Field test method for determining the relative protective effectiveness of a wood preservative in ground contact. CEN–European Committee for Standardization, Brussels

CEN (2016) EN 205, Adhesives - Wood adhesives for non-structural applications - Determination of tensile shear strength of lap joints. CEN–European Committee for Standardization, Brussels

Cristescu C (2015) Self-bonding of beech veneer. Dissertation, Luleå University of Technology

Cristescu C, Hagman O, Sandberg D et al (2015a) Could colour predict hardness of hot-pressed self-laminated beech boards. ProLigno 11(4):150–156

Cristescu C, Sandberg D, Ekevad M et al (2015b) Influence of pressing parameters on mechanical and physical properties of self-bonded laminated beech boards. Wood Mater Sci Eng 10(2):205–214

Deka M, Saikia CN (2000) Chemical modification of wood with thermosetting resin: effect on dimensional stability and strength property. Bioresour Technol 73:179–181

Devi RR, Ali I, Maji TK (2003) Chemical modification of rubber wood with styrene in combination with a crosslinker: Effect on dimensional stability and strength property. Bioresour Technol 88:185–188

DIN (1979) Resin impregnated and compressed laminated wood and insulating wood; test methods. DIN–Deutsches Institut für Normung e.V., Berlin

Dreher WA, Goldstein IS, Cramer GR (1964) Mechanical properties of acetylated wood. For Prod J 14: 66–68

Engelund Thybring E, Thygesen LG, Svensson S et al (2013) A critical discussion of the physics of wood-water interactions. Wood Sci Technol 47(1):141–161

Esteves B, Nunes L, Pereira H (2011) Properties of furfurylated wood (*Pinus pinaster*). Eur J Wood Prod 69:521–525

Fang CH, Mariotti N, Cloutier A et al (2012) Densification of wood veneers by compression combined with heat and steam. Eur J Wood Prod 70(1–3):155–163

Frihart CR, Frangi A, Gottlöber C et al (2023) Joining and reassembling of wood (Chapter 14). In: Niemz P, Teischinger A, Sandberg D (eds) Springer handbook of wood science and technology, Springer Nature, Cham

Ganne-Chedeville C, Properzi M, Pizzi A et al (2006) Parameters of wood welding: A study with infrared thermography. Holzforschung 60:434–438

Han L, Kutnar A, Sandberg D (2022a) Creep behaviour of densified European beech and Scots pine under constant climate. Wood Mat Sci Eng 17(6):475–477

Han L, Kutnar A, Couceiro J et al (2022b) Creep properties of wood in bending. Forest 13:Article ID 757

Huguenin P, Navi P (1995) Bois densifié sans résine synthétique (Wood densification without use of synthetic resin). Ingénieurs et Architectes Suisses (IAS), Geneva 13:262–268

International Thermowood Association (2021) ThermoWood handbook. Helsinki
Jones D, Sandberg D, Goli G et al (2019) Wood modification in Europe: A state-of-the-art about processes, products, applications. Firenze University Press, Florence
Kebony 2020. https://kebony.com/en (2020-03-22)
Kelley SS, Rials T, Glasser WG (1987) Relaxation behavior of the amorphous components of wood. J Mat Sci 22:617-624
Krause A (2006) Holzmodifizierung mit N-Methylolvernetzern (Wood modification with cross-linking N-methylol compounds). Dissertation, University of Göttingen
Kutnar A, Kamke AF (2008) The mechanical properties of densified VTC wood relevant for structural composites. Holz Roh-Werkstoff 66(6):439–446
Kutnar A, Kamke AF (2012) Compression of wood under saturated steam, superheated steam, and transient conditions at 150°C, 160°C, and 170°C. Wood Sci Technol 46:73-88
Kymäläinen M, Belt T, Seppäläinen H et al (2022) Decay resistance of surface carbonized wood. Mater 15:Article ID 8410
Laine K, Segerholm K, Wålinder M et al (2016a) Wood densification and thermal modification: hardness, set-recovery and micromorphology. Wood Sci Technol 50:883–894
Laine K, Segerholm K, Wålinder M et al (2016b) Surface densification of acetylated wood. Eur J Wood Pro 74: 829–835
Lande S, Westin M, Schneider M (2004) Properties of furfurylated wood. Scand J For Res 19(5):22–30
Liu H, Kamke FA, Guo K (2013) Integrated drying and thermo-hydro-mechanical modification of western hemlock veneer. Eur J Wood Prod 71:173–181
Mai C, Militz H (2023) Wood modification. (Chapter 16). In: Springer Handbook of Wood Science and Technology. Niemz P, Teischinger A, Sandberg D. (eds) Springer Nature, Cham
Militz H, Tjeerdsma B (2000) Heat treatment of wood by the PLATO-process: Proceedings of Seminar "Production and development of heat treated wood in Europe". Nov. 2000, Helsinki, Stockholm, Oslo
Minato K. Takazawa R, Ogura K (2003) Dependence of reaction kinetics and physical and mechanical properties on the reaction systems of acetylation II: physical and mechanical properties. J Wood Sci 49:519–524
Navi P, Girardet F, Heger F (2000) Thermo-hydro-mechanical post-treatment of densified wood. In: Proceedings of 5th Pacific Rim Bio-based Composites Symposium Canberra, Australia
Navi P, Sandberg D (2012) Thermo-hydro-mechanical Processing of Wood. Presses Polytechniques et Universitaires Romandes, Lausanne
Neyses B (2019) Surface densification of solid wood. Paving the way towards industrial implementation. Dissertation, Luleå University of Technology
Neyses B, Karlsson O, Sandberg D (2019) The effect of ionic liquid and superbase pre-treatment on the spring-back, set-recovery and Brinell hardness of surface-densified Scots pine. Holzforschung 74(3):303–312
Paril P, Brabec M, Manak O et al (2014) Comparison of selected physical and mechanical properties of densified beech wood plasticized by ammonia and saturated steam. Eur J Wood Prod 72:583–591
Pelit H, Budakçı M, Sönmez A (2018) Density and some mechanical properties of densified and heat post-treated Uludağ fir, linden and black poplar woods. Eur J Wood Prod 76:79–87
Rapp AO, Sailer M (2001) Oil heat treatment of wood in Germany: State of the art. In Rapp A (ed) Review of heat treatment of wood. Proc. Special seminar of COST Action E22: Environmental Optimisation of Wood Protection, February 9, Antibes
Rassam G, Ghofrani M, Taghiyari HR et al (2012). Mechanical performance and dimensional stability of nano-silver impregnated densified spruce wood. Eur J Wood Prod 70:595–600
Rautkari L, Kamke FA, Hughes M (2011) Density profile relation to hardness of viscoelastic thermal compressed (VTC) wood composite. Wood Sci Technol 45(4):693–705
Rautkari L, Laine K, Kutnar A et al (2013) Hardness and density profile of surface densified and thermally modified Scots pine in relation to degree of densification. J Mater Sci 48(6):2380–2375
Rowell RM (2012) Handbook of wood chemistry and wood composites. 2nd edn, CRC Press, Boca Raton
Rowell RM, Ibach RE, McSweeny J et al (2009) Understanding decay resistance, dimensional stability and strength changes in heat treated and acetylated wood. Wood Mat Sci Eng 4(1–2):14–22
Ryu JY, Takahashi M, Imamura Y et al (1991) Biological resistance of phenol-resin treated wood. Mokuzai Gakkaishi 37(9):852–858
Sandberg D, Kutnar A, Karlsson O et al (2021) Wood modification technologies. Principles, sustainability, and the need for innovation. CRC Press, Boca Raton
Scharf A, Lemoine A, Neyses B et al (2023) The effect of the growth ring orientation on spring-back and set-recovery in surface-densified wood. Holzforschung 77(6):394–406
Shams MI, Yano H (2004) Compressive deformation of wood impregnated with low molecular weight phenol formaldehyde (PF) resin II: effects of processing parameters. J Wood Sci 50:337–342
Stevens WC, Turner N (1970) Wood bending handbook. Her Majesty's Stationery Office, London
Strømdahl K (2000) Water sorption in wood and plant fibres. Report Serie R, Technical University of Denmark (DTU)
van Blokland J, Olsson A, Oscarsson J et al (2019) Prediction of bending strength of thermally modified timber using high-resolution scanning of fibre direction. Eur J Wood Prod 77:327–340
Vaziri M, du Plessis A Sandberg D et al (2015). Nano X-ray tomography analysis of the cell-wall density of welded beech joints. Wood Mater Sci Eng 10(4):368–372
Vaziri M, Dreimol C, Niemz P et al (2023) Water-vapour sorption process of welded European beech and Scots pine. Holzforschung 77(7):500-514
Wehsener J, Brischke C, Meyer-Veltrup L et al (2018) Physical, mechanical and biological properties of thermomechanically densified and thermally modified timber using the Vacu3-process. Eur J Wood Prod 76:809–821
Won KR, Kim TH, Hwang KK et al (2012) Effect of heat treatment on the bending strength and hardness of wood. J Korean Wood Sci Technol 40(5):303–310

Prof.Dr. Dick Sandberg received his Doctorate in Mechanical Engineering, Wood Technology and Processing from the Royal Institute of Technology (KTH), Stockholm, Sweden, in 1998. He is a Chaired Professor of Design and Fabrication of Timber Construction at the Norwegian University of Science and Technology (NTNU). He specialises in using new measurement technology, advanced information processing, and wood communication from felling to finished product. He is an Elected Fellow of the International Academy of Wood Science.

Prof.Dr. Andreja Kutnar is a full professor of wood science at the University of Primorska in Koper, Slovenia, and director of the research institute InnoRenew CoE. She graduated from the University of Ljubljana in 2003 and received her PhD in Wood Science at the University of Ljubljana in 2008. In 2006 and 2007, she was a visiting student at the Department of Wood Science and Engineering, College of Forestry, Oregon State University, Corvallis, Oregon, USA and a Research Associate from 2009 to 2010. In 2014, she was elected Acting Head of the Department of Technology at Andrej Marušič Institute, University of Primorska, Slovenia. In 2017, she was nominated as an affiliated faculty member in the Department of Wood Science and Engineering at Oregon State University, USA. From 2014 until 2020, she was an Executive Board member of InnovaWood, and as of July 2020, she is a President of the Society of Wood Science and Technology. She is also a member of the Young Academy of Europe.

4 Characteristic Values of Wood-Based Panels

Steffen Tobisch

4.1 Introduction

Characteristic values given in this chapter are 5-percentiles for strength values (*f*) and mean values for elastic properties (*E*, *G*). If necessary, the 5-percentile values of the elastic properties can be calculated as 0.85 times the mean values.

According to Wiegand (2020), indices can be given in the following way:

1st index indicates bending, tension, compression, shear or rolling shear as "m", "t", "c", "v" or "r",

2nd index indicates the orientation of the stress with respect to the grain direction of the outermost layers/veneers as "0" or"90", if relevant,

3rd index indicates weather in plane or out of plane; in plane is given as "edge", out of plane is given as "flat" only in the case of stresses perpendicular respectively to the grain and shear, and

4th index indicates characteristic "k" or mean value "mean".

The symbols for strength, modulus of elasticity and shear modulus (Wiegand 2020) and the loading conditions shown in Fig. 4.1 are hereafter used for characteristic properties of the wood-based materials.

Some characteristic values are related to conditions in so-called Service Classes (SC).

According to EN 1995-1-1 (CEN 2010a), Service Class 1 is characterised by a moisture content in the wood-based materials corresponding to a temperature of 20 °C and a relative humidity of the surrounding air which exceeds 65% only for a few weeks each year. In SC 1, the average moisture content of most softwoods does not exceed 12%.

Service Class 2 is characterised by a moisture content in the wood-based materials corresponding to a temperature of 20 °C and a relative humidity of the ambient air which exceeds only 85% for a few weeks each year. In SC 2, the average moisture content of most softwoods does not exceed 20%.

Service Class 3 covers climatic conditions which result in higher moisture content than that in Service Class 2.

The load duration and moisture content influence the strength and stiffness properties of wood and wood-based materials and must therefore be taken into account in calculations and designs for mechanical resistance and serviceability (CEN 2010a). Table 4.1 shows both load duration and examples of load exposure.

The k_{def}-values (deformation modification factor) given characterise the creep behavior of the material in certain service classes and are used to modify elastic properties as in the Equation (CEN 2010a):

$$E_{mean,fin} = \frac{Emean}{(1 + k_{def})} \tag{4.1}$$

The load duration modification factor k_{mod} modifies the characteristic value of a strength property for the design of a building material property with reference to the load duration:

$$X_d = k_{mod} \frac{X_k}{\gamma_M} \tag{4.2}$$

Equation 4.2 show the calculation of a design value X_d of a material property, X_k is the characteristic 5th percentile value of a material property, and γ_M is the partial factor for a material property (CEN 2010a).

4.2 Fibreboard

4.2.1 Density, Strength, and Stiffness

See Tables 4.2, 4.3, 4.4, 4.5, 4.6 and 4.7.

S. Tobisch (✉)
Institut Für Holztechnologie Dresden, Dresden, Germany
e-mail: steffen.tobisch@ihd-dresden.de

P. Niemz et al. (eds.), *Wood Material and Processing Data*, https://doi.org/10.1007/978-3-031-38329-8_4

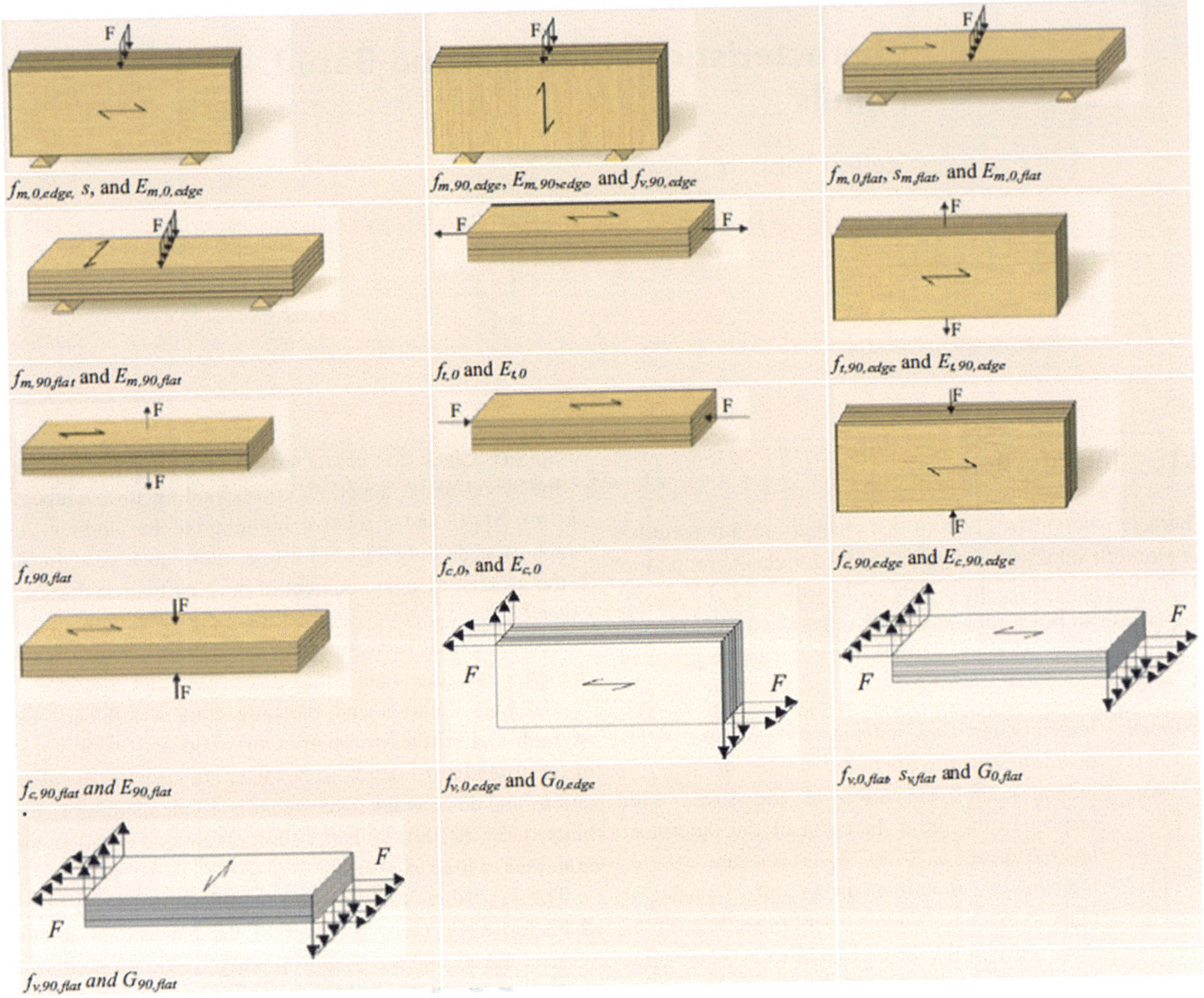

Fig. 4.1 Symbols for strengths, modulus of elasticity and shear modulus (Hakkarainen et al. 2019)

Table 4.1 Load duration classes and examples of load exposure (CEN 2010a)

Load Duration Classes	Length of the accumulated duration of the characteristic load duration	Examples of the load exposure
Permanent	Longer than 10 years	Own-weight
Long-term	6 month – 10 years	Stock materials
Medium-term	1 week – 6 months	Working load, snow
Short-term	Shorter than one week	Snow, wind
Instantaneous		Wind and extraordinary impacts

4.2.2 Creep and Duration of Load

See Table 4.8.

4.2.3 Moisture, Vapor, and Thermal Conductivity

See Table 4.9.

Table 4.2 Characteristic strengths and densities of fibreboard HB.HLA2 and MBH.LA2 according to the EN 622-2 (CEN 2004a), EN 622-3 (CEN 2004b) and EN 1995-1-1/NA (DIN 2013) standards

Product Type	Nominal thickness range (mm)	Characteristic strength (MPa) and density (kg/m³)							
		Characteristic density	Bending		Tension	Compression		Shear	
			perpendicular to plane	in plane		perpendicular to plane	perpendicular to plane	perpendicular to plane	perpendicular to plane
			$f_{m,k}$	$f_{m,edge,k}$	$f_{t,k}$	$f_{c,flat,k}$	$f_{c,edge,k}$	$f_{v,edge,k}$	$f_{v,flat,k}$
HB.HLA2	> 3.5 – 5.5	850	35	26	26	12	27	18.0	3.00
	> 5.5	800	32	23	23	12	24	16.0	2.50
MBH.LA2	≤ 10	650	17	9	9	8	9	5.5	0.30
	> 10	600	15	8	8	8	8	4.5	0.25

Table 4.3 Mean modulus of elasticity and densities of fibreboard HB.HLA2 and MBH.LA2 according to EN 622–2 (CEN 2004a), EN 622–3 (CEN 2004b) and DIN EN 1995–1-1/NA (DIN 2013) standards

Product Type	Nominal thickness range (mm)	Mean modulus of elasticity (MPa) and density (kg/m³)				
		Minimum density	Bending		Shear	
			perpendicular to plane	in plane	perpendicular to plane	in plane
			$E_{m,flat,mean}$	$E_{m,edge,mean}$	$G_{edge,mean}$	$G_{flat,mean}$
HB.HLA2	>3.5 – 5.5	850	4800	4800	2000	200
	>5.5	800	4600	4600	1900	200
MBH.LA2	≤10	650	3100	3100	1300	100
	>10	600	2900	2900	1200	100

Table 4.4 Characteristic strengths and densities of medium-density fibreboard MDF.LA according to the EN 12369–1 standard (CEN 2001)

Nominal thickness range (mm)	Characteristic strength (MPa) and density (kg/m³)				
	Minimum density	Bending	Tension	Compression	Shear
		perpendicular to plane		in plane	in plane
		$f_{m,k}$	$f_{t,k}$	$f_{c,edge,k}$	$f_{v,flat,k}$
>1.8 – 12	650	21	13.0	13.0	6.5
>12 – 19	600	21	12.5	12.5	6.5
>19 – 30	550	21	12.0	12.0	6.5
>30	500	19	10.0	10.0	5.0

Table 4.5 Mean modulus of elasticity and densities of medium-density fibrerboard MDF.HLS according to the EN 12369-1 standard (CEN 2001)

Nominal thickness range (mm)	Mean Modulus of elasticity (MPa) and density (kg/m³)				
	Minimum density	Bending	Tension	Compression	Shear
		perpendicular to plane		in plane	perpendicular to plane
		$E_{m,flat,mean}$	$E_{t,mean}$	$E_{c,edge,mean}$	$G_{edge,mean}$
>1.8 – 12	650	3700	2900	2900	800
>12 – 19	600	3000	2700	2700	800
>19 – 30	550	2900	2000	2000	800
>30	500	2700	1600	1600	600

Table 4.6 Characteristic strengths and densities of medium-density fiberboards MDF.HLS according to the (CEN 2001) standard

Nominal thickness range (mm)	Characteristic strength (MPa) and density (kg/m^3 ▷)				
	Minimum density	Bending	Tension	Compression	Shear
		perpendicular to plane		in plane	in plane
		$f_{m,k}$	$f_{t,k}$	$f_{c,edge,k}$	$f_{v,flat,k}$
>1.8 – 12	650	22.0	18.0	18.0	8.5
>12 – 19	600	22.0	16.5	16.5	8.5
>19 – 30	550	21.0	16.0	16.0	8.5
>30	500	18.0	13.0	13.0	7.0

Table 4.7 Mean modulus of elasticity and densities of medium-density fibreboard MDF.HLS according to the (CEN 2001) standard

Nominal thickness range (mm)	Mean modulus of elasticity (MPa) and density (kg/m^3)				
	Minimum density	Bending	Tension	Compression	Shear
		perpendicular to plane		in plane	perpendicular to plane
		$E_{m,flat,mean}$	$E_{t,mean}$	$E_{c,edge,mean}$	$G_{edge,mean}$
>1.8 – 12	650	3700	3100	3100	1000
>12 – 19	600	3200	2800	2800	1000
>19 – 30	550	3100	2700	2700	1000
>30	500	2800	2400	2400	800

Table 4.8 Values of k_{mod} and k_{def} for medium-density fibreboard according to the EN 622-5 (CEN 2010b) and EN 1995-1-1 (CEN 2010a) standards

Product	Service Class	k_{mod}-values with regard to Load-Duration Classes				
		Permanent	Long-term	Medium-term	Short-term	Instantaneous
MDF.LA	1	0.20	0.40	0.60	0.80	1.10
MDF.HLS	2				0.45	0.80
	3					
		k_{def}-values in				
		Service Class 1	Service Class 2	Service Class 3		
MDF.LA		2.25				
MDF.HLS		2.25	3.00			

Table 4.9 Equilibrium moisture content, shrinkage and swelling values, water vapor resistance and thermal conductivity of medium-density fibreboard (MDF) (Niemz et al. 1998[1], Schwab et al. 1997[2], Sattler 1999[3], Sonderegger and Niemz 2006[4])

1	2	3	4	5	6	7	8	9	10	11	12
Specification of wood-based material	Standard	**Equilibrium moisture content, u_{gl} (%) in relative humidity**			**Differential shrinkage** **V in % per % of change in wood moisture content in the direction of:**						
		ϕ=30 %	**ϕ=65 %**	**ϕ=85 %**	**thickness**	**length width**	**3**	**4**	**5**	**6**	**7**
MDF MDF MDF.H	EN 622-5 (CEN 2009a	6.4 (5.4–7.5)	7.7 (6.6–9.5)	14.1 (12.3–15.8)	0.840 (0.710–0.930)	0.021 (0.017–0.037)	[1, 4]				[1, 2, 3]
		Water vapor resistance factor μ in relation to density (kg/m³)			**Thermal conductivity λ (W/mK) in relation to density (kg/m³)**		No. in the literature [No.] refers to column				
		Density	**Wet cup μ**	**Dry cup μ**	**Density**	**λ**	**3**	**4**	**5**	**6**	**7**
MDF MDF	EN 622-5 (CEN 2009a)	250	2	5	250	0.05	EN 13986 (CEN 2015)				EN 13986 (CEN 2015)
		400	5	10	400	0.07					
		600	12	20	600	0.10					
		800	20	30	800	0.14					

Table 4.10 Characteristic strengths and densities of particleboard P4 according to the EN 12369-1 standard (CEN 2001)

Nominal thickness range (mm)	Characteristic strength (MPa) and density (kg/m³)					
	Minimum density	Bending	Tension	Compression	Shear	
		perpendicular to plane		in plane	perpendicular to plane	in plane
		$f_{m.k}$	$f_{t.k}$	$f_{c.edge.k}$	$f_{v.edge.k}$	$f_{v.flat.k}$
>6 – 13	650	14.2	8.9	12.0	6.6	1.8
>13 – 20	600	12.5	7.9	11.1	6.1	1.6
>20 – 25	550	10.8	6.9	9.6	5.5	1.4
>25 – 32	550	9.2	6.1	9.0	4.8	1.2
>32 – 40	500	7.5	5.0	7.6	4.4	1.1

Table 4.11 Mean modulus of elasticity and densities of particleboard P4 according to the EN 12369-1 standard (CEN 2001)

Nominal thickness range (mm)	Mean modulus of elasticity (MPa) and density (kg/m³)				
	Minimum density	Bending	Tension	Compression	Shear
		perpendicular to plane		in plane	perpendicular to plane
		$E_{m.flat.mean}$	$E_{t.mean}$	$E_{c.edge.mean}$	$G_{edge.mean}$
>6 – 13	650	3200	1800	1800	860
>13 – 20	600	2900	1700	1700	830
>20 – 25	550	2700	1600	1600	770
>25 – 32	550	2400	1400	1400	680
>32 – 40	500	2100	1200	1200	600

Table 4.12 Characteristic strengths and densities of particleboard P5 according to the EN 12369-1 standard (CEN 2001)

Nominal thickness range (mm)	Characteristic strength (MPa) and density (kg/m³)					
	Minimum density	Bending	Tension	Compression	Shear	
		perpendicular to plane		in plane	perpendicular to plane	in plane
		$f_{m.k}$	$f_{t.k}$	$f_{c.edge.k}$	$f_{v.edge.k}$	$f_{v.flat.k}$
>6 – 13	650	15.0	9.4	12.7	7.0	1.9
>13 – 20	600	13.3	8.5	11.8	6.5	1.7
>20 – 25	550	11.7	7.4	10.3	5.9	1.5
>25 – 32	550	10.0	6.6	9.8	5.2	1.3
>32 – 40	500	8.3	5.6	8.5	4.8	1.2

4.3 Organic-Bonded Particleboard

According to the EN 312 standard (CEN 2010), seven types of particleboard are classified: P1 - General purpose boards for use in dry conditions to P7 - Heavy duty load-bearing boards for use in humid conditions, which are specified by differentmechanical properties and are also related to the types of adhesives used.

4.3.1 Density, Strength, and Stiffness

See Tables 4.10, 4.11, 4.12, 4.13, 4.14, 4.15.

4.3.2 Creep and Duration of Load

See Table 4.16.

Table 4.13 Mean modulus of elasticity and densities of particleboard P5 according to the EN 12369-1 standard (CEN 2001)

Nominal thickness range (mm)	Mean modulus of elasticity (MPa) and density (kg/m³)				
	Minimum density	Bending	Tension	Compression	Shear
		perpendicular to plane		in plane	perpendicular to plane
		$E_{m.flat.mean}$	$E_{t.mean}$	$E_{c.edge.mean}$	$G_{edge.mean}$
>6 – 13	650	3500	2000	2000	960
>13 – 20	600	3300	1900	1900	930
>20 – 25	550	3000	1800	1800	860
>25 – 32	550	2600	1500	1500	750
>32 – 40	500	2400	1400	1400	690

Table 4.14 Characteristic strengths and densities of particleboard P7 according to the EN 12369–1 standard (CEN 2001)

Nominal thickness range (mm)	Characteristic strength (MPa) and density (kg/m³)					
	Minimum density	Bending	Tension	Compression	Shear	
		perpendicular to plane		in plane	perpendicular to plane	in plane
		$f_{m.k}$	$f_{t.k}$	$f_{c.edge.k}$	$f_{v.edge.k}$	$f_{v.flat.k}$
>6 – 13	650	18.3	11.5	15.5	8.6	2.4
>13 – 20	600	16.7	10.6	14.7	8.1	2.2
>20 – 25	550	15.4	9.8	13.7	7.9	2.0
>25 – 32	550	14.2	9.4	13.5	7.4	1.9
>32 – 40	500	13.3	9.0	13.2	7.2	1.9

Table 4.15 Mean modulus of elasticity and densities of particleboard P7 according to the EN 12369-1 standard (CEN 2001)

Nominal thickness range (mm)	Mean modulus of elasticity (MPa) and density (kg/m³)				
	Minimum density	Bending	Tension	Compression	Shear
		perpendicular to plane		in plane	perpendicular to plane
		$E_{m.flat.mean}$	$E_{t.mean}$	$E_{c.edge.mean}$	$G_{edge.mean}$
>6 – 13	650	4600	2600	2600	1250
>13 – 20	600	4200	2500	2500	1200
>20 – 25	550	4000	2400	2400	1150
>25 – 32	550	3900	2300	2300	1100
>32 – 40	500	3500	2100	2100	1050

Table 4.16 Values of k_{mod} and k_{def} for particleboard according to the EN 312 (CEN 2010c) and EN 1995-1-1 (CEN 2010a) standards

Product	Service Class	k_{mod}-value with regard to Load-Duration Class				
		Permanent	Long-term	Medium-term	Short-term	Instantaneous
P4, P5	1	0.30	0.45	0.65	0.85	1.10
P5	2	0.20	0.30	0.45	0.60	0.80
P6, P7	1	0.40	0.50	0.70	0.90	1.10
P7	2	0.30	0.40	0.55	0.70	0.90
		k_{def}-values				
		Service Class 1	**Service Class 2**	**Service Class 3**		
P4		2.25				
P5		2.25	3.00			
P6		1.50				
P7		1.50	2.25			

4.3.3 Moisture, Vapor, and Thermal Conductivity

See Table 4.17.

4.4 Inorganic-Bonded Particleboard

4.4.1 Density, Strength, and Stiffness

See Tables 4.18 and 4.19.

4.4.2 Creep and Duration of Load

See Table 4.20.

4.4.3 Moisture, Vapor, and Thermal Conductivity

See Table 4.21.

4.5 Oriented Strand Board (OSB)

4.5.1 Density, Strength, and Stiffness

See Tables 4.22, 4.23, 4.24 and 4.25.

4.5.2 Creep and Duration of Load

See Table 4.26.

4.5.3 Moisture, Vapor and Thermal Conductivity

See Table 4.27.

4.6 Plywood

4.6.1 Density, Strength, and Stiffness

There is only one Document, prEN 12,369 (CEN 1996), which gives characteristic values for Swedish, Finnish, US-American, Canadian, German, Italian and French plywood depending on wood species, hardwood and softwood, and layer thicknesses in different panel thickness ranges. Unfortunately, this major document was never officially approved.

Currently, plywood is characterised like laminated veneer lumber (LVL) only in strength classes showing bending strength and bending elasticity both parallel to and perpendicular to the fibre orientation of the surface layer. F20/10 E40/20 indicates a board showing characteristic bending strengths of 20 MPa parallel to and 10 MPa perpendicular to the fibre orientation and mean bending modulus (MOE) of 4000 MPa and 2000 MPa, respectively.

The following values apply only to the Strength Classes listed. Characteristic values for other Strength Classes not listed herein shall be determined on medium-sized test specimens (Tables 4.28 and 4.29).

4.6.2 Creep and Duration of Load

See Table 4.30.

Table 4.17 Equilibrium moisture content, shrinkage and swelling values, water vapor resistance and thermal conductivity of particleboard (PB) (Niemz et al. 1998[1], Sonderegger and Niemz 2006[2], Sattler 1999[3], Schwab et al. 1997[4], Niemz and Kucera 2000[5])

1	2	3	4	5	6	7	8	9	10	11	12
Specification of wood-based material	Standard	**Equilibrium moisture content, u_{gl} (%) in relative humidity**			**Differential shrinkage** **V in % per % of change in wood moisture content in the direction of**						
		ϕ=30 %	**ϕ=65 %**	**ϕ=85 %**	**thickness**	**length** **width**	3	4	5	6	7
PB P3, P5	EN 312 (CEN 2010c)	7.3 (6.5–8.1)	9.9 (9.1–10.9)	16.2 (14.5–19.3)	0.69 (0.45–0.93)	0.037 (0.030–0.050) 0.028 (0.026–0.031)	[1, 2, 5]			[1]	[1, 3, 4]
		Water vapor resistance factor μ in relation to density (kg/m³)			**Thermal conductivity λ (W/mK) in relation to density (kg/m³)**		No. in the literature [No.] refers to column				
		Density	**Wet cup μ**	**Dry cup μ**	**Density**	**λ**	3	4	5	6	7
PB	*EN 312* (CEN 2010c)	300	10	50	300	0.07	EN 13986 (CEN 2015)				EN 13986 (CEN 2015)
		600	15	50	600	0.12					
		900	20	50	900	0.18					

Table 4.18 Characteristic strengths and densities of cement-bonded particleboard according to the DiBt (1995, 2000, 2001, 2004a,b, 2006, 2011) structural design type approvals

Nominal thickness range (mm)	Characteristic strength (MPa) and density (kg/m³)							
	Minimum density	Bending		Tension	Compression		Shear	
		perpendicular to plane	in plane		perpendicular to plane	in plane	perpendicular to plane	in plane
		$f_{m.k}$	$f_{m.edge.k}$	$f_{t.k}$	$f_{c.flat.k}$	$f_{c.edge.k}$	$f_{v.edge.k}$	$f_{v.flat.k}$
8–13	1160	10.2	10.0	4.3	10.0	17.0	9.3	3.0
>13–20		9.2	8.7	3.9	10.0	14.6	7.5	2.5
>20–40		8.3	8.1	3.6	10.0	14.2	7.2	2.5

Table 4.19 Mean modulus of elasticity and densities of cement-bonded particleboard according to the DiBt (1995, 2000, 2001, 2004a,b, 2011) structural design type approvals

Nominal thickness range (mm)	Mean modulus of elasticity (MPa) and density (kg/m³)							
	Minimum density	Bending		Tension	Compression		Shear	
		perpendicular to plane	in plane		perpendicular to plane	in plane	perpendicular to plane	in plane
		$E_{m,flat,mean}$	$E_{m,edge,mean}$	$E_{t,mean}$	$E_{c,flat,mean}$	$E_{c,edge,mean}$	$G_{edge,mean}$	$G_{flat,mean}$
8–13	1160	5200	5100	4900		2250		2200
>13–20		5400	4950	4800		3036		2040
>20–40		5400	4950	4800		3036		2040

Table 4.20 Values of k_{mod} and k_{def} for cement-bonded particleboard (CBPB) according to DIN EN 1995-1-1/NA standard (DIN 2013)

Product	Service Class	k_{mod}-values with regard to Load-Duration Classes				
		Permanent	Long-term	Medium-term	Short-term	Instantaneous
CBPB	1	0.30	0.45	0.65	0.85	1.10
	2	0.20	0.30	0.45	0.60	0.80
	3					
		k_{def}-values in				
		Service Class 1	**Service Class 2**	**Service Class 3**		
CBPB		2.25	3.00			

Table 4.21 Equilibrium moisture content, shrinkage and swelling values, water vapor resistance and thermal conductivity of cement-bonded particleboard (DiBt 1995[1], 2000[2], 2001[3], 2004a,b[4–5], 2006[6], 2011[7], CEN 2007[8], Anon. 2004[9], CEN 2015[10])

1	2	3	4	5	6	7		8	9	10	11	12
Specification of wood-based material	Standard	**Equilibrium moisture content, u_{gl} (%)**			**Differential shrinkage**							
		in relative humidity			***V* in % per % of change in wood moisture content in the direction of**							
		ϕ=30 %	**ϕ=65 %**	**ϕ=85 %**	**thickness**	**length**	**width**	3	4	5	6	7
CBPB	[1-7]	5	10	14	0.04	0.03	0.03	[5]			[1-9]	
		Water vapor resistance factor *μ*			**Thermal conductivity λ(W/mK)**			No. in the literature [No.] refers to column				
		in relation to density (kg/m3)			**in relation to density (kg/m3)**							
		Density	**Wet cup μ**	**Dry cup μ**	**Density**	**λ**		3	4	5	6	7
CBPB	EN 634-2 [8]	1200	30	50	1200	0.23 … 0.35		[10]				[1-7, 10]

Table 4.22 Characteristic strengths and minimum densities of OSB/3 according to prEN 12369-3 (CEN 2020)

Nominal thickness range (mm)	Characteristic strength (MPa) and density (kg/m³)											
	Minimum density	Bending				Tension		Compression			Shear	
		perpendicular to plane		in plane				perpen-dicular to plane	in plane		perpen-dicular to plane	in plane
		$f_{m,0,k}$	$f_{m,90,k}$	$f_{m,0,edge,k}$	$f_{m,90,edge,k}$	$f_{t,0,k}$	$f_{t,90,edge,k}$	$f_{c,90,flat,k}$	$f_{c,0,k}$	$f_{c,90,edge,k}$	$f_{v,0,90,edge,k}$	$f_{v,0,90,flat,k}$
>6–10	550	18.0	9.0			9.9	7.2		15.9	12.9	6.8	1.0
>10–18		16.4	8.2			9.4	7.0		15.4	12.7	6.8	1.0
>18–25		14.8	7.4			9.0	6.8		14.8	12.4	6.8	1.0

4.6.3 Moisture, Vapor, and Thermal Conductivity

See Table 4.31.

4.7 Laminated Veneer Lumber (LVL)

4.7.1 Density, Strength, and Stiffness

Characteristic values for laminated veneer lumber (LVL) are related not to thickness ranges, but regulated according to strength classes, related to the strength in bending perpendicular to plane. The usual thickness of common LVL ranges of 24 mm to 75 mm.

The following values apply only to the strength classes listed. Characteristic values for other strength classes not listed herein shall be determined on medium-sized test specimens (Tables 4.32, 4.33, 4.34 and 4.35).

4.7.2 Creep and Duration of Load

See Table 4.36.

4.7.3 Moisture, Vapor, and Thermal Conductivity

See Table 4.37.

4.8 Solid-Wood Panels (SWP)

4.8.1 Density, Strength, and Stiffness

Table 4.23 Mean modulus of elasticity and minimum densities of OSB/3 according to the prEN 12369-3 standard (CEN 2020)

Nominal thickness range (mm)	Mean modulus of elasticity (MPa) and density (kg/m³)													
	Minimum density	Bending				Tension		Compression			Shear			
		perpendicular to plane		in plane				perpendicular to plane	in plane		perpendicular to plane		in plane	
		$E_{m.0.flat,mean}$	$E_{m.90.flat,mean}$	$E_{m.0.edge,mean}$	$E_{m.90.edge,mean}$	$E_{t.0,mean}$	$E_{t.90.edge,mean}$	$E_{c.90.flat,mean}$	$E_{c.0,mean}$	$E_{c.90.edge,mean}$	$G_{0.edge,mean}$	$G_{90.edge,mean}$	$G_{0.flat,mean}$	$G_{90.flat,mean}$
>6–10	550	4930	1980			3800	3000		3800	3000	1080		50	
>10–18														
>18–25														

Table 4.24 Characteristic strengths and minimum densities of OSB/4 according to the prEN 12369-3 standard (CEN 2020). Due to a lack of values in prEN 12369-3, characteristic values for bending in plane of OSB/4 were calculated from characteristic values taken from DiBt structural design type approvals (DiBt 2013, 2016)

Nominal thickness range (mm)	Characteristic strength (MPa) and density (kg/m³)												
	Minimum density	Bending				Tension		Compression				Shear	
		perpendicular to plane		in plane				perpendicular to plane	in plane			perpendicular to plane	in plane
		$f_{m.0.k}$	$f_{m.90.k}$	$f_{m.0.edge.k}$	$f_{m.90.edge.k}$	$f_{t.0.k}$	$f_{t.90.edge.k}$	$f_{c.0.90.flat.k}$	$f_{c.0.k}$	$f_{c.90.edge.k}$		$f_{v.0.90.edge.k}$	$f_{v.0.90.flat.k}$
>6–10	550	24.5	13.0	19.3	15.8	11.9	8.5	10	18.1	14.3		6.9	1.1
>10–18		23.0	12.2	19.3	15.8	11.4	8.2	10	17.6	14.0		6.9	1.1
>18–25		21.0	11.4	18.3	15.8	10.9	8.0	10	17.0	13.7		6.9	1.1

Table 4.25 Mean modulus of elasticity and minimum densities of OSB/4 according to the prEN 12369-3 standard (CEN 2020). Due to a lack of values in prEN 12369-3, characteristic values for bending in plane of OSB/4 were calculated from characteristic values taken from DiBt structural design type approvals (DiBt 2013, 2016)

Nominal thickness range (mm)	Mean modulus of elasticity (MPa) and density (kg/m³)											
	Minimum density	Bending				Tension		Compression			Shear	
		perpendicular to plane		in plane				perpendicular to plane	in plane		perpendicular to plane	in plane
		$E_{m,0,flat,mean}$	$E_{m,90,flat,mean}$	$E_{m,0,edge,mean}$	$E_{m,90,edge,mean}$	$E_{t,0,mean}$	$E_{t,90,edge,mean}$	$E_{c,90,flat,mean}$	$E_{c,0,mean}$	$E_{c,90,edge,mean}$	$G_{0,90,edge,mean}$	$G_{0,90,flat,mean}$
>6–10	550	6780	2680	3900	3100	4300	3200		4300	3200	1090	60
>10–18												
>18–25												

Table 4.26 Values of k_{mod} and k_{def} for oriented strand board (OSB) according to the EN 300 (CEN 2006) and the EN 1995-1-1 (CEN 2010a) standards

Product	Service Class	k_{mod} -values with regard to load duration classes				
		Permanent	Long-term	Medium-term	Short-term	Instantaneous
OSB/2	1	0.30	0.45	0.65	0.85	1.10
OSB/3	1	0.40	0.50	0.70	0.90	1.10
OSB/4	2	0.30	0.40	0.55	0.70	0.90
		k_{def} -values in				
		Service Class 1	**Service Class 2**	**Service Class 3**		
OSB/2		2.25				
OSB/3, OSB/4		1.50	2.25			

Table 4.27 Equilibrium moisture content, shrinkage and swelling values, water vapor resistance and thermal conductivity of oriented strand board (OSB) (Schwab et al. 1997[1], Niemz et al. 1998[2], Sattler 1999[3], Sonderegger and Niemz 2006[4])

1	2	3	4	5	6	7		8	9	10	11	12
Specification of wood-based material	Standard	**Equilibrium moisture content, u_{gl} (%) in relative humidity**			**Differential shrinkage** ***V* in %/% of change in moisture content in the direction of**							
		ϕ=30 %	**ϕ=65 %**	**ϕ=85 %**	**thickness**	**length**	**width**	**3**	**4**	**5**	**6**	**7**
OSB OSB/3, OSB/4	EN 300 (CEN 2006)	6.9	9.5	15.8	0.99	0.046	0.025	[3, 4]			[2]	[2–4]
		(6.6–7.5)	(8.9–9.8)	(15.4–16.2)	(0.74–1.20)	(0.037–0.067)	(0.015–0.032)					
		Water vapor resistance factor *μ* in relation to density (kg/m3)			**Thermal conductivity λ (W/mK) in relation to density (kg/m3)**			No. in the literature [No.] refers to column				
		Density	**Wet cup μ**	**Dry cup μ**	**Density**	**λ**		**3**	**4**	**5**	**6**	**7**
OSB		650			650	0.13		EN 13,986 (CEN 2015)				EN 13,986 (CEN 2015)

Table 4.28 Characteristic strengths and characteristic densities of plywood according to the DIN 20000–1 standard (DIN 2017)

Strength Class	Characteristic strength (MPa) and characteristic density (kg/m³)												
	Characteristic density	Bending				Tension		Compression			Shear		
		perpendicular to plane		in plane				perpen-dicular to plane	in plane		perpen-dicular to plane	in plane	
	ρ_k	$f_{m,0,k}$	$f_{m,90,k}$	$f_{m,0,edge,k}$	$f_{m,90,edge,k}$	$f_{t,0,k}$	$f_{t,90,edge,k}$	$f_{c,0,90,flat,k}$	$f_{c,0,k}$	$f_{c,90,edge,k}$	$f_{v,0,90,edge,k}$	$f_{v,0,flat,k}$	$f_{v,90,flat,k}$
F20/10 E40/20	350	20.0	10.0	9.0	7.0	9.0	7.0	4.0	15.0	10.0	3.5	0.9	0.6
F20/15 E30/25		20.0	15.0	8.0	7.0	8.0	7.0	4.0	13.0	13.0	4.0	1.0	0.7
F40/30 E60/40	600	40.0	30.0	29.0	31.0	29.0	31.0	9.0	21.0	22.0	2.2	1.1	1.1
F50/25 E70/25		50.0	25.0	36.0	24.0	36.0	24.0	10.0	36.0	17.0	2.5	1.1	1.1
F60/10 E90/10		60.0	10.0	36.0	24.0	36.0	24.0	10.0	26.0	18.0	2.5	1.1	1.1

Table 4.29 Mean modulus of elasticity and characteristic densities of plywood according to the DIN 20000-1 standard (DIN 2017)

Strength Class	Mean modulus of elasticity (MPa) and characteristic density (kg/m³)								
	Characteristic density	Bending				Shear			
		perpendicular to plane		in plane		perpendicular to plane		in plane	
	ρ_k	$E_{m,0,flat,mean}$	$E_{m,90,flat,mean}$	$E_{m,0,edge,mean}$	$E_{m,90,edge,mean}$	$G_{0,edge,mean}$	$G_{90,edge,mean}$	$G_{0,flat,mean}$	$G_{90,flat,mean}$
F20/10 E40/20	350	4000	2000	4000	3000	35	25	350	350
F20/15 E30/25		3000	2500	4000	3000	35	25	350	350
F40/30 E60/40	600	6000	4000	4400	4700	150	150	600	600
F50/25 E70/25		7000	2500	5500	3650	200	200	700	700
F60/10 E90/10		9000	1000	5500	3700	200	200	700	700

Table 4.30 Values of k_{mod} and k_{def} for plywood according to the EN 636 (CEN 2012) and EN 1995-1-1 (CEN 2010a) standards

Product	Service Class	k_{mod}-values with regard to Load Duration Classes				
		Permanent	Long-term	Medium-term	Short-term	Instantaneous
EN 636-1	1	0.60	0.70	0.80	0.90	1.10
EN 636-2	2	0.60	0.70	0.80	0.90	1.10
EN 636-3	3	0.50	0.55	0.65	0.70	0.90
		k_{def}-values in				
		Service Class 1	**Service Class 2**	**Service Class 3**		
EN 636-1		0.80				
EN 636-2		0.80	1.00			
EN 636-3		0.80	1.00	2.50		

Table 4.31 Equilibrium moisture content, shrinkage and swelling values, water vapor resistance and thermal conductivity of plywood

1	2	3	4	5	6	7		8	9	10	11	12
Specification of wood-based material	Standard	**Equilibrium moisture content, u_{gl} (%)**			**Differential shrinkage**							
		in relative humidity			***V* in % per % of change in wood moisture content in the direction of**							
		ϕ=30 %	**ϕ=65 %**	**ϕ=85 %**	**thickness**	**length**	**width**	**3**	**4**	**5**	**6**	**7**
Plywood	EN 636 (CEN 2015)					0.02	0.32					(DIN 2013)
		Water vapor resistance factor μ in relation to density (kg/m3)			**Thermal conductivity λ (W/mK) in relation to density (kg/m3)**			No. in the literature [No.] refers to column				
		Density	**Wet cup μ**	**Dry cup μ**	**Density**	**λ**		**3**	**4**	**5**	**6**	**7**
Plywood	EN 636 (CEN 2012)	300	50	150	300	0.09		EN 13986 (CEN 2015)			EN 13986 (CEN 2015)	EN 13986 (CEN 2015)
		500	70	200	500	0.13						
		700	90	220	700	0.17						
		1000	110	250	1000	0.24						

Table 4.32 Characteristic strengths and characteristic densities of LVL-P (all veneers laid in the same direction) according to Hakkarainen et al. (2019)

Strength Class	Characteristic strength (MPa) and characteristic density (kg/m³)													
	Characteristic density	Bending				Tension		Compression			Shear			
		perpendicular to plane		in plane				perpen-dicular to plane	in plane		perpendicular to plane		in plane	
		$f_{m,0,k}$	$f_{m,90,k}$	$f_{m,0,edge,k}$	$f_{m,90,edge,k}$	$f_{t,0,k}$	$f_{t,90,edge,k}$	$f_{c,90,flat,k}$	$f_{c,0,k}$	$f_{c,90,edge,k}$	$f_{v,0,edge,k}$	$f_{v,90,edge,k}$	$f_{v,0,flat,k}$	$f_{v,90,flat,k}$
LVL 32 P	410	32		27		22	0.5	0.8	26	4	3.2		2.0	
LVL 48 P	480	48		44		35	0.8	2.2	35	6	4.2		2.3	
LVL 80 P	730	80		75		60	1.5	12.0	69	14	8.0		8.0	

Table 4.33 Mean modulus of elasticity and characteristic densities of LVL-P (all veneers laid in the same direction) according to Hakkarainen et al. (2019)

Strength Class	Mean modulus of elasticity (MPa) and characteristic density (kg/m³)													
	Characteristic density	Bending				Tension		Compression			Shear			
		Perpendicular to plane		in plane				perpendicular to plane	in plane		perpendicular to plane		in plane	
		$E_{m,0,flat,mean}$	$E_{m,90,flat,mean}$	$E_{m,0,edge,mean}$	$E_{m,90,edge,mean}$	$E_{t,0,mean}$	$E_{t,90,edge,mean}$	$E_{c,90,flat,mean}$	$E_{c,0,mean}$	$E_{c,90,edge,mean}$	$G_{0,edge,mean}$	$G_{90,edge,mean}$	$G_{0,flat,mean}$	$G_{90,flat,mean}$
LVL 32 P	410	9600		9600		9600			9600		500		320	
LVL 48 P	480	13800		13800		13800	430		13800	430	600		380	
LVL 80 P	730	16800		16800		16800	470		16800	470	760		850	

Table 4.34 Characteristic strengths and characteristic densities of LVL-C (20% of the veneers glued in a cross-wise direction) according to Hakkarainen et al. (2019)

Strength Class	Characteristic strength (MPa) and characteristic density (kg/m³)													
	Characteristic density	Bending				Tension		Compression			Shear			
		perpendicular to plane		in plane				perpendicular to plane	in plane		perpendicular to plane		in plane	
		$f_{m,0,k}$	$f_{m,90,k}$	$f_{m,0,edge,k}$	$f_{m,90,edge,k}$	$f_{t,0,k}$	$f_{t,90,edge,k}$	$f_{c,90,flat,k}$	$f_{c,0,k}$	$f_{c,90,edge,k}$	$f_{v,0,edge,k}$	$f_{v,90,edge,k}$	$f_{v,0,flat,k}$	$f_{v,90,flat,k}$
LVL 22 C	410	22		19		14	4	1.0	18	8	3.6		1.1	
LVL 32 C	480	32	7	28		18	5	2.2	18	9	4.5		1.3	0.6
LVL 75 C	730	75	20	60		51	8	16.0	64	23	7.8		3.8	

Table 4.35 Mean modulus of elasticity and characteristic densities of LVL-C (20% of the veneers glued in a cross-wise direction) according to Hakkarainen et al. (2019)

Strength Class	Mean modulus of elasticity (MPa) and characteristic density (kg/m³)													
	Characteristic density	Bending				Tension		Compression			Shear			
		perpendicular to plane		in plane				perpendicular to plane	in plane		perpendicular to plane		in plane	
		$E_{m,0,flat,mean}$	$E_{m,90,flat,mean}$	$E_{m,0,edge,mean}$	$E_{m,90,edge,mean}$	$E_{t,0,mean}$	$E_{t,90,edge,mean}$	$E_{c,90,flat,mean}$	$E_{c,0,mean}$	$E_{c,90,edge,mean}$	$G_{0,edge,mean}$	$G_{90,edge,mean}$	$G_{0,flat,mean}$	$G_{90,flat,mean}$
LVL 22 C	410	6700		6700		6700			6700		500		70	
LVL 32 C	480	10000	1200	10000	2400	10000	2400		10000	2400	600		80	22
LVL 75 C	730	13200		13200		13200			13200		820		430	

Table 4.36 Values of k_{mod} and k_{def} for LVL according to the EN 1995-1-1 standard (CEN 2010a) and Hakkarainen et al. (2019)

Product	Service Class	k_{mod}-values with regard to Load Duration Classes				
		Permanent	Long-term	Medium-term	Short-term	Instantaneous
LVL-P	1	0.60	0.70	0.80	0.90	1.10
LVL-C	2	0.60	0.70	0.80	0.90	1.10
	3	0.50	0.55	0.65	0.70	0.90
		k_{def}-values in				
		Service Class 1	**Service Class 2**	**Service Class 3**		
LVL-P		0.60	0.80	2.00		
LVL-C		0.80	1.00	2.50		

Table 4.37 Equilibrium moisture content, shrinkage and swelling values, water vapor resistance and thermal conductivity of LVL

1		2	3	4	5	6	7		8	9	10	11	12
Specification of wood-based material		Standard	**Equilibrium moisture content, u_{gl} (%) in relative humidity**			**Differential shrinkage** *V* **in % per % of change in wood moisture content in the direction of**							
			$\phi = 30$ %	**$\phi = 65$ %**	**$\phi = 85$ %**	**thickness**	**length**	**width**	3	4	5	6	7
LVL	LVL-P LVL-C	Hakkarainen et al. (2019)	6.40	11.40	15.90	0.32	0.01	0.32 0.03	Hakkarainen et al. (2019)			Hakkarainen et al. (2019)	Hakkarainen et al. (2019)
			Water vapor resistance factor μ in relation to density (kg/m³)			**Thermal conductivity λ (W/mK) in relation to density (kg/m³)**			No. in the literature [No.] refers to column				
			Density	**Wet cup µ**	**Dry cup µ**	**Density**	**λ**		3	4	5	6	7
LVL		EN 14279 (CEN 2009b)	300	30	150	300	0.09		EN 13986 (CEN 2015), Hakkarainen et al. (2019)				EN 13986 (CEN 2015), Hakkarainen et al. (2019)
			440	65	180	500	0.13						
			510	70	200	700	0.17						
			700	90	220	1000	0.24						
			1000	110	250								

Table 4.38 Characteristic strengths and densities of solid-wood panels according to the prEN 12369-3 standard (CEN 2020)

Nominal thickness range (mm)	Characteristic strength (MPa) and density (kg/m³)													
	Minimum density	Bending				Tension		Compression			Shear			
		perpendicular to plane		in plane				perpendicular to plane	in plane		perpendicular to plane		in plane	
		$f_{m,0,k}$	$f_{m,90,k}$	$f_{m,0,edge,k}$	$f_{m,90,edge,k}$	$f_{t,0,k}$	$f_{t,90,edge,k}$	$f_{c,90,flat,k}$	$f_{c,0,k}$	$f_{c,90,edge,k}$	$f_{v,0,edge,k}$	$f_{v,90,edge,k}$	$f_{v,0,flat,k}$	$f_{v,90,flat,k}$
12–20	410	30.0	5.0	25.0	12.0	12.0	3.0		18.0	12.0	4.0	4.0	1.0	1.0
>20–30		27.0	5.0	18.0	12.0	9.0	3.0		16.0	10.0	4.0	4.0	1.0	1.0
>30–80		20.0	10.0	12.0	12.0	6.0	3.0		10.0	10.0	2.5	2.5	1.0	1.0

Table 4.39 Mean moduli of elasticity and densities of solid-wood panels according to the prEN 12369-3 standard (CEN 2020)

Nominal thickness range (mm)	Mean modulus of elasticity (MPa) and density (kg/m³)													
	Minimum density	Bending				Tension		Compression			Shear			
		perpendicular to plane		in plane				perpen-dicular to plane	in plane		perpendicular to plane		in plane	
		$E_{m,0,flat,mean}$	$E_{m,90,flat,mean}$	$E_{m,0,edge,mean}$	$E_{m,90,edge,mean}$	$E_{t,0,mean}$	$E_{t,90,edge,\ mean}$	$E_{c,90,flat,mean}$	$E_{c,0,mean}$	$E_{c,90,edge,mean}$	$G_{0,edge,mean}$	$G_{90,edge,mean}$	$G_{0,flat,mean}$	$G_{90,flat,mean}$
12–20	410	10,000	650	6000	4000	6000	4000		6000	4000	450	450	50	50
>20–30		10000	800	5000	4000	5000	4000		3500	2500	450	450	50	50
>30–80		8000	1800	4000	4000	4000	4000		2500	2500	450	450	50	50

See Tables 4.38 and 4.39.

4.8.2 Creep and Duration of Load

See Table 4.40.

4.8.3 Moisture, Vapor, and Thermal Conductivity

See Table 4.41.

Table 4.40 Values of k_{mod} and k_{def} for solid-wood panels (SWP) according to the EN 13,353 (CEN 2011) and DIN EN 1995-1-1/NA standards (DIN 2013)

Product	Service Class	k_{mod}-values with regard to load duration classes				
		Permanent	Long-term	Medium-term	Short-term	Instantaneous
SWP/1	1	0.6	0.7	0.8	0.9	1.1
SWP/2	2	0.6	0.7	0.8	0.9	1.1
SWP/3	3					
		k_{def}-values in				
		Service Class 1	**Service Class 2**	**Service Class 3**		
SWP/1, SWP/2, SWP/3		0.6	0.8			

Table 4.41 Equilibrium moisture content, shrinkage and swelling values of solid-wood panels (SWP)

1	2	3	4	5	6	7		8	9	10	11	12
Specification of wood-based material	Standard	**Equilibrium moisture content (%) in relative humidity**			**Differential shrinkage** ***V*** **in %/% of change in moisture content in the direction of**							
		$\phi=30$ %	$\phi=65$ %	$\phi=85$ %	**thickness**	**length**	**width**	**3**	**4**	**5**	**6**	**7**
SWP	SWP/1 SWP/2 SWP/3	EN 13353 (CEN 2011)	8.8 (8.7-8.9)	12.5 (12.2-12.8)	17.8 (7.4-18.1)	0.62	0.02 (0.008-0.035)	0.042 (0.024-0.067)	Niemz et al. (1998), Tobisch (2006)			Niemz et al. (1998)	Schwab et al. (1997), Niemz et al. (1998), Tobisch (2006)
		Water vapor resistance factor μ			**Thermal conductivity** λ **(W/mK)**			No. in the literature refers to column				
		Density (kg/m³)	**Wet cup** μ	**Dry cup** μ	**Density (kg/m³)**	λ		**3**	**4**	**5**	**6**	**7**
SWP	SWP/1	EN 13353	300	50	150	300	0.09	EN 13986 (CEN 2015)			EN 13986 (CEN 2015)	EN 13986 (CEN 2015)
	SWP/2	(CEN 2011)	500	70	200	500	0.13					
	SWP/3		700	90	220	700	0.17					
			1000	110	250	1000	0.24					

References

Anon (2004) Cement bonded particleboard (CBPB). Tech. Rep. Building Research Establishment (BRE), London

CEN (1996) prEN 12369, Wood-based panels - Characteristic values for established products. CEN–European Committee for Standardization, Brussels

CEN (2001) EN 12369-1, Wood-based panels - Characteristic values for structural design - Part 1: OSB, particleboards and fibreboards. CEN–European Committee for Standardization, Brussels

CEN (2004a) EN 314, Plywood – Bonding quality. CEN-European Committee for Standardization, Brussels

CEN (2004b) EN 622-3, Fibreboards - Specifications - Part 3: Requirements for medium boards. CEN–European Committee for Standardization, Brussels

CEN (2006) EN 300, Oriented strand boards (OSB) - Definitions, classification and specifications. CEN–European Committee for Standardization, Brussels

CEN (2007) EN 634-2, Cement-bonded particleboards - Specifications - Part 2: Requirements for OPC bonded particleboards for use in dry, humid and external conditions. CEN–European Committee for Standardization, Brussels

CEN (2009a) EN 622-5, Fibreboards - Specifications - Part 5: Requirements for dry process boards (MDF). CEN–European Committee for Standardization, Brussels

CEN (2009b) EN 14279, Laminated veneer lumber (LVL) - Definitions, classification and specifications. CEN–European Committee for Standardization, Brussels

CEN (2010a) EN 1995-1-1, Eurocode 5: Design of timber structures - Part 1-1: General - Common rules and rules for buildings. CEN–European Committee for Standardization, Brussels

CEN (2010b) EN 1534, Wood flooring - Determination of resistance to indentation: Test method. CEN–European Committee for Standardization, Brussels

CEN (2010c) EN 312, Particleboards – Specifications. CEN–European Committee for Standardization, Brussels

CEN (2011) EN 13353, Solid wood panels (SWP) – Requirements. CEN–European Committee for Standardization, Brussels

CEN (2012) EN 636, Plywood – Specifications. CEN–European Committee for Standardization, Brussels

CEN (2020) prEN 12369-3, Wood-based panels – Characteristic values for structural design - Part 3: Solid wood panels. CEN–European Committee for Standardization, Brussels

DiBt (1995) Z-9.1–324, Agepan zementgebundene Spanplatte für die Holztafelbauart. (Agepan cement-bonded chipboard for wooden panel construction). Technical Report, Deutsches Institut für Bautechnik, Berlin

DiBt (2000) Z-9.1–173, Mineralisch gebundene Flachpressplatten "Isopanel" für die Holztafelbauart (Mineral bonded flat pressed panels "Isopanel" for wooden panel construction). Technical Report, Deutsches Institut für Bautechnik, Berlin

DiBt (2001) Z-9.1–120, Mineralisch gebundene Flachpressplatten (Holzzementplatten) "Duripanel" (Mineral bonded flat pressed panels (Wood Cement Board) "Duripanel"). Technical Report, Deutsches Institut für Bautechnik, Berlin

DiBt (2004a). Z-9.1–267, Mineralisch gebundene Flachpressplatten "CETRIS". (Mineral-bonded flat pressed boards "CETRIS"). Technical Report, Deutsches Institut für Bautechnik, Berlin

DiBt (2004b) DiBt: Z-9.1–89, Mineralisch gebundene Flachpressplatten "Betontyp" (Mineral-bound flat pressed panels "concrete type") Technical Report, Deutsches Institut für Bautechnik, Berlin

DiBt (2006) Z-9.1–340, Spezialbauplatte "Cospanel" (Special construction board "Cospanel). Technical Report, Deutsches Institut für Bautechnik, Berlin

DiBt (2011a) Z-9.1–328, Cospan-Massivbauplatte (Cospan Solid Board). Technical Report, Deutsches Institut für Bautechnik, Berlin

DiBt (2013) Z-9.1–566, EUROSTRAND OSB 4 TOP. Technical Report, Deutsches Institut für Bautechnik, Berlin

DiBt (2016) Z-9.1–503, SWISS KRONO OSB/4. Technical Report, Deutsches Institut für Bautechnik.

DIN (2013) DIN EN 1995-1-1/NA, Nationaler Anhang - National festgelegte Parameter - Eurocode 5: Bemessung und Konstruktion von Holzbauten - Teil 1-1: Allgemeines - Allgemeine Regeln und Regeln für den Hochbau (National Annex - Nationally determined parameters - Eurocode 5: Design of timber structures - Part 1-1: General - Common rules and rules for buildings). DIN–Deutsches Institut für Normung e.V., Berlin

CEN (2015) EN 13986, Wood-based panels for use in construction - Characteristics, evaluation of conformity and marking. CEN–European Committee for Standardization, Brussels

DIN (2017) 20000-1, Anwendung von Bauprodukten in Bauwerken – Teil 1: Holzwerkstoffe (Use of building products in buildings - Part 1: Wood-based materials). DIN–Deutsches Institut für Normung e.V., Berlin

Hakkarainen J, Linkosalmi L, Huovinen A et al (2019) LVL handbook europe. Federation of the Finnish Woodworking Industries, Helsinki

Niemz P, Kucera LJ, Hebeisen S (1998) Quellverhalten von Holzwerkstoffen (Swelling behavior of wood based materials). Holz-Zent.bl 121:1772

Niemz P, Kucera LJ (2000) Quellung von Spanplatten und Buchensperrholz, Untersuchungen zur Gleichgewichtsfeuchtigkeit sowie Längen- und Dickenquellung (Swelling of particleboard and beech plywood, investigations on equilibrium moisture content as well as swelling in length and thickness). Holz-Zent.bl 27:368

Sattler H (1999) Hygrisch bedingte Dimensionsänderungen, Diskussion von Ergebnissen aus den Untersuchungen von Holzwerkstoffen (Schluss) (Hygric dimensional changes, discussion of results from the investigations of wood-based materials (conclusion)) Holz Kunstst.verarb. 34(12):36, 38–40

Schwab E, Steffen A, Korte C (1997) Feuchtebedingte Längenänderung von Holzwerkstoffen in Plattenebene (Humidity-related change in length of wood-based materials in panel level). Holz Roh- Werkstoff 55:227–233

Sonderegger W, Niemz P (2006) Untersuchungen zur Quellung und Wärmedehnung von Faser-, Span- und Sperrholzplatten (Investigation of swelling and thermal expansion of fibreboard, particleboard and plywood). Holz Roh- Werkstoff 64:11–20

Tobisch S (2006) Methoden zur Beeinflussung ausgewählter Eigenschaften von dreilagigen Massivholzplatten aus Nadelholz (Methods for influencing selected properties of three-layer solid wood panels made of softwood). Dissertation, Universität Hamburg

Wiegand T (2020) Comparison of symbols for Cross laminated timber, LVL, GLVL, plywood and SWP. CEN/TC 250 SC5 WG1. Technical Report (CEN/TC 250 SC5 WG1), CEN–European Committee for Standardization, Brussels

Prof. Dr. Steffen Tobisch studied Process Engineering at TU Dresden, Germany, specialising in Wood Technology and Fibre Materials Technology. Since 1993, he has been working on the non-destructive testing of materials, since 1998 at the Institut für Holztechnologie Dresden gemeinnützige GmbH (IHD). In 2000, he was appointed head of the Materials Department; in 2004, he became head of the IHD. He received his doctorate from the University of Hamburg, Germany, in 2006 and the honorary professorship at the Technische Universität Dresden in 2016. His tasks included standardising European wood-based materials as chairman of CEN/TC 112 and manufacturing technology. He is one of the publishers of the scientific journal Holztechnologie.

5 Selected Material Parameters

Peter Niemz, Dick Sandberg and Rico Emmler

5.1 Introduction

Peter Niemz, Dick Sandberg

This chapter provides information on the properties of first-hand wood-based panels, with a focus on aspects and relationships not covered in the relevant standards—specifically, information that is not included in Chapter 4 of this handbook. It also presents valuable data regarding adhesives and coatings. The chapter covers the rigidity of various timber connection types, such as glued, nailed, and dowelled joints, as well as the impact of load duration in timber construction. Additionally, correction factors for both static and dynamic flexural behaviour are discussed. Special attention is given to the ageing of wood and glued wood. Finally, the chapter includes information on other natural fibres beyond wood.

5.2 Solid Wood

Peter Niemz, Dick Sandberg

5.2.1 Maximum Tangential Swelling in Different Liquids

See Table 5.1

P. Niemz (✉)
Department Wood Physics (today, Wood Materials Science), Institute of Building Materials, ETH Zürich, Zurich, Switzerland
e-mail: niemzp@retired.ethz.ch

D. Sandberg
Department of Manufacturing and Civil Engineering, Norwegian University of Science and Technology (NTNU), Gjøvik, Norway
e-mail: dick.sandberg@ntnu.no

R. Emmler
Entwicklungs- und Prüflabor Holztechnologie GmbH, Dresden, Germany
e-mail: rico.emmler@eph-dresden.de

5.3 Wood-Based Panels

Peter Niemz, Dick Sandberg

5.3.1 Moisture-Related Properties

See Table 5.2, 5.3 and 5.4

5.3.2 Fibreboard

See Fig. 5.1, 5.2

5.3.3 Organic-Bonded Particleboard

See Fig. 5.3, 5.4 and 5.5

5.3.4 Inorganic-Bonded Particleboard

See Table 5.5 and 5.6

5.3.5 Oriented Strand Board (OSB)

See Table 5.7

5.3.6 Plywood

See Table 5.8

5.3.7 Laminated Veneer Lumber (LVL)

See Table 5.9 and 5.10

P. Niemz et al. (eds.), *Wood Material and Processing Data*, https://doi.org/10.1007/978-3-031-38329-8_5

Table 5.1 Maximum tangential swelling in various solvents relative to that in water at 23 °C (Mantanis et al. 1994a, b). The organic solvents were selected so they represented various solvent chemical classes (Class 1 to 5), i.e. the solvents have different properties (hydrogen-bonding parameter, molar volume, cohesive energy density, basicity, dipole moment, dielectric constant, volubility parameter, etc.) known to influence the wood swelling phenomenon in various ways

Classes	Medium	Douglas-fir I	Douglas-fir II	Quaking aspen	Sitka spruce	Sugar maple
REF	Water at 23°C	100	100	100	100	100
Class 1	Formamide	109.1	113.0	134.8	133.3	158.5
	Ethylene glycol	103.4	116.0	113.5	113.1	98.1
	Methanol	83.0	92.0	94.4	97.6	82.1
	Ethanol	71.6	73.0	86.5	83.3	65.1
	Propanol	58.0	66.0	67.4	58.3	50.0
Class 2	DMF	130.7	147.0	148.3	140.5	124.5
	Pyridine	129.5	150.0	136.0	145.2	125.5
	Acetone	52.3	72.0	84.3	67.9	67.0
	2-Butanone	46.6	56.0	67.4	51.2	48.1
	Dioxane	85.2	89.0	102.2	67.9	79.2
	Ethyl acetate	30.7	26.0	61.8	31.0	34.9
	Butyraldehyde	7.5	8.4	38.2	8.3	11.4
	Methyl acetate	53.4	52.0	64.0	59.5	52.8
	Propyl acetate	15.9	17.0	55.1	26.2	29.2
Class 3	Nitromethane	45.5	46.0	67.4	53.6	50.9
	Butyrolacetone	85.2	86.0	103.4	85.7	90.6
	Furfural	61.4	57.0	87.6	65.5	71.7
Class 4	Chloroform	17.0	17.0	61.8	16.7	36.8
	Ethylene dichloride	23.9	24.0	56.2	25.0	43.4
	Carbon tetrachloride	14.8	14.0	13.5	14.3	10.4
Class 5	Benzaldehyde	10.2	12.0	68.0	11.9	19.8
	Nitrobenzene	4.5	5.6	23.6	6.0	6.8
	Toluene	17.0	17.0	18.0	19.0	14.2
	Isopropyl ether	14.8	15.0	14.6	17.9	9.4
	Benzyl alcohol	29.5	22.0	107.9	34.5	84.0
	Quinoline	3.3	5.6	27.0	4.4	5.3
	Octane	8.0	7.0	9.0	10.0	5.5
Others	Pyrrole	78.4	78.0	113.5	90.5	113.2
	DMSO	155.7	167.0	161.8	165.5	136.8
	Benzyl benzoate	10.2	11.0	10.1	11.9	9.4
Acids	Formic acid	197.7	214.0	191.0	177.4	191.5
	Acetic acid	86.4	93.0	130.3	103.6	96.2
	Propionic acid	71.6	72.0	89.9	76.2	94.3
Amines	2,6-Dimethylpyridine	11.0	11.0	91.0	15.0	59.0
	Diethylamine	118.2	124.0	119.1	120.2	103.8
	Butylamine	186.4	211.0	204.5	172.6	182.1
	Dibetylamine	11.4	11.0	10.1	15.5	7.5
	2-Methylpyridine	131.8	128.0	131.5	128.6	112.3
	Piperidine	7.0	6.0	138.2	10.7	40.6

5.3.8 Solid-Wood Panels

According to Muszynski (2021), the annual global output of cross-laminated timber (CLT) in 2020 was approx. 1.56 million m^3 produced in 76 production lines. The global annual capacity per-shift was then approx. 1.22 million cubic metre. The Alpine Region accounted for over 70% of the output volume and almost 65% of the annual per-shift capacity (Fig. 5.6). It should be noted that in North America, a difficult to assess volume is produced as non-structural panels for the industrial use.

Other mass-timber-panel products like the Massiv-Holz-Mauer (MHM) system (solid-timber wall system), dowel-joined CLT, and other similar products, probably contributed another 75,000 m^3 of panel products to the mass-timber panel market in 2019.

5.4 Adhesives

Peter Niemz, Dick Sandberg

5.4.1 General

See Table 5.14, 5.15, 5.16 and 5.17 and Fig. 5.8 also Table 5.18

Table 5.2 Diffusion resistance factors of wood-based materials according to the EN ISO 10456 standard (CEN 2007)

Material	Density	Diffusion resistance factor, μ (-)	
	(kg/m³)	dry	wet
Solid wood	450–500	50	20
	700	200	50
Plywood, LVL	300	150	50
	500	200	70
	700	220	90
	1000	250	110
Particleboard	300	50	10
	600	50	15
	900	50	20
OSB	650	50	30
Cement-bonded particleboard	1200	50	30
Insulation fibreboard	40–250	5	3
Fibreboard, MDF	400	10	5
	600	20	12
	800	30	20

Table 5.3 Shinkage and swelling of wood-based materials (Halász and Scheer 1986). EMC – equilibrium moisture content, MDF/HDF – medium/high-density fibreboard, PF - phenolformaldehyde adhesive, UF – urea–formaldehyde adhesive

Material	Swelling/shrinking (% per % change in EMC)	
	Perpendicular to the plane	In plane
Plywood	0.30	0.020
Particleboard bonded with PF	0.45	0.025
Particleboard bonded with UF	0.70	0.035
MDF, HDF	0.80	0.040

Table 5.4 Equilibrium moisture content (EMC) of wood and wood-based materials under different climatic conditions (Halász and Scheer 1995). MDF/HDF – medium/high-density fibreboard, PF – phenol–formaldehyde adhesive, UF – urea–formaldehyde adhesive

Material	EMC at 20 °C and relative humidisty		
	30%	65%	85%
Cement-bonded boards		2–4	
Gypsum-bonded panels		2–3	
MDF, HDF	3–5	6–8	10–14
Particleboard bonded with PF	4–6	10–12	15–23
Particleboard bonded with other adhesives	4–8	9–12	13–18
Plywood	4–6	8–12	12–18
Veneer		6–14	
Wood		12	

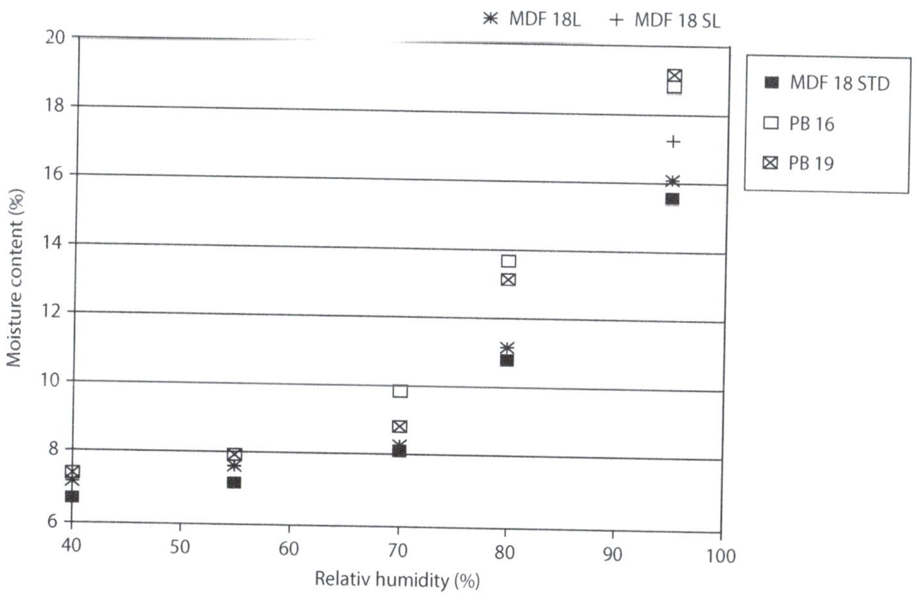

Fig. 5.1 Sorption isotherms of particleboard (PB) and medium-density fibreboard (MDF). Material codes: 16, 18 and 19 are the thickness in mm, STD is a density of 700 kg/m³, L is 600 kg/m³, and SL is 500 kg/m³ (Niemz et al. 1999)

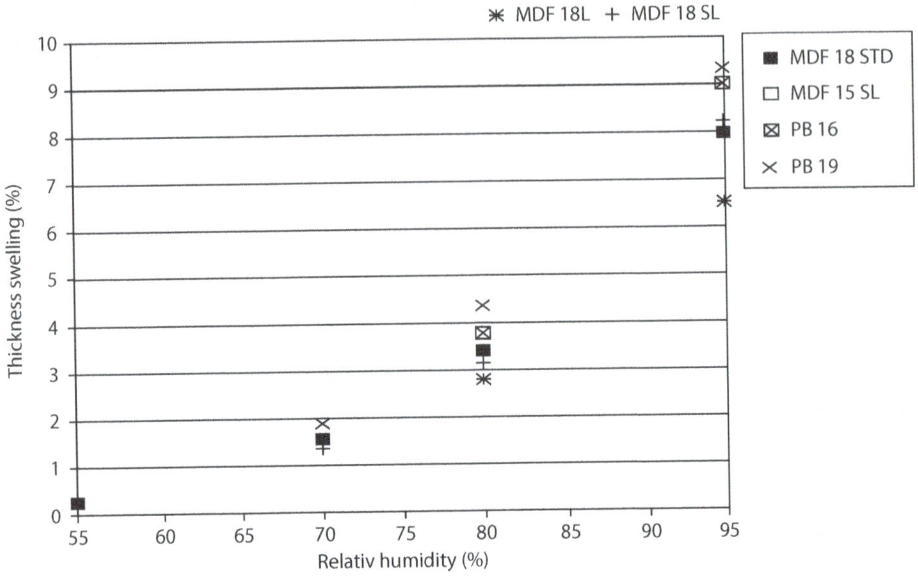

Fig. 5.2 Thickness swelling of different particleboard (PB) and medium-density fibreboard (MDF) as a function of relative humidity. Material codes: 15, 16, 18 and 19 are the thickness in mm, STD is a density of 700 kg/m^3, L is 600 kg/m^3, and SL is 500 kg/m^3 (Niemz et al. 1999)

Manufacturers tend to select local structural grade softwoods for the manufacture of CLT, although some manufacturers sometimes import sawn timber from distant

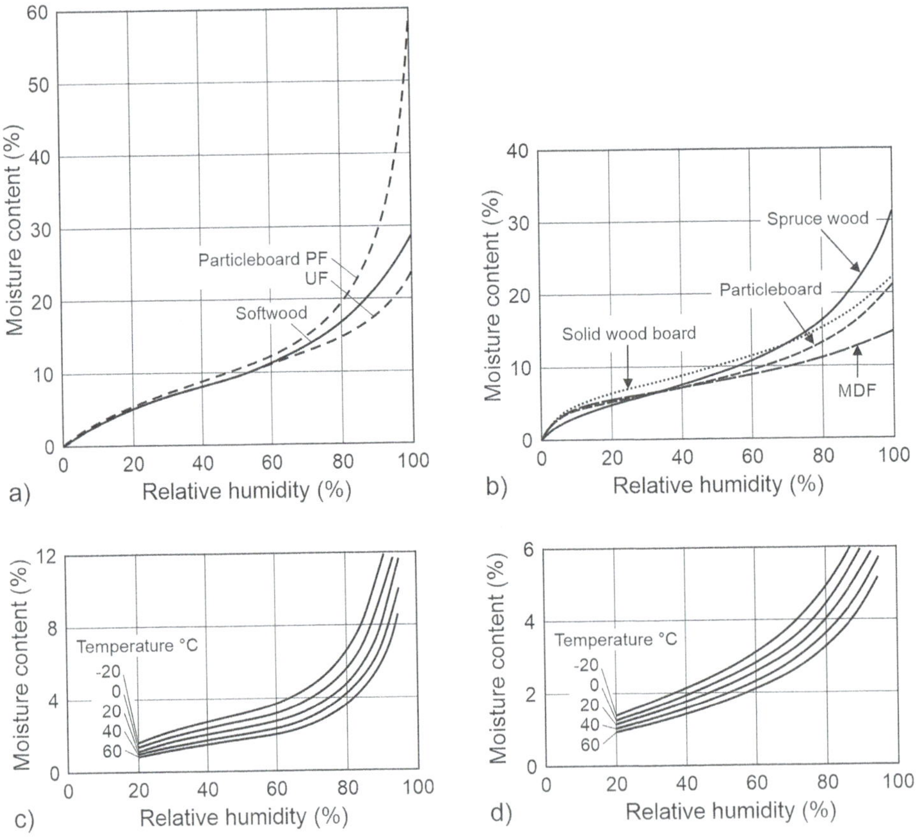

Fig. 5.3 Sorption of particleboard bonded with phenolformaldehyde (PF) or urea–formaldehyde (UF) adhesive: a) sorption of different wood-based materials and softwoods (Anon. 1975), b) solid wood, particleboard, MDF, solid-wood panels (Dunky 2002), (c) gypsum-bonded particleboard (Greubel and Drewes 1987), and (d) cement-bonded particleboard (Greubel and Drewes 1987)

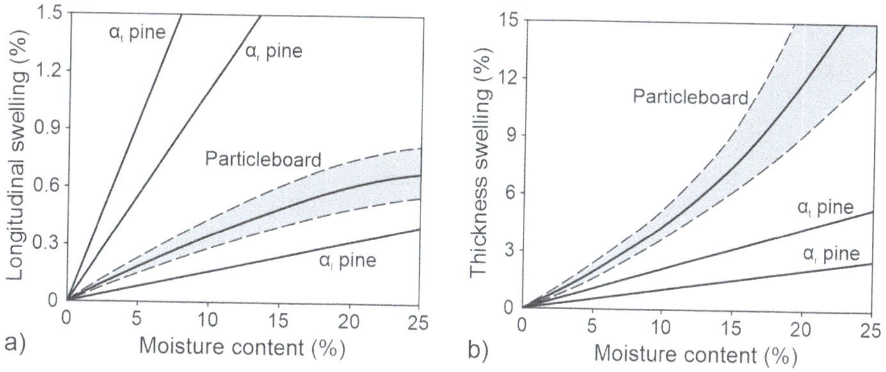

Fig. 5.4 Swelling in (a) the longitudinal and (b) the thickness directions of particleboard and Scots pine in the longitudinal (α_l), radial (α_r) and tangential (α_t) directions. After Mörath and Teischgraeber in Anon. (1975)

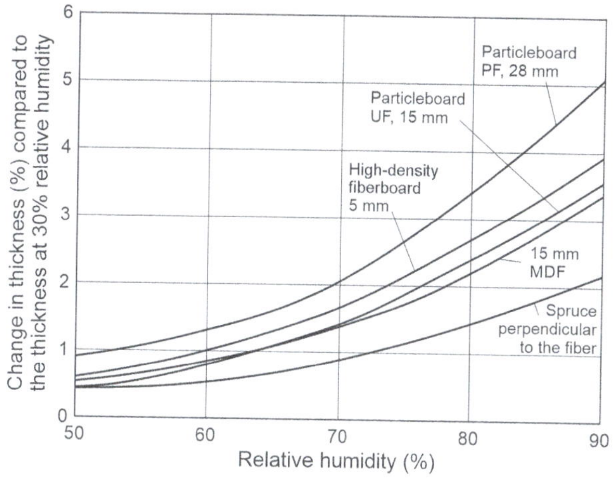

Fig. 5.5 Thickness swelling of wood-based materials as a function of the relative humidity (Niemz and Sonderegger 2021)

Table 5.5 Properties of cement-bonded particleboard (Neroth and Vollenschaar 2011). MOE – modulus of elasticity

Property	Unit	
Thickness	(mm)	8–32
Density	(kg/m³)	1000–1350
MOE in bending	(MPa)	4500–5000
Bending strength	(MPa)	9–18
Internal bond strength	(MPa)	0.4–0.6
Thickness swelling after 2 h in water	(%)	max. 1.5
Thickness swelling after 24 h in water	(%)	max. 2.0

markets if the prices are favourable. It is different to find precise information on the volumes of individual species used by the CLT industry, but Fig. 5.7 give a general idea of the diversity in species (Table 5.11, 5.12 and 5.13).

Table 5.6 Characteristic values of light-weight panels of wood wool (Neroth and Vollenschaar 2011)

Thickness (mm)	Density (kg/m³)	Bending strength (MPa)	Thermal conductivity (W/mK)
10–15	570–800	Up to 1.7	0.150
25–100	360–460	0.4–1.0	0.093

5.4.2 Properties of adhesive films and bond-lines

See Fig. 5.9, 5.10, 5.11 and Table 5.19, also Fig. 5.12, 5.13 and 5.14

5.5 Coatings

Rico Emmler

5.5.1 Properties of coating materials

See Table 5.20, 5.21, 5.22, 5.23, 5.24, 5.25 and 5.26

5.6 Loading of Materials

Peter Niemz, Dick Sandberg

5.6.1 Relation Between Strength and Density

See Fig. 5.15

Table 5.7 Properties of OSB (board thickness 10–18 mm) according to the EN 300 standard (CEN 2006)

OSB type	Bending strength		Modulus of elasticity (MOE)	
	Parallel to the particle orientation (MPa)	Perpendicular to particle orientation (MPa)	Parallel to the particle orientation (MPa)	Perpendicular to particle orientation (MPa)
OSB 1	18	9	2500	1200
OSB 2	20	10	3500	1400
OSB 3	20	10	3500	1400
OSB 4	28	15	4800	1900

Table 5.8 Properties of plywood (Niemz and Sonderegger 2021)

Property	Unit	Direction of load	Thickness (mm)		
			≤8	>8–15	15–29
Bending strength perpendicular to the plane	(MPa)	// ⊥	65–130 7.5–19	58–110 25–70	50–75 30–60
Bending strength in the plane direction	(MPa)	// ⊥	27 18	27 18	27 18
MOE in the plane direction	(MPa)	// ⊥	7000–14 000 3000–5000	6000–1200 2000–7000	5000–10 000 3500–6500
Shear modulus in the plane direction	(MPa)		600	600	600
Shear modulus perpendicular to the plane	(MPa)		500–1000	500–1000	500–1000
Compression strength in the plane direction	(MPa)	// ⊥	25–50 10–25	22–45 14–30	20–40 17–35
Tensile strength in the plane direction	(MPa)	// ⊥	45–80 20–45	40–75 25–53	37–70 30–60
Shear strength in the plane direction	(MPa)	//	3–5	3–5	3–5
Shear strength perpendicular to the plane		⊥	10–18	10–18	10–18
Thermal conductivity	(W/mK)	⊥	0.15	0.15	0.15
Diffusion resistance factor[a]			50/400	50/400	50/400

// - parallel to the fibre direction/board plane in surface layer
⊥ - perpendicular to the fibre direction/board plane in the surface layer
a) Diffusion resistance factor: 50 is wet cup, and 400 is dry cup

Table 5.9 Characteristic strength values for LVL made of Norway spruce (Kerto from FinnForest Oy, Finland) Characteristic density according to the Z-9.1–100 structural design type approval (DiBt 2011). For symbols for strengths, modulus of elasticity and shear modulus, see Chap. 4, Fig. 4.1

Thickness (t):	Unit	Kerto S	Kerto Q	
	(mm)	$20 \leq t \leq 75$	$21 \leq t \leq 24$	$27 \leq t \leq 69$
Strength when loaded as a plate				
Bending[a)] $f_{m,0,k}$	(MPa)	50	32	36
Bending at right angles[a)] $f_{m,0,k}$	(MPa)		9[b]	9
Compression $f_{c,90,k}$	(MPa)	2	2	
Rolling shear $f_{v,k}$	(MPa)	2.3	1.5	
Strength when loaded as a disc				
Bending[a)] $f_{m,0,k}$	(MPa)	48	32	36
Tension parallel to the fibre $f_{t,0,k}$	(MPa)	38	20	27
Tension perpendicular to the fibre $f_{t,90,k}$	(MPa)	0.8	6	
Compressionparalle to the fibre $f_{c,0,k}$	(MPa)	38	20	27
Compression perpendicular to the fibre $f_{c,90,k}$	(MPa)	6	9	

a) Values apply to $h \leq 300$ mm. For $300 < h \leq 1000$ the characteristic strength value must be multiplied by the factor $k_h = (300/h)^{0.12}$, h is the dimension of the cross-section in mm that is decisive for the entire bending stress
b) for $t = 21$ mm the structure I-III-I may $f_{m,90,k}$ 16 MPa or 2500 MPa be accepted

Table 5.10 Characteristics of beech LVL according to the DIBT Z-9.1.-837 structural design type approval (DiBt 2013)

Property	Unit	Length wise	Transverse
Thickness	(mm)	$20 \leq t \leq 120$	$20 \leq t \leq 100$
Strength when loaded as a plate			
Bending $f_{m,0,k}$	(MPa)	66	40
Compression $f_{c,90,k}$	(MPa)	10	10
Rolling shear $f_{v,k}$	(MPa)	3.3	
Strength when loaded as a disc			
Bending[a)] $f_{m,0,k}$	(MPa)	70	60
Tension parallel to fibre $f_{t,0,k}$	(MPa)	70	40
Tension perpendicular to fibre $f_{t,90,k}$	(MPa)	1.5	17
Compression parallel to fibre $f_{c,0,k}$	(MPa)	41.6	24.2
Compression perpendicular to fibre $f_{c,90,k}$	(MPa)	14	14
Shearing $f_{v,k}$	(MPa)	9	
Stiffness parameters			
Modulus of elasticity (MOE) $E_{0,\,mean}$	(MPa)	16 800	11 800
Modulus of elasticity (MOE) $E_{0,05}$	(MPa)	14 900	10 700
Modulus of elasticity (MOE) $E_{90,mean}$	(MPa)	470	3700
Shear modulus upright G_{mean}	(MPa)	760	890
Shear modulus flat G_{mean}	(MPa)	850	430
Density ρ_k	(kg/m³)	680	

a) Values apply for $h \leq 300$ mm. For $300 < h \leq 100$ the characteristic strength value must be multiplied by the coefficient $k_h = (300/h)^{0.12}$; h is the dimension of the cross-section in mm that is decisive for the entire bending stress

5.6.2 Load Slip for for Adhesive Bonding and Mechanical Connectors

See Fig. 5.16

5.6.3 Factors Affecting Duration of Load

See Table 5.27

5.6.4 Load-Duration Classes

See Table 5.28

5.6.5 Deformation Factor k_{def} for sawn Timber and Wood-based Materials

See Table 5.29

5.6.6 Fatigue Factors

See Table 5.30

5.7 Ageing of Material

Peter Niemz, Dick Sandberg

5.7.1 Ageing of Wood

See Fig. 5.17, 5.18 and 5.19

Extensive studies on the use of wood in mechanical engineering and the development of a database were summarised by the TU Dresden as part of the joint project "Wood in Mechanical Engineering" (participation of all major German higher education institutions in the field of wood science and technology 2022). Studies on adhesives, veneers, plywood, etc. were for example documented.

A database is available (www.holz-im-maschinenbau.de) at the TU Chemnitz, Germany (https://www.tu-chemnitz.de/mb/FoerdTech/aew/aew_start.php) (Wagenführ 2022) (Table 5.31).

5.7.2 Ageing of Adhesives and Adhesive Films

See Fig. 5.20, 5.21, 5.22 and 5.23 also Table 5.32

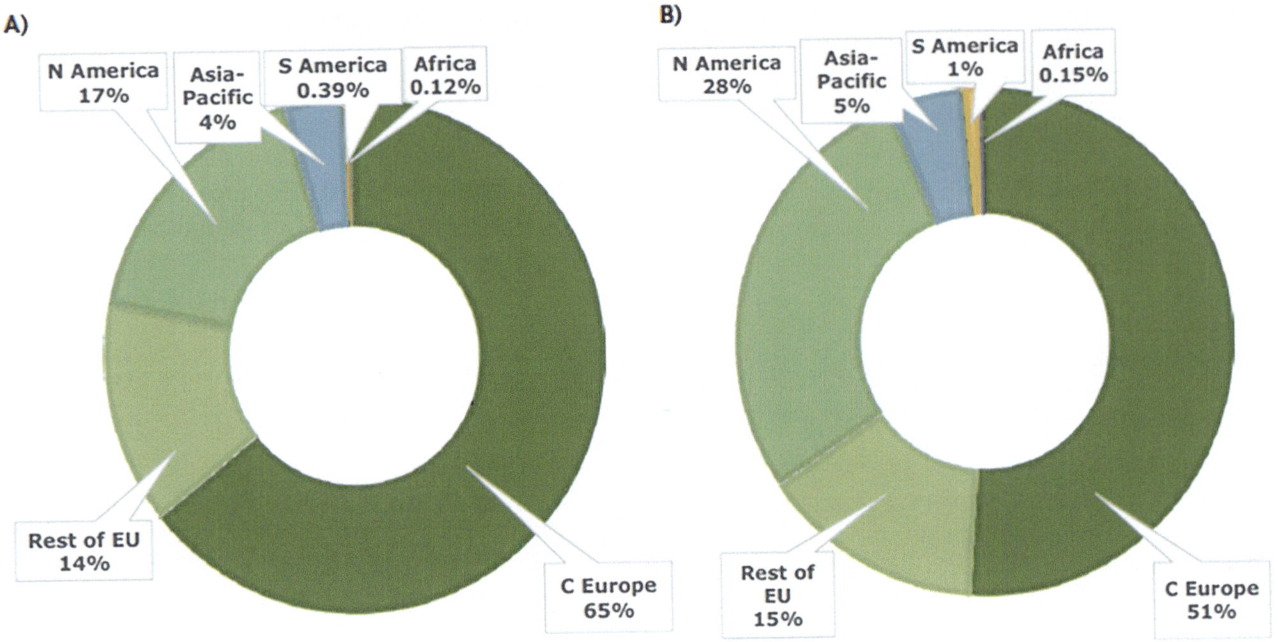

Fig. 5.6 Overview of CLT production worldwide: a) regional distribution of the total global output volume, and b) shift capacity (Muszynski 2021)

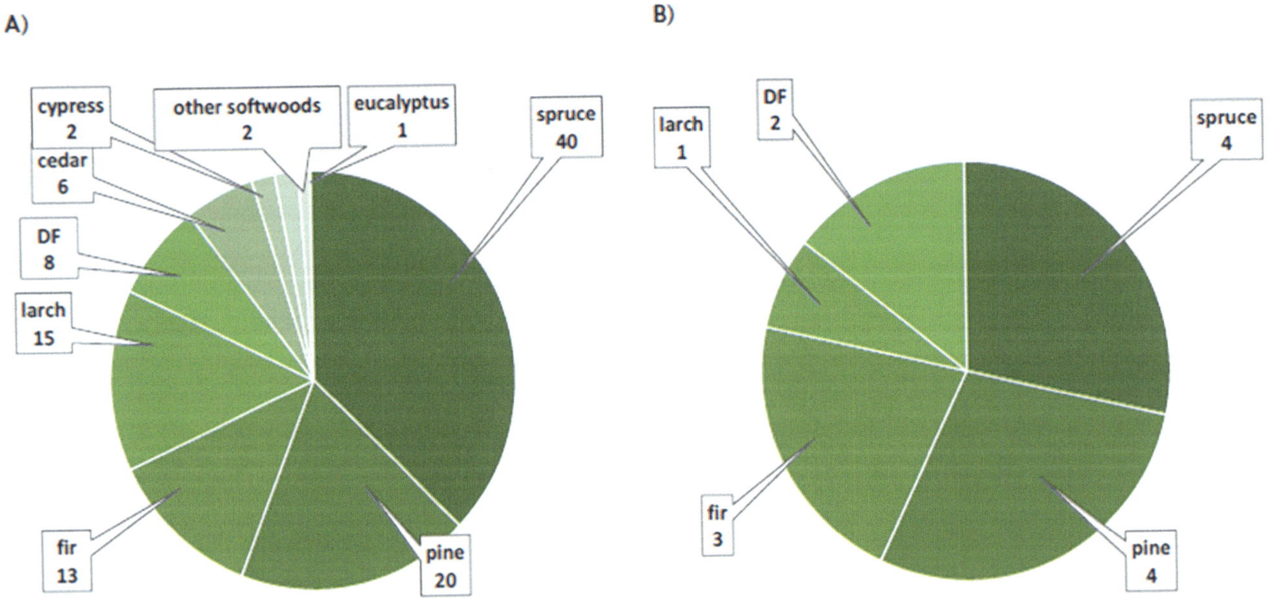

Fig. 5.7 Estimate of species use in CLT manufacture as species groups rather than specific species such as spruces, pines and firs etc.: (a) globally, and (b) in the USA (Muszynski 2021). DF - Douglas fir

Table 5.11 Properties of single-layer solid-wood panels (edge-glued panels) with a thickness of 20–30 mm according to the EN 13353 standard (CEN 2011)

Property	Unit	
Density	(kg/m³)	410
Bending strength	(MPa)	40
Modulus of elasticity (MOE)	(MPa)	8500

Table 5.12 Properties of multi-layer solid-wood panels according to the EN 13353 standard (CEN 2011)

Property	Unit	Thickness (mm)			
		12–20	>20–30	>30–42	>42
Density	(kg/m³)	410	410	410	410
Bending strength:					
perpendicular to the panel plane	(MPa)				
parallel to the fibre direction of the top layer	(MPa)	35	30	16	12
perpendicular to the fibre direction of the top layer	(MPa)	5	5	9	9
Modulus of elasticity (MOE) in bending:					
perpendicular to the panel plane	(MPa)				
parallel to the fibre direction of the top layer	(MPa)	8500	7000	6500	6000
perpendicular to the fibre direction of the top layer	(MPa)	470	470	1300	1300

Table 5.13 Physical properties of three-layer solid-wood panels (Niemz and Sonderegger 2021)

Panel structure[a]	Density	Moisture content	Thermal conductivity	Diffusion resistance factor	Swelling in relation to the fibre direction in the surface:	
					//	⊥
(mm)	(kg/m³)	(%)	(W/m·K)		(%/%)	
10 / 10 / 10	417	11.7	0.091	134.2	0.012	0.016
10 / 20 / 10	436	12.3	0.096	122.4	0.018	0.011
10 / 40 / 10	432	12.6	0.099	148.3	0.014	0.010
7 / 14_G / 7	435	11.6	0.095	109.2	0.010	0.009
20 / 30_G / 20	394	12.6	0.094	95.7	0.013	0.015
10_{st} / 10 / 10_{st}	424	11.6	0.088	120.7	0.015	0.016
14_{st} / 14_{st} / 14_{st}	467	10.8	0.104	165.7	0.011	0.020
10 / 19_{MDF} / 10	596	9.2	0.105	109.0	0.014	0.046
10 / 17_{OSB} / 10	544	10.5	0.107	159.1	0.011	0.031
10 / 10_{A0} / 10	446	12.0	0.101	124.9	0.018	0.017
10 / 10_{A5} / 10	443	12.6	0.100	117.6	0.014	
10 / 10_{A10} / 10	418	12.4	0.094	81.1	0.012	
10 / 10_{A30} / 10	397	11.7	0.091	69.8	0.015	
10_{li} / 10 / 10_{li}	441	12.0	0.096	91.5	0.010	0.024

a) Layer thickness, G – slotted, st – vertical growth rings (quarter sawn), li – horizontal growth rings (flat sawn), Ax – nonglued lamellae (with spacing x in mm)

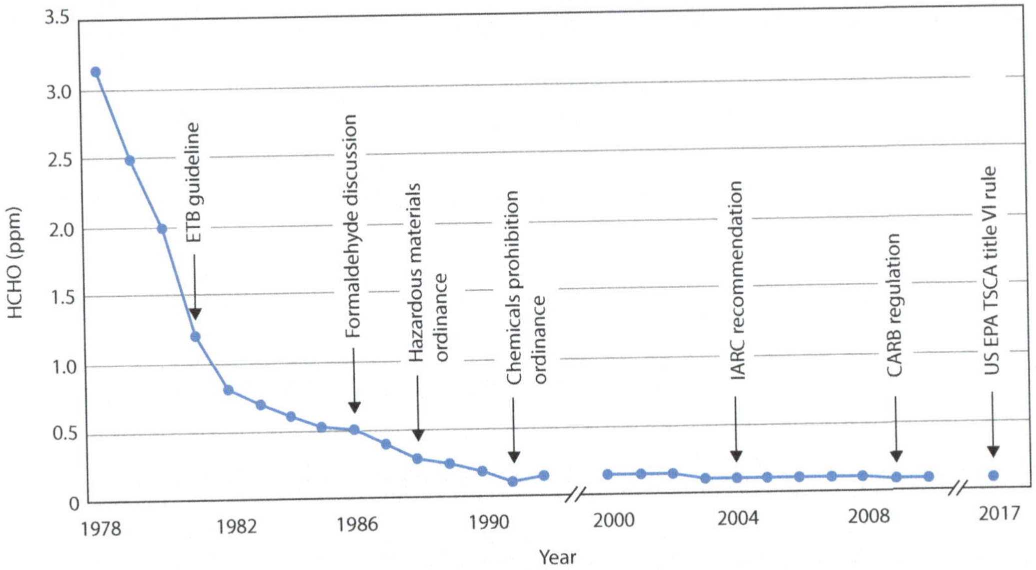

Fig. 5.8 Reduction in formaldehyde (HCHO) emission from wood-based panels between 1978 and 2017 (Dunky 2023)

Table 5.14 Adhesive consumption in 2022 for wood-based materials worldwide (updated from Dunky 2023). PB – particleboard, OSB – oriented strand board, HDF/MDF/LDF – high/medium/low-density fibreboard

Region	All types	Plywood	PB	OSB	MDF/HDF	Hardboard	LDF	Fibreboards Total
Million m³/year (liquid form)								
Africa	3.2	1.3	1.2	0	0.4	0.3	0	0.7
Americas	65.2	17.0	11.8	21.5	10.8	0.8	3.3	14.9
North A.	45.2	10.9	6.0	20.9	3.8	0.3	3.3	7.4
Central A. and Caribean	1.3	0.3	0.9	0	0.03	0.1	0	0.1
South A.	18.7	5.8	4.9	0.6	7.0	0.4	0.03	7.4
Asia	215.2	81.8	51.4	7.1	69.4	4.4	1.1	74.9
Europe	88.8	8.4	44.8	9.6	20.0	2.0	4.0	26.0
Oceania	3.0	0.6	1.1	0	1.3	0	0	1.3
Total worldwide:	375.4	109.1	110.3	38.2	101.9	7.5	8.4	117.8

Table 5.15 Demand for adhesives in a million tons delivered from adhesive producers and based on production volumes of the wood-based panels in 2022 (updated from Dunky 2023). UF may be slightly fortified with melamine. PB – particleboard, OSB – oriented strand board, HDF/MDF/LDF – high/medium/low-density fibreboard

Adhesive type	All types	Plywood	PB	OSB	Hardboard/LDF	HDF, MDF
	Million tonnes/year (liquid form)					
Urea-formaldehyde (UF) a)	21.6	4.0	6.5	0.1	0	11.0
Melamine-urea formaldehyde (MUF)	2.7	1.0	1.0	0.2	0	0.5
Phenol-formaldehyde (PF)	2.2	1.5	0.3	0.3	0.1	0
Polymeric methyl diphenyl diisocyanate (PMDI)	1.2	0	0.1	1.0	0	0.1
Total:	27.7	6.5	7.9	1.6	0.1	11.6

a) Including melamine-modified UF-resins

Table 5.16 Adhesive comsumption in solid wood-based materials 2021 (Henkel company). EWP – engineered wood product, KVH – Konstruktionsvollholz, i.e. industrially produced solid-wood components for construction with defined product characteristics, developed to meet the high requirements in timber constructions

Type of EWP	Production volume of EWP	Adhesive consumption		Country code
	(m^3)	(kg/m^3 EWP)	(Tonnes)	
Glued-laminated timber	2 907 000	7	20 349	AT, CH, DE
Cross-laminated timber	1 139 000	5	5695	AT, CH, CZ, DE, IT
KVH	3 489 000	1	3489	AT, CH, DE

Table 5.17 Development of the European wood-based panel industry (Dunky 2022). PB – particleboard, OSB – oriented strand board, HDF/MDF – high/medium/low-density fibreboard. (Source: FAOSTAT)

Parameter	1982	2022
Recycled wood material	Mainly virgin wood	Mainly recycled wood, but wide variation up to 100%
Requirements for physical and mechanical properties	Partly still national standards, as in Germany DIN 68761 and DIN 68763 ("V20" and "V100")	European standards: PB: EN 312 (CEN 2010a), EN 13986 (CEN 2015) MDF: EN 622–5 (CEN 2009) OSB: EN 300 (CEN 2006) The requirements themselves remained the same, despite the strong decrease in emission.
Production volume in Europe (million tones)	PB + OSB: 27.4 Total fibreboard: 7.0 Plywood: 5.3	PB: 40.5 OSB: 10.1 MDF/HDF: 20.6 Plywood: 9.1

Table 5.18 Adhesives used in EU, North Amerika und Asia/Pacific (Henkel company 2021). Adhesives: 1C PUR – one-component polyurethane, MUF/MF – melamine-urea formaldehyde, PRF/PF – formaldehyde-phenol-resorcinol, EPI – emulsion-polymer-isocyanate, Glulam – glued-laminated timber, CLT – cross-laminated timber

Type of EWP	1C PUR (%)	MUF/MF (%)	PRF/PF (%)	EPI (%)
CLT	69	18		13
Glulam	8	48	9	35
I-Joist	1	6	78	15

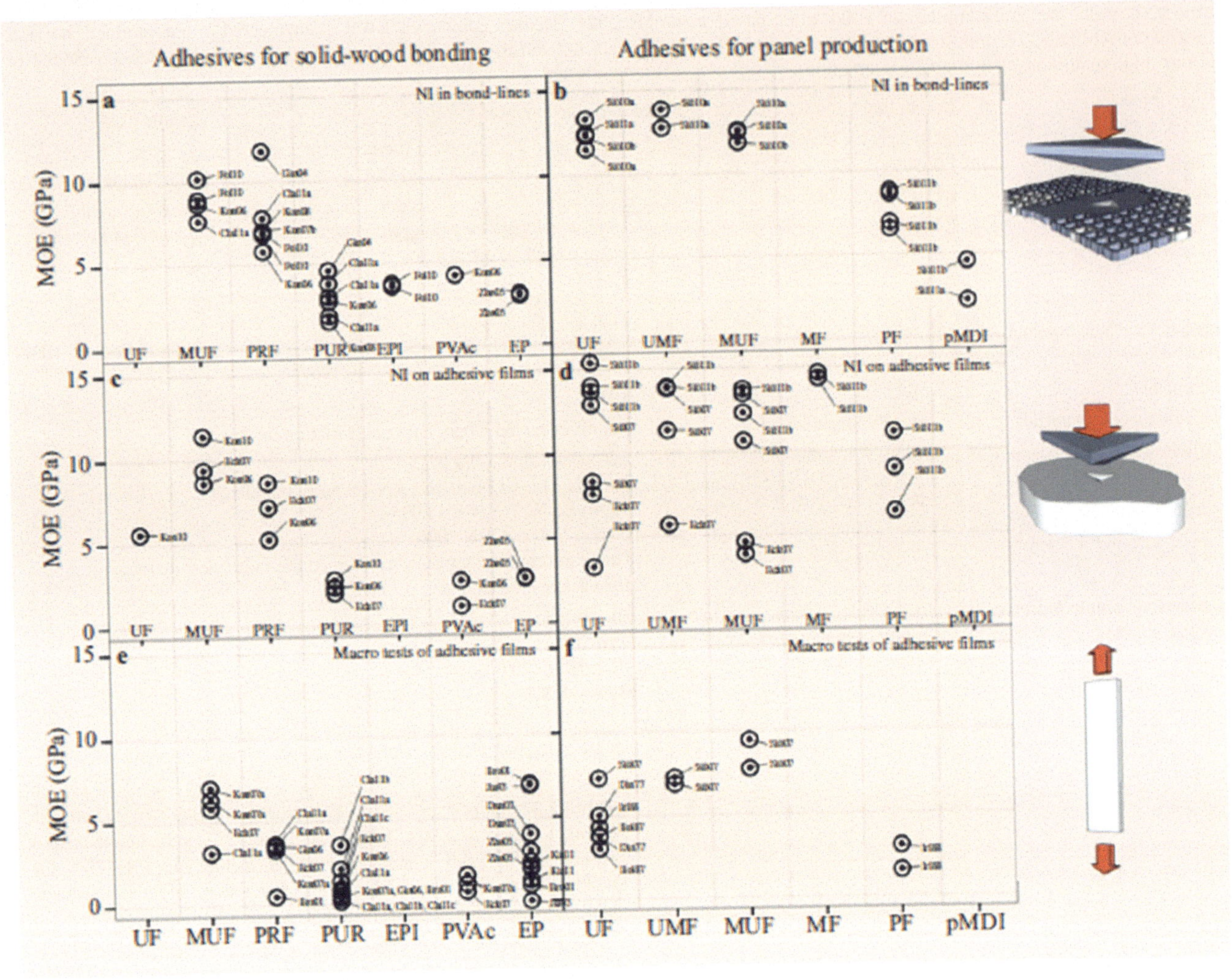

Fig. 5.9 Elastic modulus of wood adhesives designed for solid-wood bonding (left collumn) and for panel production (right collumn) measured by nanoidentation inside bondlines (top) and on adhesive films (middle) and by macroscopic testing (bottom) (Stöckel et al. 2013)

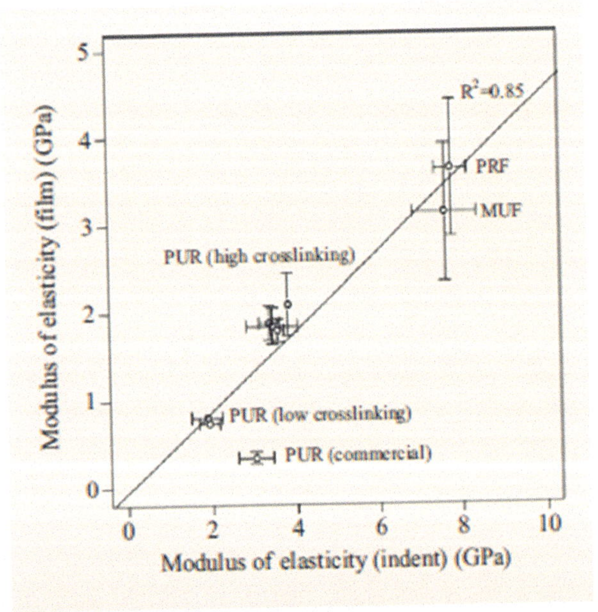

Fig. 5.10 Nanoidentation results of bondline interphase region (Clauss 2011)

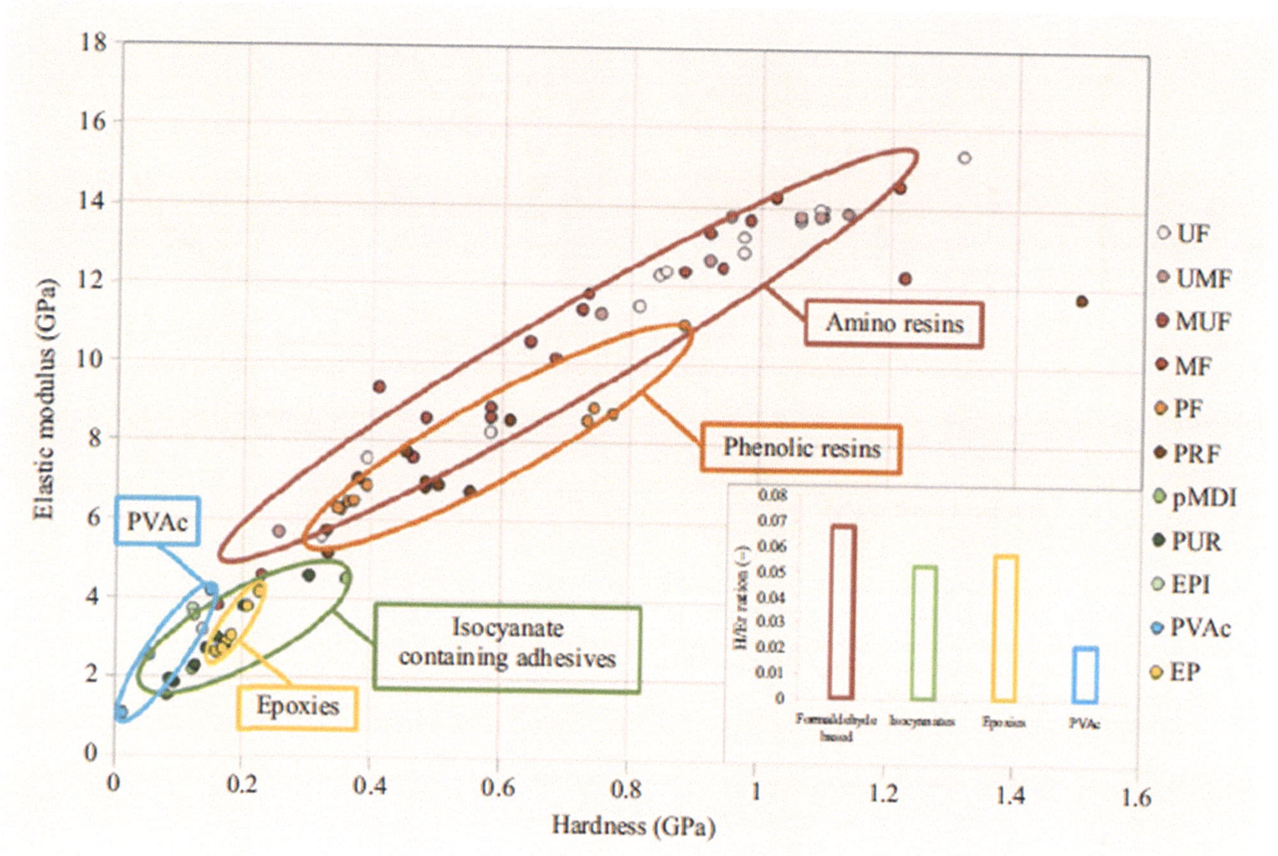

Fig. 5.11 Mechanical properties of wood adhesives by means of nanoidentation as synonym for reduced modulus of elasticity (E_r) (Stöckl et al. 2013)

Table 5.19 Properties of adhesive films conditioned at 20 °C and various relative humidities (RH) (Bachtiar 2017). 1C PUR – one-component polyurethane adhesive, MOE – modulus of elasticity, σ_u – ultimate strength, ν - Poisson's ratio, Coefficient of variance in parentheses (%)

Adhesive	Property		Properties after conditioning at 20 °C and 5, 35, 50, 65 and 85% RH				
		RH/MC[3]:	5%/3%	35%/12%	50%/12%	65%/20%	85%/26%
Hide	MOE	(MPa)	4160 (10%)	4040 (8%)	4350 (4%)	3450 (14%)	1890 (18%)
	σ_u	(MPa)	76 (14%)	66 (5%)	57 (7%)	46 (9%)	28 (27%)
	ν	(-)				0.36 (3%)	
			5%/3%	**35%/12%**	**50%/15%**	**65%/19%**	**85%/29%**
Bone	MOE	(MPa)	5420 (4%)	4390 (5%)	4020 (14%)	3890 (20%)	130 (21%)
	σ_u	(MPa)	82 (6%)	74 (3%)	59 (12%)	47 (13%)	14 (8%)
	ν	(-)				0.37 (6%)	
			5%/2%	**35%/9%**	**50%/12%**	**65%/16%**	**85%/23%**
Fish	MOE	(MPa)	4990 (8%)	4530 (7%)	3710 (15%)	1780 (25%)	720 (16%)
	σ_u	(MPa)	67 (13%)	57 (1%)	50 (1%)	41 (6%)	14 (13%)
	ν	(-)				0.37 (8%)	
			5%/-	**35%/-**	**50%/-**	**65%/-**	**85%/-**
1C-PUR[1]	MOE	(MPa)	900 (5%)	800 (3%)	830 (35%)	730 (12%)	560 (7%)
	σ_u	(MPa)	22 (7%)	19 (5%)	20 (1%)	18 (6%)	17 (5%)
	ν	(-)				0.39 (5%)	
			5%/0.1	**35%/0.8**		**65%/1.7**	**85%/3.1**
1C-PUR[2]	MOE	(MPa)	1300 (6%)	1100 (7%)		1000 (7%)	900 (3%)
	σ_u	(MPa)	33 (1%)	28 (3%)		27 (5%)	23 (1%)
	ν	(-)					

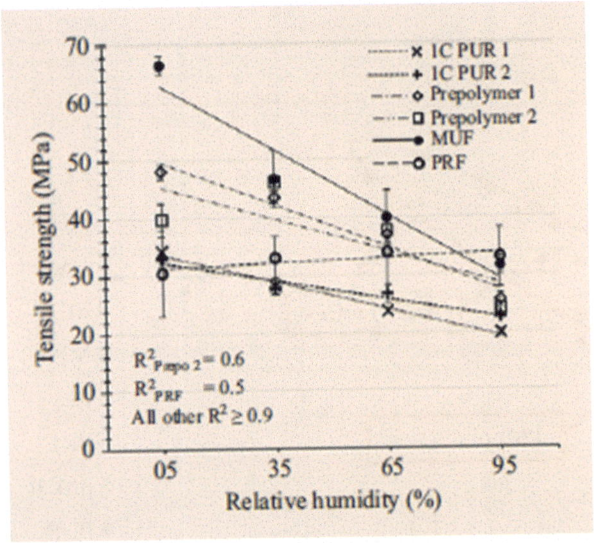

Fig. 5.12 Tensile strength of polymer films as a function of the relative air humidity. Lines: Linear regressions. Data points: Average values. Olid lines are linear regressions, data points are average values, error bars show 95% confidence intervals of average values, and R^2 is the coefficient of determination (Kläusler 2014)

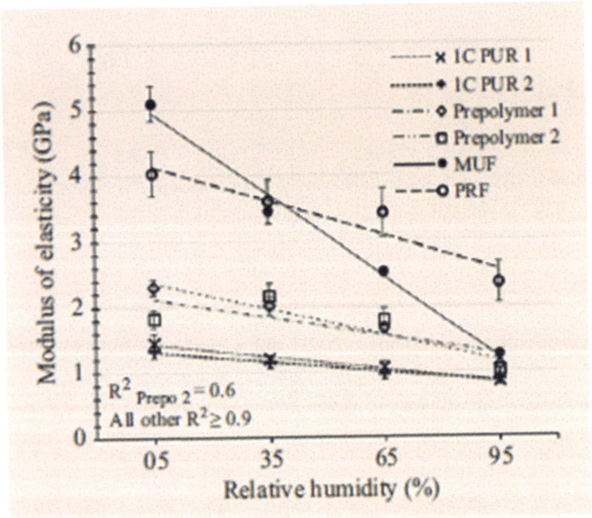

Fig. 5.13 Modulus of elasticity of polymer films as a function of the relative air humidity. Lines are linear regressions, data points are average values, error bars show 95% confidence intervals of average values, and R^2 is the coefficient of determination (Kläusler 2014)

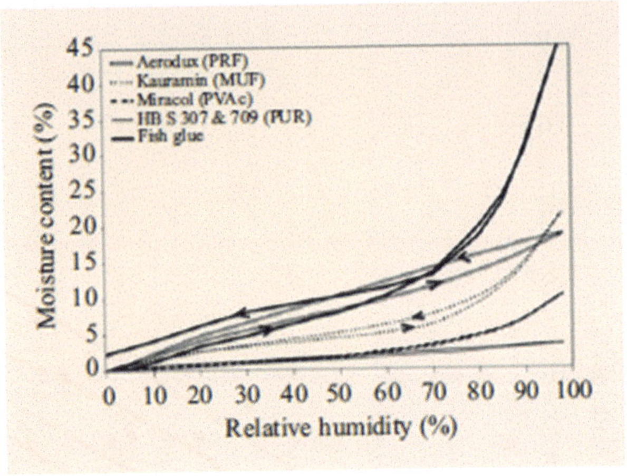

Fig. 5.14 Sorption of moisture by adhesive films (Wimmer et al. 2013)

Table 5.20 Classification system and typical applications of high-pressure laminates (HPL) less than 2 mm thick intended for bonding to supporting substrates. Material type see Table 5.21, and classification index numbers, see Table 5.22

Performance category	Material type	Classification index numbers			Equivalent alphabetical classification	Examples of typical applications
		Wear resistance	Impact resistance	Scratch resistance		
Very high resistance to surface wear Very high resistance to impact Very high resistance to scratching	S, F or P	4	4	4	HDS (horizontal heavy-duty standard) HDF (horizontal heavy-duty flame-retardant) HDP (horizontal heavy- duty postforming)	Counter tops, institutional applications (prisons, military barracks, etc.)
High resistance to surface wear High resistance to impact High resistance to scratching	S, F or P	3	3	3	HGS (horizontal general-purpose standard) HGF (horizontal general-purpose flame-retardant) HGP (horizontal general-purpose postforming)	Kitchen and office working surfaces, restaurant and hotel tables, doors and wall coverings in public areas, interior walls of public transport vehicles
Medium resistance to surface wear Medium resistance to impact Medium resistance to scratching	S, F or P	2	2	2	VGS (vertical general-purpose standard), VGF (vertical general- purpose flame-retardant) VGP (vertical general-purpose postforming)	Front panels for kitchen, office and bathroom furniture, wall coverings, ceiling panels, shelves, and furniture elements

Table 5.21 Classification of high-pressure laminates (HPL) less than 2 mm thick intended for bonding to supporting substrates

First letter	Second letter	Third letter
H (horizontal grade) or V (vertical grade)	G (general purpose) or D (heavy duty)	S (standard grade) or P (postformable grade) or F (flame-retardant grade)

Table 5.22 Numerical classification of high-pressure laminates (HPL) less than 2 mm thick intended for bonding to supporting substrates

Initial point (revs)	**First Index number - Wear resistance** [a]		
	2	**3**	**4**
	≥ 50	≥ 150	≥ 350
Small diametre ball (N)	**Second Index number - Impact resistance** [a]		
	2	**3**	**4**
	≥ 15	≥ 20	≥ 25
Scratch resistance (rating)	**Third Index number - Scratch resistance** [a]		
	2	**3**	**4**
	2	3	4

a) Index numbers 2, 3, and 4 are specified to maintain consistency with EN 438–1 (CEN 1991). Index number 1 represents a lower quality level that does not apply to HPL covered by the scope of this part of the standard

Table 5.23 General requirements of high-pressure laminates (HPL) less than 2 mm thick intended for bonding to supporting substrates

Property	Test method EN 438–2 (CEN 2016a) Clause No. unless otherwise stated	Property or attribute	Unit (max or min)	Laminate grade		
				HDS HDF HDP	HGS HGF HGP	VGS VGF VGP
				444	333	222
Resistance to surface wear	10	Wear resistance	revolutions (min) initial point	350	150	50
Resistance to immersion in boiling water	12	Appearance	Rating (min) gloss finish other finishes	3 4	3 4	3 4
Resistance to water vapour	14	Appearance	Rating (min) gloss finish other finishes	3 4	3 4	3 4
Resistance to dry heat (160 °C)	16	Appearance	Rating (min) gloss finish other finishes	3 4	3 4	3 4
Dimensional stability at elevated temperature	17	Cumulative dimensional change	% (max) L [a)] T [b)]	0.45 0.90	0.55 1.05	0.75 1.25
Resistance to wet heat (100 °C)	18	Appearance	Rating (min) gloss finish other finishes	3 4	3 4	3 4
Resistance to impact by small diametre ball	20	Spring force	N (min)	25	20	15
Resistance to impact by large diametre ball (optional)	21	Drop height Indent diameter	mm (min) mm (max)	1000 10	800 10	600 10
Resistance to cracking under stress (optional)	23	Appearance	Rating (min)	4	4	4
Resistance to scratching	25	Force	Rating (min) smooth finishes textured finishes	3 4	2 3	1 2
Resistance to staining	26	Appearance	Rating (min) groups 1 and 2 group 3	5 4	5 4	5 4
Light fastness (xenon arc)	27	Contrast	Greyscale rating	4 to 5	4 to 5	4 to 5
Density	EN ISO 1183–1 (CEN 2019)	Density	kg/m^3 (min)	1350	1350	1350

a) L – in the longitudinal (or machine) direction of the fibrous sheet material (normally the direction of the longest dimension of the laminate)
b) T – in the cross-longitudinal (cross-machine) direction of the fibrous sheet material (at right angles to direction L)

Table 5.24 General requirements for surfaces of melamine-faced boards (MFB) for interior use according to the EN 14323 standard (CEN 2017a)

Property	Test method	Units	Minimum requirement
Surface defects	EN 14323	(mm^2/m^2) (mm/m^2)	points ≤ 2 length ≤ 20
Resistance to scratching	EN 14323	(N)	≥ 1.5
Resistance to staining	EN 14323	Rating	≥ 3
Resistance to cracking	EN 14323	Rating	≥ 3

Table 5.25 Classification of abrasion resistance of surfaces of melamine-faced boards (MFB) for interior use according to the EN 14322 standard (CEN 2017b). IP – initial point, i.e. the initial abrasion point (start of grinding through)

Class	IP in revolutions
1	< 50
2	≥ 50
3A	≥ 150
3B	≥ 250
4	≥ 350

5.8 Natural Fibres

Peter Niemz, Dick Sandberg

Natural fibres from wood and various plants, as are the chemical and mechanical properties of different fibre materials, are becoming increasingly important. Natural fibres offer a sustainable and economically viable alternative to synthetic fibres. Unlike their synthetic counterparts, plant fibres are celebrated for their beneficial properties, including low cost, favourable mechanical characteristics, and a production process that demands relatively low energy. Fibres can be broadly categorised into natural and man-made fibres, with the former further divided into animal, mineral, and plant-based fibres. Within plant-derived fibres, a distinction is made between monocotyledonous and dicotyledonous types, each possessing unique attributes. These fibres are classified according to their plant origin, with key groups including bast fibres, leaf fibres, seed hair fibres, fruit fibres, grass fibres, and wood fibres. As of 2018, global fibre production was reported to be approximately 107 million tons, with synthetic fibres dominating the market since the mid-1990s, particularly in the textile industry. Despite this, natural fibres, including plant-based materials, play a crucial role in global fibre production, accounting for a substantial portion of the industry (Table 5.33).

Table 5.26 Classification requirements for surfaces of direct-pressed-laminate floor (DPL) coverings according to the EN 13329 standard (CEN 2016b)

Class	Levels of use							Test method
	Domestic			Commercial				
	Moderate	General	Heavy	Moderate	General	Heavy	Very heavy	
	21	22	23	31	32	33	34	
Abrasion resistance	AC1	AC2	AC3		AC4	AC5	AC 6	Annex E
Impact resistance Small ball	≥ 8 N				≥ 12 N	≥ 15 N	≥ 20 N	Annex H
Big ball	≥ 500 mm				≥ 750 mm	≥ 1000 mm	≥ 1600 mm	
Resistance to staining	4, (groups 1 and 2) 3, (group 3)		5, (groups 1 and 2) 4, (group 3)				5, (groups 1, 2 and 3)	EN 438 series (CEN 2016c)
Effect of a furniture leg	–		No damage shall be visible, when tested with foot type 0					EN 424 (CEN 2001)
Effect of a castor chair[a]	–		25 000 cycles, No damage[a)]				25 000 cycles No damage with type H wheels[a)]	EN 425 (CEN 2002a,)
Surface soundness	≥ 1.0 MPa				≥ 1.25 MPa		≥ 1.50 MPa	Annex D

a) No visible damage on the surface of the assembled test area caused by detachment of layers, opening of joints, or crazing. Ignore any flattening or change in appearance, e.g. change in gloss

b) Using soft castor wheels W PU (95 ± 5) Shore A except for class 34 wheels H PA (95 ± 5) Shore A

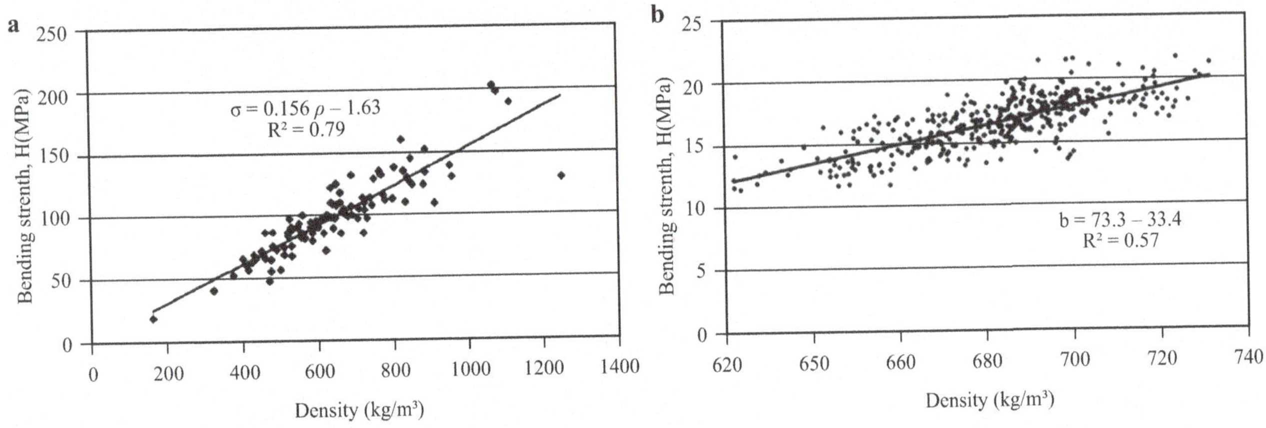

Fig. 5.15 Relation between (a) bending strength (σ_b) of solid wood and density (ρ) at 12% moisture content using material parameters from the literature for 103 wood species, and (b) bending strength of particleboard and density (ρ) at 12% moisture content (Niemz and Sonderegger 2017)

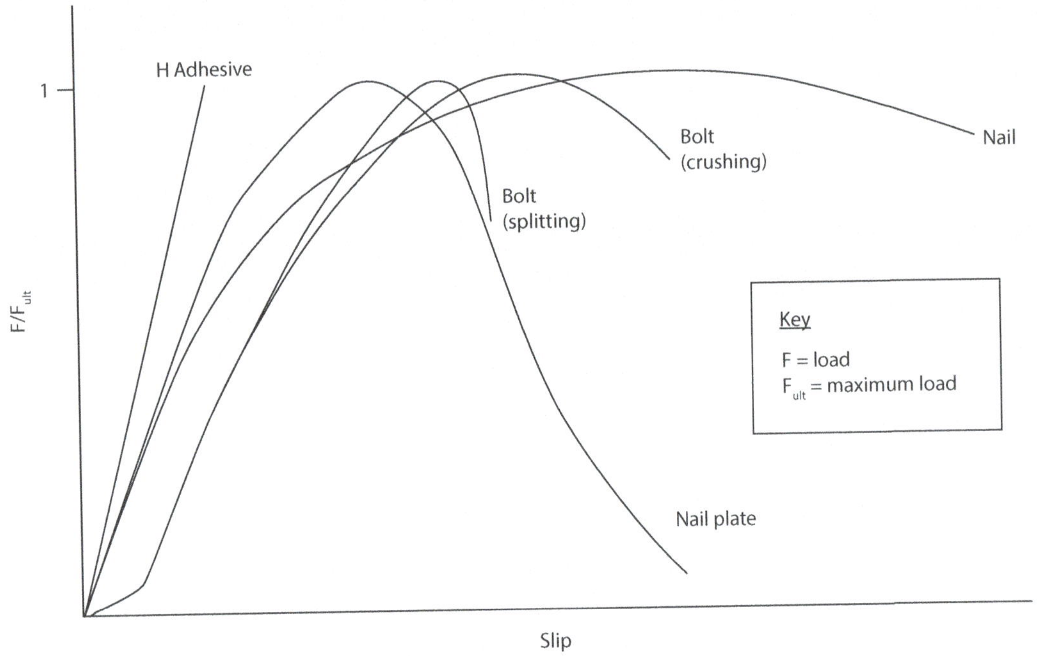

Fig. 5.16 Normalised load-slip curves for adhesive bonding and mechanical connectors (Larsen and Jensen 2000)

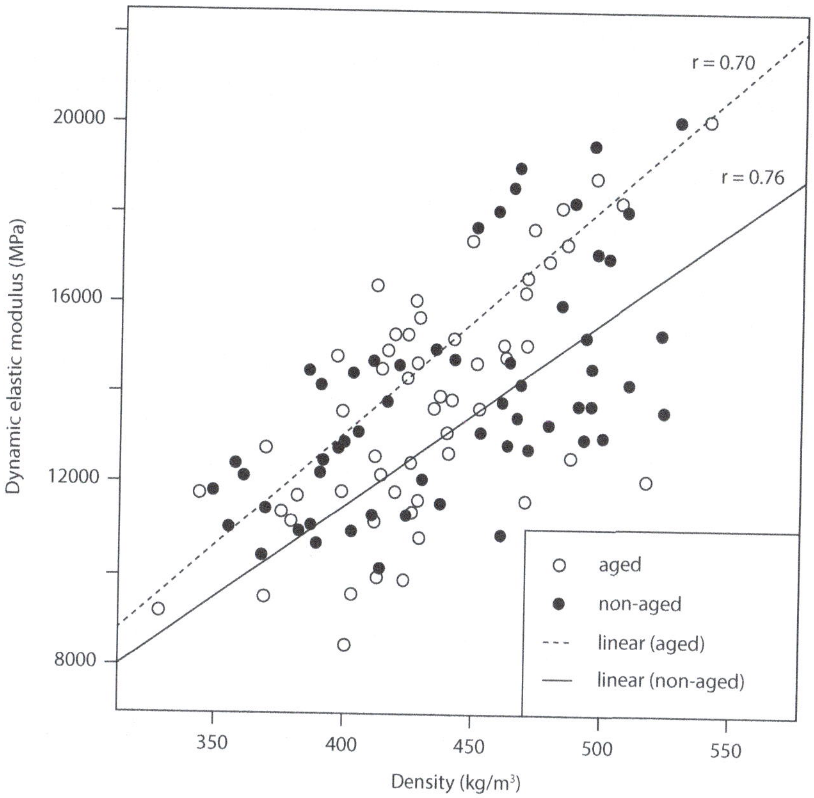

Fig. 5.17 Dynamic elastic modulus as a function of density tested on sawn timber (Kránitz 2014; Kránitz et al. 2014, 2016)

Table 5.27 Modification factors for Load Duration Class according the EN 1995–1 and 1995–2 standards (CEN 2010b)

Material	Standard	Service Class	Load Duration Class (type of action)				
			Permanent	Long-term	Medium-term	Short-term	Instantaneous
Sawn timber	EN 14081–1	1	0.60	0.70	0.80	0.90	1.10
		2	0.60	0.70	0.80	0.90	1.10
		3	0.50	0.55	0.65	0.70	0.90
Glued-laminated timber	EN 14080	1	0.60	0.70	0.80	0.90	1.10
		2	0.60	0.70	0.80	0.90	1.10
		3	0.50	0.55	0.65	0.70	0.90
LVL	EN 14279, EN 14374	1	0.60	0.70	0.80	0.90	1.10
	EN 14374	2	0.60	0.70	0.80	0.90	1.10
		3	0.50	0.55	0.65	0.70	0.90
Plywood	EN 636 Parts 1–3	1	0.60	0.70	0.80	0.90	1.10
	Parts 2–3	2	0.60	0.70	0.80	0.90	1.10
	Part 3	3	0.50	0.55	0.65	0.70	0.90
OSB	EN 300 OSB/2	1	0.30	0.45	0.65	0.85	1.10
	OSB/3, OSB/4	1	0.40	0.50	0.70	0.90	1.10
	OSB/3, OSB/4	2	0.30	0.40	0.55	0.70	0.90
Particleboard	EN 312, Parts 4–5	1	0.30	0.45	0.65	0.85	1.10
	Part 5	2	0.20	0.30	0.45	0.60	0.80
	Parts 6–7	1	0.40	0.50	0.70	0.90	1.10
	Part 7	2	0.30	0.40	0.55	0.70	0.90
Fibreboard (hardboards)	EN 622–2						
	HB.LA, HB.HLA 1 or 2	1	0.30	0.45	0.65	0.85	1.10
	HB.HLA 1 or 2	2	0.20	0.30	0.45	0.60	0.80
Fibreboard (medium boards)	EN 622–3						
	MBH.LA 1 or 2	1	0.20	0.40	0.60	0.80	1.10
	MBH.HLS 1 or 2	1	0.20	0.40	0.60	0.80	1.10
	MBH.HLS 1 or 2	2				0.45	0.80
Fibreboard (MDF)	EN 622–5						
	MDF.LA, MDF.HLS	1	0.20	0.40	0.60	0.80	1.10
	MDF.HLS	2				0.45	0.80

Table 5.28 Load Duration Classes according to the EN 1995–1 and 1995–2 standards (CEN 2010b)

Load Duration Class	Accumulated duration of characteristic load
Permanent	More than 10 years
Long-term	6 month – 10 years
Medium-term	1 week – 6 months
Short-term	Less than 1 week
Instantaneous	

Table 5.30 Fatigue factors ($k_{fat\infty}$) under cycling dynamic loads according to SIA 265/1 (SNV 2003)

Type of Lload	$k_{fat\infty}$
Compression	1.0
Tension	
Bending	0.5
Tension–compression cyclic loading	
Shear	0.3

Table 5.29 Deformation factor k_{def} for sawn timber and wood-based materials (correction load duration = creep) according to the EN 1995–1 and 1995–2 standards (CEN 2010b)

Material	Standard	Service Class		
		1	2	3
Sawn timber	EN 14081–1 (CEN 2016d)	0.60	0.80	2.00
Glued-laminated timber	EN 14080 (CEN 2013b)	0.60	0.80	2.00
LVL	EN 14279 (CEN 2004c), EN 14374 (CEN 2004d)	0.60	0.80	2.00
Plywood	EN 636 (CEN 2012) Part 1	0.80		
	Part 2	0.80	1.00	
	Part 3	0.80	1.00	2.50
OSB	EN 300 (CEN 2006) OSB/2	2.25		
	OSB/3, OSB/4	1.50	2.25	
Particleboard	EN 312 (CEN 2010a), Part 4	2.25		
	Part 5	2.25	3.00	
	Part 6	1.50		
	Part 7	1.50	2.25	
Fibreboard (hard)	EN 622–2 (CEN 2004a)			
	HB.LA,	2.25		
	HB.HLA 1, HB.HLA 2	2.25	3.00	
Fibreboard (medium)	EN 622–3 (CEN 2004b)			
	MBH.LA 1, MBH.LA 2	3.00		
	MBH.HLS 1, MBH.HLS 2	3.00	4.00	
Fibreboard (MDF)	EN 622–5 (CEN 2009)			
	MDF.LA	2.25		
	MDF.HLS	2.25	3.00	

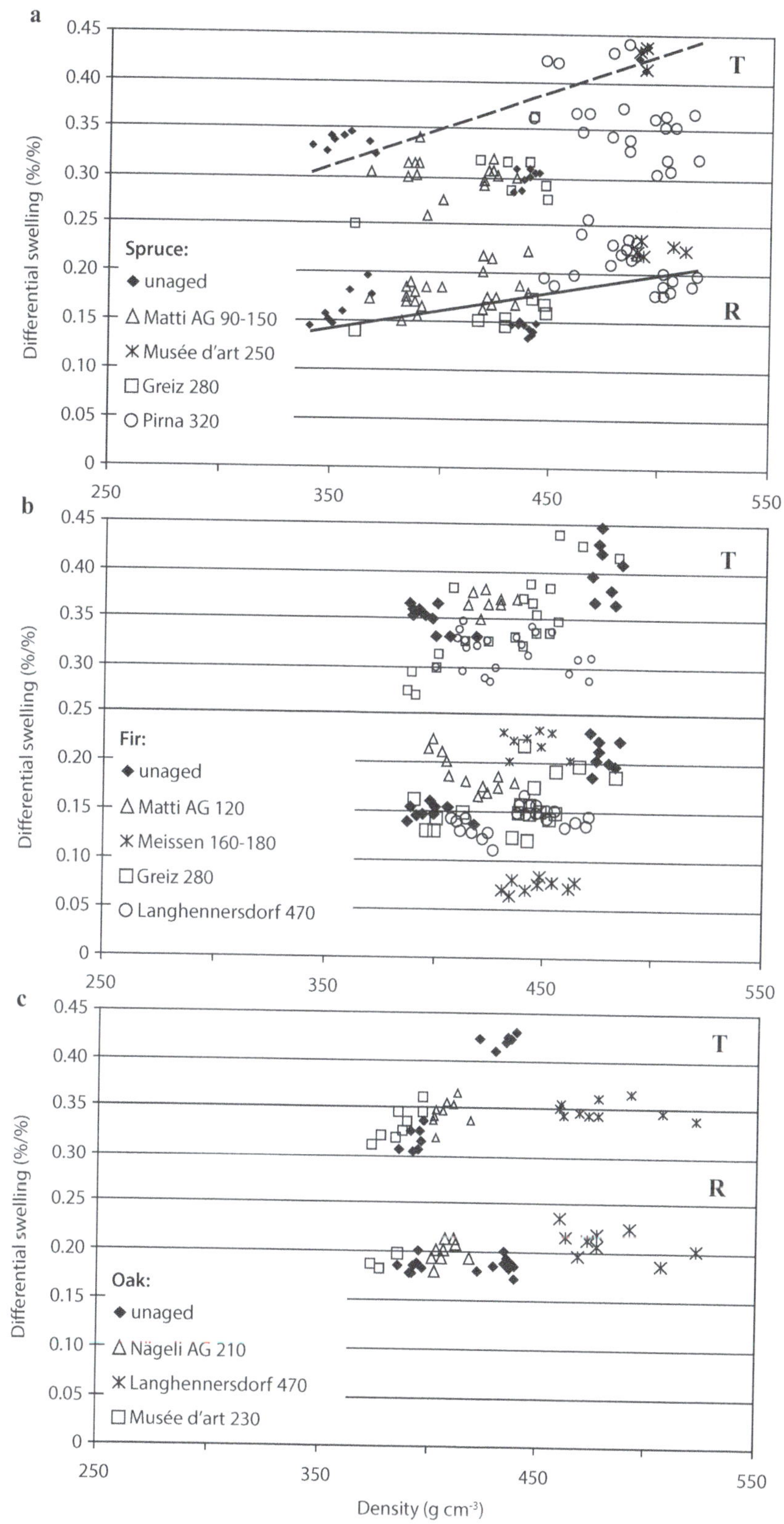

Fig. 5.18 The difference in swelling between non-aged and aged wood is a function of wood density. Aged wood has been collected in different regions of Switzerland (Sonderegger et al. 2015) Its age ranges between 90 and 470 years: (a) Norway spruce with linear trends according to. Sonderegger et al. (2008), where prominent symbols and the solid line in the figure indicate radial swelling, small symbols and the dashed line indicates tangential (T) swelling; (b) European silver fir, and c) European oak

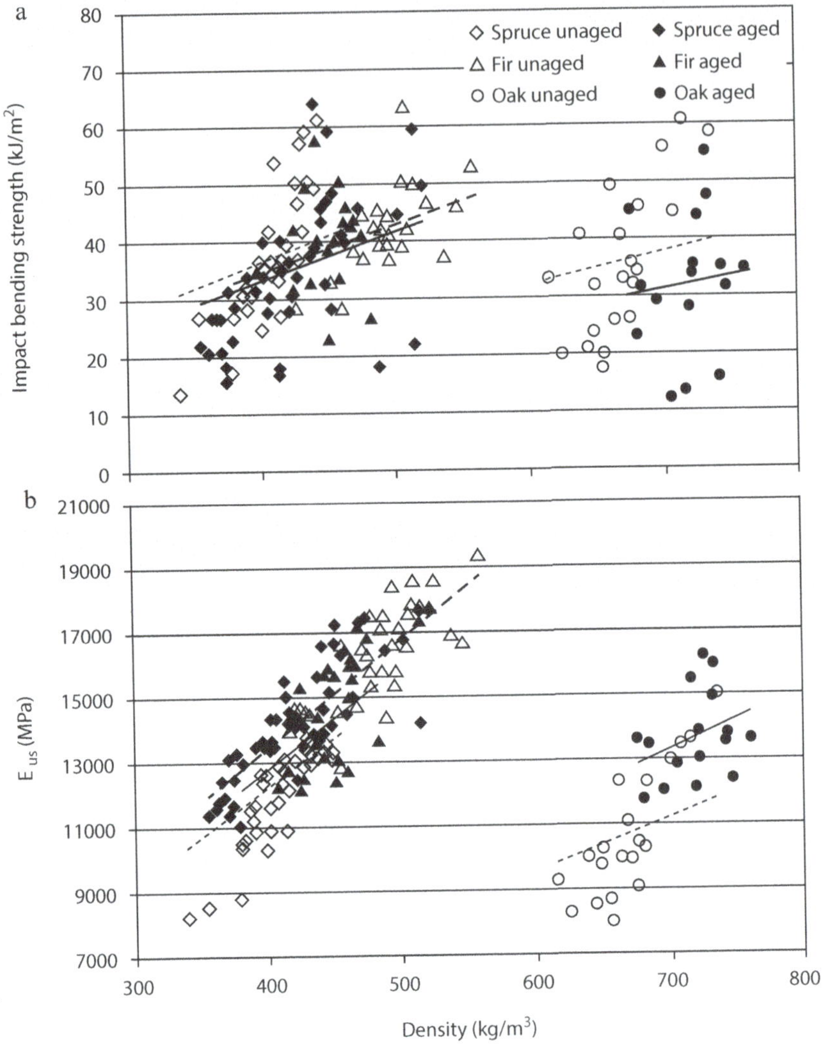

Fig. 5.19 Change in mechanical properties of aged European oak, European silver fir, and Norway spruce wood (Sonderegger et al. 2015): (a) relationship between impact bending strength and density, and (b) modulus of elasticity, determined by ultrasound (E_{US}), as a function of density prior to the impact bending test. The regression curves pass through the zero-point (0,0), dashed lines show non-aged wood, solid lines show aged wood

Table 5.31 Dynamic elastic modulus and shear modulus of aged wood (Kránitz et al. 2014). SA – aged Norway spruce, SR – non-aged Norway spruce, FA – aged fir, E BPV-E-ultrasound (BPV), E_{ef} – E-eigenfrequency, E_B – E-bending test. grw – growth-ring width, and ρ_b – bulk density. Mean – mean value, SD – standard deviation, COV – covariance

Specimen		Age (years)	n*		ρ_b (kg/m^3)	grw (mm)	E_{BPV} (MPa)	E_{ef} (MPa)	E_B (MPa)
Norway spruce	SA06	150	45173	Mean	410	1.14	13 472	11 742	10 385
				SD	0.02	0.1	585	635	619
				COV	4.0%	8.7%	4.3%	5.4%	6.0%
	SA07	150	45090	Mean	460	1.6	15 911	13 279	12 233
				SD	0.01	0.57	968	902	1275
				COV	3.0%	3.6%	6.1%	6.8%	10.4%
	SA08	120	45059	Mean	440	2.47	14 794	12 643	12 900
				SD	0.01	0.43	1139	1090	686
				COV	2.5%	17.5%	7.7%	8.6%	5.3%
	SA09	150-150	62/29	Mean	420	2.63	14 343	12 084	11 936
				SD	0.06	1.28	1974	1830	1584
				COV	13.6%	48.8%	13.8%	15.1%	13.3%
	aged all	120-150	97/44	Mean	420	2.33	14 533	12 283	11 945
				SD	0.05	1.17	1783	1622	1496
				COV	11.6%	50.4%	12.3%	13.2%	12.5%
	SR01	Non-aged	87/44	Mean	410	2.62	12 583	10 719	11 002
				SD	0.02	0.97	1809	1551	2194
				COV	5.6%	36.9%	14.4%	14.5%	19.9%
Fir	FA08	120-150	45173	Mean	430	2.27	14 804	12 334	14 325
				SD	0.04	0.6	1338	849	2382
				COV	9.2%	26.4%	9.0%	6.9%	16.6%
	FA06	120	45266	Mean	450	2.12	15 110	12 344	14 117
				SD	0.02	0.32	1244	1280	1903
				COV	3.4%	14.9%	8.2%	10.4%	13.5%
	FA07	120	45173	Mean	450	1.61	13 445	10 784	12 975
				SD	0.01	0.24	890	961	2425
				COV	2.8%	14.6%	6.6%	8.9%	18.7%
	aged all	120-150	30/14	Mean	440	2	14 519	11 857	13 850
				SD	0.03	0.42	1352	1262	2097
				COV	5.8%	21.1%	9.3%	10.6%	15.1%
	FR01	Non-aged	62/27	Mean	480	2.4	16 249	13 224	14 759
				SD	0.03	0.83	1533	1313	1664
				COV	6.1%	34.6%	9.4%	9.9%	11.3%

* number of specimens (n) by measurement of BPV and eigenfrequency / n by bending test

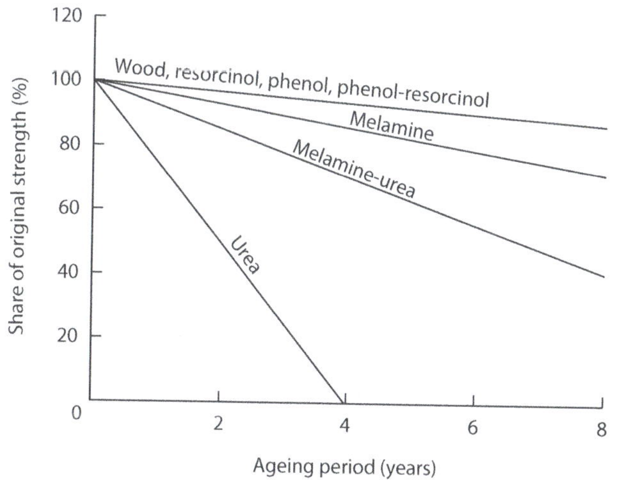

Fig. 5.20 Relative rates of deterioration of bond strength of small specimens exposed directly to weathering (Forest Products Laboratory 2021)

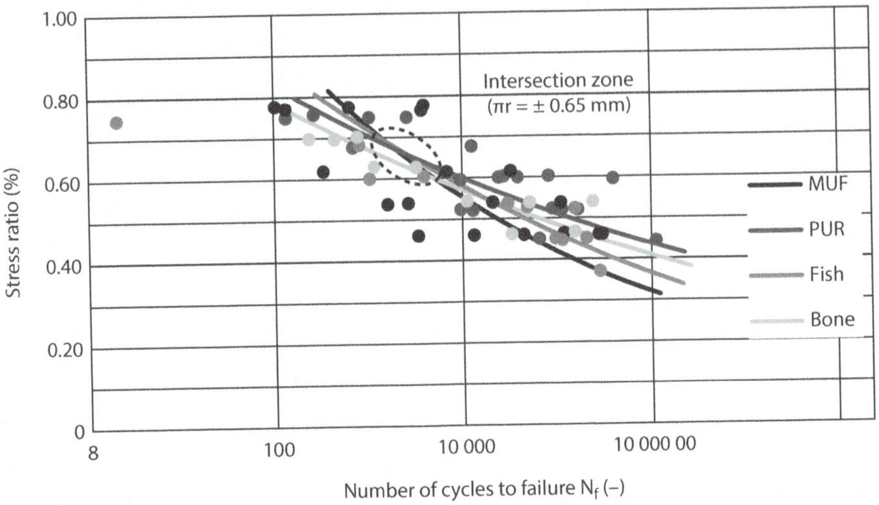

Fig. 5.21 Stress ratio (τ_r) in cyclic loading in tensile shear test as a function of number of cycles to failure (N_f) for different wood-adhesive bond-lines (Bachtiar et al. 2017). The ratio is between the maximum applied shear stress and the shear strength. Bone – adhesive based on animal bone, Fish – adhesive based on fish bone or skin, MUF - melamine-urea formaldehyde adhesives, and PUR – polyurethane adhesive

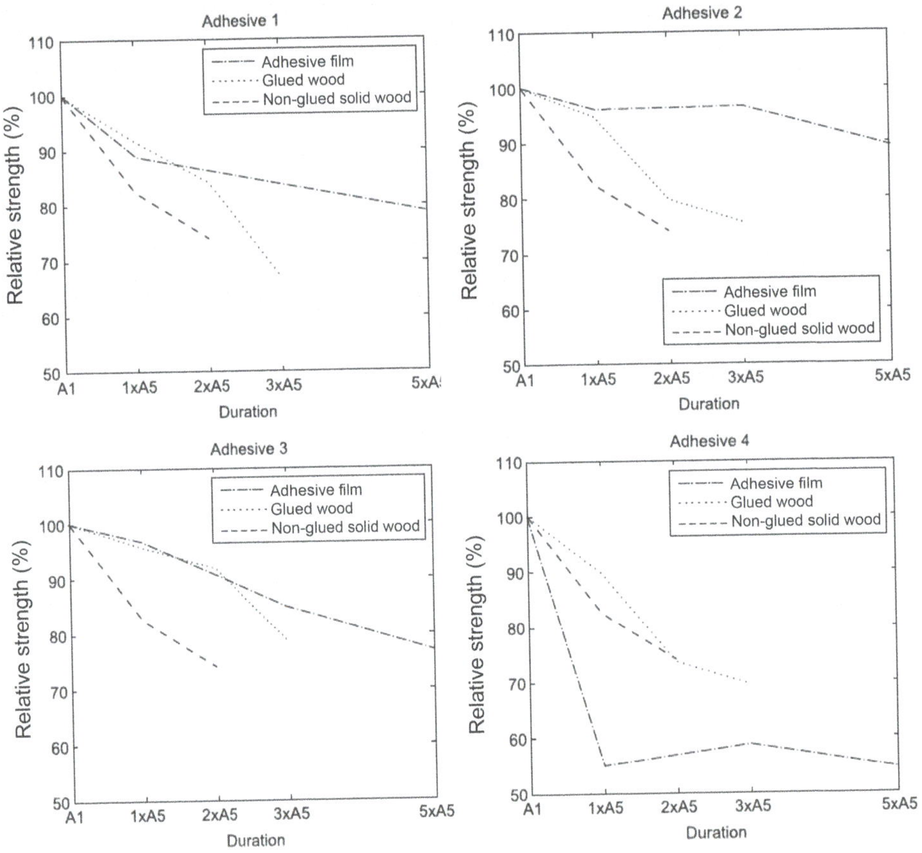

Fig. 5.22 The relative strength of artificial weathered adhesive film (1C-PUR adhesive) and glued (1C-PUR adhesive) and non-glued European beech wood after 1, 2, 3 and 5 cycles of A5 treatment according to CEN (2004e). A1 = Before the test, all material was dried at 70 °C for 5 months and conditioned at 20 °C and 65% RH. The A5 treatment means: (1) specimens are placed in boiling water for 6 hours, (2) after that placed in 20 °C water, and finally (3) conditioned at 20 °C and 65% RH (Clerc et al. 2017; Clerc 2020)

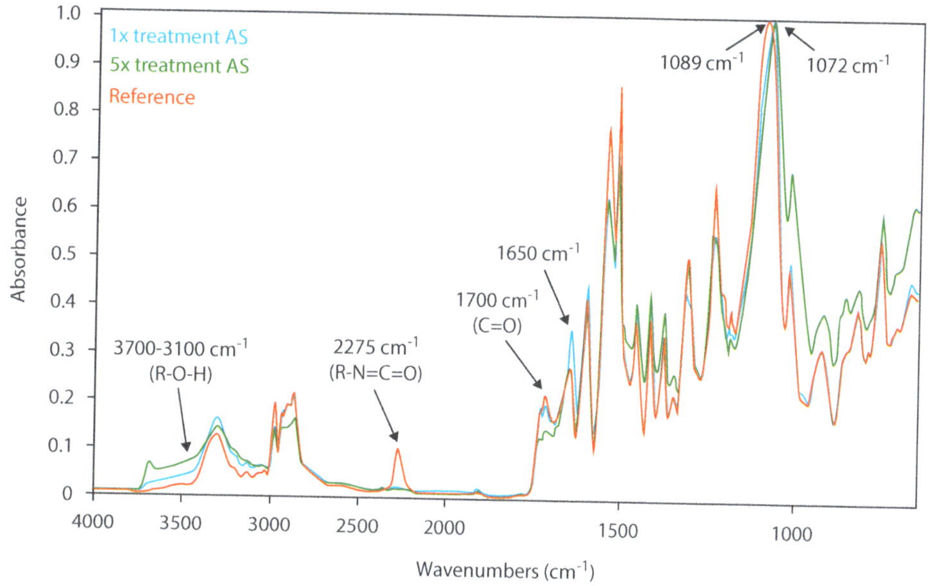

Fig. 5.23 Hydrolysis of adhesive films after 1 and 5 cycles of A5 treatments according the to the EN 302–1 standard (CEN 2004e, 2013a, Clerc et al. 2017). See also explanation of Figure 5.22

Table 5.32 Ageing of various adhesives used in structural timber engineering (Egner and Kolb 1966). The strength of the solid wood is reference (100%). RH – relative humidity, TM – trade mark

Adhesive	Type of exposure over time	Transverse tensile strength of adhesive bond-lines for different species									
		European beech		European oak		Norway spruce		Scots pine		Silver fir	
		(MPa)	(%)	(MPa)	(%)	(MPa)	(%)	(MPa)	(%)	(MPa)	(%)
Solid wood	50 days at 20 °C/65% RH	5.40		6.26	100	3.20	100	3.32	100	1.77	
	5 years in a damp cellar	6.50	100	7.21		2.52		4.10		2.11	100
	1 year in a damp cellar	5.10		6.72		3.40		3.56		2.42	
Phenol-resorcinol-formaldehyde (PRF) adhesive (TM: Aerodux 187)	50 days at 20 °C/65% RH	5.82	102	6.30	94	2.89	95	3.13	86	2.11	106
	5 years in damp cellar	5.24	92	8.43	123	3.40	112	3.72	102	2.30	115
	10 years in a damp cellar	5.82	102	6.62	98	3.09	102	3.09	101	2.86	144
Phenol-resorcinol-formaldehyde (PRF) adhesive (TM: Aerodux 185B)	50 days at 20 °C/65% RH					3.00	99				
	5 years in a damp cellar	6.63	117	6.35	95	3.08	101	3.42	94	25.80	130
Urea formaldehyde (UF) (TM: Kaurit WHK)	50 days at 20 °C/65% RH	5.40	93	6.42	91	3.12	103	3.33	91	2.22	112
	Then 5 years in a damp cellar followed by 6 years at 20 °C/65% RH	7.06	124	30.90	46	3.20	105	3.60	98	2.60	131
Casein adhesive, cold curing (TM: Keimelin)	50 days at 20 °C/65% RH	6.04	106	63.80	95	2.98	98	3.71	104	2.07	104
	5 years in a damp cellar	0	0	0	0	0	0	0	0	0	0

Table 5.33 Primary commercial fibre sources

Fibre source	World production (10^3 tonnes)	Reference
Wood	1 750 000	Rowell (1998)
Straw (barley, flax, grass, oat, rice, rye, wheat)	1 145 000	Rowell (1998)
Stalks (corn, cotton, sorghum)	970 000	Rowell (1998)
Sugar cane bagasse	75 000	Faruk et al. (2012), Rowell (1998)
Reeds	30 000	Rowell (1998)
Bamboo	30 000	Faruk et al. (2012), Rowell (1998)
Cotton staple	15 000	Rowell (1998)
Core (hemp, jute, kenaf)	8000	Rowell (1998)
Papyrus	5000	Rowell (1998)
Bast (hemp, jute, kenaf)	2900	Rowell (1998)
Jute	2300	Faruk et al. (2012)
Cotton linters	1000	Rowell (1998)
Kenaf	970	Faruk et al. (2012)
Flax	830	Faruk et al. (2012)
Grass	700	Faruk et al. (2012)
Esparto grass	500	Rowell (1998)
Leaf (abaca, henequen, sisal)	480	Rowell (1998)
Sisal	378	Faruk et al. (2012)
Hemp	214	Faruk et al. (2012)
Sabai grass	200	Rowell (1998)
Coir	100	Faruk et al. (2012)
Ramie	100	Faruk et al. (2012)
Abaca	70	Faruk et al. (2012)

Table 5.34 Chemical composition of some natural fibres (Faruk et al. 2012)

Fibre type	Cellulose (wt%)	Hemicelluloses (wt%)	Lignin (wt%)	Waxes (wt%)
Abaca	56–63	20–25	7–9	3
Bagasse	55.2	16.8	25.3	
Bamboo	26–43	30	21–31	
Caraua	73.6	9.9	7.5	
Coir	32–43	0.15–0.25	40–45	
Flax	71	18.6–20.6	2.2	1.5
Hemp	68	15	10	0.8
Jute	61–71	14–20	12–13	0.5
Kenaf	72	20.3	9	
Oil palm	65		29	
Pineapple	81		12.7	
Ramie	68.6–76.2	13–16	0.6–0.7	0.3
Rice husk	35–45	19–25	20	14–17
Rice straw	41–57	33	8–19	8–38
Sisal	65	12	9.9	2
Wheat straw	38–45	15–31	12–20	

Table 5.35 Mechanical properties of selected natural fibres (Faruk et al. 2012)

Fibre type	Tensile strength	Modulus of elasticity	Strain at failure	Density
	(MPa)	(GPa)	(%)	(kg/m^3)
Abaca	400	12	3–10	1500
Bagasse	290	17		1250
Bamboo	140–230	11–17		600–1100
Caraua	500–1150	11.8	3.7–4.3	1400
Coir	175	4–6	30	1200
Flax	345–1035	27.6	2.7–3.2	1500
Hemp	690	70	1.6	1480
Jute	393–773	26.5	1.5–1.8	1300
Kenaf	930	53	1.6	
Oil palm	248	3.2	25	700–1550
Pineapple	400–627	1.44	14.5	800–1600
Ramie	560	24.5	2.5	1500
Sisal	511–635	9.4–22	2.0–2.5	1500

Table 5.36 Average tensile properties of humidified flax fibres (Yan et al. 2014). RH – relative humidity, STD – standard deviation. Duralin treatment of flax consists of steam- or water-heating of the rippled straw-flax at a temperature above 160 °C for 30 min in an autoclave, followed by a drying step and a heating (curing) step above 150 °C for 2 hours.

RH	Flax	Fibre length	Tensile strength	STD
(%)	-	(mm)	(MPa)	(MPa)
30	Green	3.5	677	425
	Duralin		809	134
66	Green	3.5	799	398
	Duralin		1080	368
90	Green	3.5	818	318
	Duralin		642	344
30	Green	8.0	619	461
	Duralin		651	176
66	Green	8.0	760	390
	Duralin		913	250
90	Green	8.0	761	369
	Duralin		884	180

5.8.1 Chemical Composition

See Table 5.34

5.8.2 Mechanical Properties

See Table 5.35, 5.36, 5.37 and 5.38

5.8.3 Equilibrium Moisture Content

See Table 5.39

Table 5.37 Properties of flax fibres (Yan et al. 2014)

Diameter (μm)	Density (kg/m^3)	Tensile strength (MPa)	Modulus of elasticity (GPa)	Strain at failure (%)
12–600	1400–1500	343–2000	27.6–103	1.2–3.3
10–60	1520	840	100	1.8
10–60	1520	1500	50	
76.0 ± 16		470 ± 165	37 ± 15	1.4 ± 0.5
17.8 ± 5.8	1530	1339 ± 486	58 ± 15	3.27 ± 0.4
		621 ± 295	51.7 ± 18.2	1.33 ± 0.56
		600–2000	12–85	1–4
		600–1500	50–80	1.4
	–1400	800–1500	60–80	1.2–1.6
	1400–1500	600–1100	45–100	1.5–2.4
12–34		1100	89 ± 35	
		1111 ± 544	71.7 ± 23.3	1.7 ± 0.6
		733 ± 271	49.5 ± 3.2	1.7 ± 0.6
		741 ± 400	45.6 ± 16.7	1.7 ± 0.6
		863 ± 447	48.0 ± 20.3	2.1 ± 0.8
		899 ± 461	55.5 ± 20.9	1.7 ± 0.6
		808 ± 442	51.1 ± 15.0	1.6 ± 0.4
		365–1060	36.8–61.9	0.9–2.1
15 ± 0.6	1.53	1381 ± 419	71 ± 25	2.1 ± 0.8

Table 5.38 Properties of various fibres (Yan et al. 2014)

Fibre type	Diameter (μm)	Density (kg/m^3)	Tensile strength (MPa)	Modulus of elasticity (GPa)	Strain at failure (%)
E-glass	< 17	2500–2600	2000–3500	70–76	1.8–4.8
Abaca		1500	400–980	6.2–20.0	1.0–10
Alfa		890	35	22	5.8
Bagasse	10–34	1250	222–290	17.0–27.1	1.1
Bamboo	25–40	600–1100	140–800	11.0–32	2.5–3.7
Banana	12–30	1350	500	12.0	1.5–9.0
Coir	10–460	1150–1460	95–230	2.8–6.0	15.0–51.4
Cotton	10–45	1500–1600	287–800	5.5–12.6	3–10
Curaua	7–10	1400	87–1150	11.8–96.0	1.3–4.9
Flax	12–600	1400–1500	343–2000	27.6–103.0	1.2–3.3
Hemp	25–600	1400–1500	270–900	23.5–90.0	1.0–3.5
Henequen		1200	430–570	10.1–16.3	3.7–5.9
Isora		1200 1300	500–600		5–6
Jute	20–200	1300–1490	320–800	30	1.0–1.8
Kenaf		1400	223–930	14.5–53.0	1.5–2.7
Nettle			650	38	1.7
Oil palm		700–1550	150–500	80–248	17–25
Piassava		1400	134–143	1.07–4.59	7.8–21.9
PALF	20–80	800–1600	180–1627	1.44–82.50	1.6–14.5
Ramie	20–80	1000–1550	400–1000	24.5–128.0	1.2–4.0
Sisal	8–200	1330–1500	363–700	9.0–38.0	2.0–7.0

Table 5.39 Equilibrium moisture content (EMC) of various fibres at 21 °C and 65% relative humidity (Faruk et al. 2012)

Fiber type	Equilibrium moisture content (%)
Abaca	15
Bagasse	9
Bamboo	9
Coir	10
Flax	7
Hemp	9
Jute	12
Pineapple	13
Ramie	9
Sisal	11

References

Anon (1975) Werkstoffe aus Holz und andere Werkstoffe der Holzindustrie (Materials made of wood and other materials used in the wood industry). Fachbuchverlag, Leipzig

Bachtiar EV (2017) Material characterization of wood, adhesive and coating of cultural heritage under various climatic condition. Dissertation, ETH Zurich

Bachtiar EV, Clerc G, Brunner A et al (2017) Static and dynamic tensile shear test of glued lap wooden joint with four different types of adhesives. Holzforschung 71:391–396

CEN (1991) EN 438-1 High-pressure decorative laminates (HPL) - Sheets based on thermosetting resins (usually called laminates) - Part 1: Introduction and general information. CEN–European Committee for Standardization, Brussels

CEN (2001) EN 424, Resilient floor coverings - Determination of the effect of simulated movement of a furniture leg. CEN–European Committee for Standardization, Brussels

CEN (2002) EN 425, Resilient and laminate floor coverings - Castor chair test. CEN–European Committee for Standardization, Brussels

CEN (2004a) EN 622-2, Fibreboards - Specifications - Part 2: Requirements for hardboards. CEN–European Committee for Standardization, Brussels

CEN (2004b) EN 622-3, Fibreboards - Specifications - Part 3: Requirements for medium boards. CEN–European Committee for Standardization, Brussels

CEN (2004c) EN 14279, Laminated veneer lumber (LVL) - Definitions, classification and specifications. CEN–European Committee for Standardization, Brussels

CEN (2004d) EN 14374, Timber structures - Structural laminated veneer lumber - Requirements. CEN–European Committee for Standardization, Brussels

CEN (2004e) EN 302-1, Adhesives for load-bearing timber structures - Test methods - Part 1: Determination of bond strength in longitudinal tensile shear strength. CEN-European Committee for Standardization, Brussels

CEN (2006) EN 300, Oriented strand boards (OSB) - Definitions, classification and specifications. CEN–European Committee for Standardization, Brussels

CEN (2007) EN ISO 10456, Building materials and products - Hygrothermal properties - Tabulated design values and procedures for determining declared and design thermal values. CEN–European Committee for Standardization, Brussels

CEN (2009) EN 622-5, Fibreboards - Specifications - Part 5: Requirements for dry process boards (MDF). CEN–European Committee for Standardization, Brussels

CEN (2010a) EN 312, Particleboards – Specifications. CEN–European Committee for Standardization, Brussels

CEN (2010b) EN 1995-1-1, Eurocode 5: Design of timber structures - Part 1-1: General - Common rules and rules for buildings. CEN–European Committee for Standardization, Brussels

CEN (2011) EN 13353, Solid wood panels (SWP) – Requirements. CEN–European Committee for Standardization, Brussels

CEN (2012) EN 636, Plywood – Specifications. CEN–European Committee for Standardization, Brussels

CEN (2013a) EN 302-1, Adhesives for load-bearing timber structures - Test methods - Part 1: Determination of longitudinal tensile shear strength. CEN–European Committee for Standardization, Brussels

CEN (2013b) EN 14080, Timber structures - Glued laminated timber and glued solid timber - Requirements. CEN–European Committee for Standardization, Brussels

CEN (2015) EN 13986, Wood-based panels for use in construction - Characteristics, evaluation of conformity and marking. CEN–European Committee for Standardization, Brussels

CEN (2016a) EN 438-2, High-pressure decorative laminates (HPL) - Sheets based on thermosetting resins (usually called laminates) - Part 2: Determination of properties. CEN–European Committee for Standardization, Brussels

CEN (2016b) EN 13329:2016+A1:2017: Laminate floor coverings – Elements with a surface layer based on aminoplastic thermosetting resins –Specifications, requirements and test methods. CEN–European Committee for Standardization, Brussels

CEN (2016c) EN 438-3, High-pressure decorative laminates (HPL) – Sheets based on thermosetting resins (usually called laminates) – Part 3: Classification and specifications for laminates less than 2 mm thick intended for bonding to supporting substrates. CEN–European Committee for Standardization, Brussels

CEN (2016d) EN 14081-1, Timber structures - Strength graded structural timber with rectangular cross section - Part 1: General requirements. CEN–European Committee for Standardization, Brussels

CEN (2017a) EN 14323, Wood-based panels - Melamine faced boards for interior uses - Test methods. CEN–European Committee for Standardization, Brussels

CEN (2017b) EN 14322, Wood-based panels – Melamine faced boards for interior uses –Definition, requirements and classification. CEN–European Committee for Standardization, Brussels

CEN (2019) EN ISO 1183-1, Plastics - Methods for determining the density of non-cellular plastics - Part 1: Immersion method, liquid pycnometer method and titration method. CEN–European Committee for Standardization, Brussels

Clauss S (2011) Structure-property relationships of one-component moisture-curing polyurethane adhesives under thermal load. Dissertation, ETH Zürich

Clerc G (2020) Analysis of the fatigue performance of adhesively bonded wood joints. Dissertation, TU München

Clerc G, Brülisauer M, Affolter S et al (2017) Characterization of the ageing process of one-component polyurethane moisture curing wood adhesive. Int J Adhes Adhes 72:130–138

DiBt (2011) Z-9.1–100, Zusammengesetzte Bauteile aus Furnierschichtholz "Kerto S" und "Kerto Q" (Composite components made of laminated veneer lumber "Kerto S" and "Kerto Q"). Technical Report, Deutsches Institut für Bautechnik, Berlin

DiBt (2013) Z-9.1–566, EUROSTRAND OSB 4 TOP. Technical Report, Deutsches Institut für Bautechnik, Berlin

Dunky MNP (2002) Holzwerkstoffe und Leime: Technologie und Einflussfaktoren (Wood-based materials and glues: technology and influencing factors). Springer, Berlin

Dunky MNP (2023) Production from wood based materials and adhesive consumption. Internal information, Vienna

Egner K, Kolb H (1966) Versuche über das Alterungsverhalten von Leimen für tragende Holzbauteile. (Experiments on the aging behavior of glues for load-bearing wooden components) Holz Roh- Werkst 24(10):439–442

Faruk O, Bledzki A, Fink H-P et al (2012) Biocomposites reinforced with natural fibers: 2000–2010. Prog Polym Sci 1552–1596

Forest Products Laboratory (2021) Wood handbook. Wood as an engineering material. General Technical Report FPL-GTR-282. Department of Agriculture, Forest Service, Madison WI

Greubel D, Drewes H (1987) Ermittlung der Sorptionsisothermen von Holzwerkstoffen bei verschiedenen Temperaturen mit einem neuen Messverfahren (Determination of the sorption isotherms of wood-based materials at different temperatures using a new measuring method). Holz Roh Werk 45(7):289–295

Halász von EhR, Scheer C (1986) Holzbau-Taschenbuch Band 1. Grundlagen, Entwurf und Konstruktionen (Timber construction paperback Volume 1. Basics, design and construction). 8th edn. Ernst & Sohn Verlag fur Architektur und technische Wissenschaften, Berlin

Halász von Eh., Scheer C (1995) Holzbau-Taschenbuch Band 1. Grundlagen, Entwurf und Konstruktionen (Timber construction paperback Volume 1. Basics, design and construction). 9th edn. Ernst & Sohn Verlag fur Architektur und technische Wissenschaften, Berlin

Kläusler OF (2014) Improvement of one-component polyurethane bonded wooden joints under wet conditions. Dissertation, ETH Zürich

Kránitz K (2014) Effect of natural aging on wood. Dissertation, ETH Zürich

Kránitz K, Deublein M, Niemz P (2014) Determination of dynamic elastic moduli and shear moduly of aged wood by means of ultrasonic devices. Mater Struct 47:925-936

Kránitz K, Sonderegger W, Bues C-T et al (2016) Effects of aging on wood: a literature review. Wood Sci Technol 50(1):7-22

Larsen HJ, Jensen JL (2000) Influence of semi-rigidity of joints on the behaviour of timber structures. Prog Struct Eng 2(3):267–277

Mantanis GI, Young RA, Rowell RM (1994a) Swelling of Wood. Part 1. Swelling in water. Wood Sci Technol 28(2):119–134

Mantanis GI, Young RA, Rowell RM (1994b) Swelling of wood. Part 2. Swelling in organic liquids. Holzforschung 48(6):480–490

Muszynski L (2021) The global mass timber panel industry. In: Anderso, A, Dawson E, Muszynski L (eds) International Mass Timber Report 2021, Forest Business Network

Neroth G, Vollenschaar D (eds) (2011) Wendehorst Baustoffkunde: Grundlagen - Baustoffe - Oberflächenschutz (Wendehorst building materials: basics - building materials - surface protection Springer-Verlag, Berlin

Niemz P, Sonderegger W (2017) Holzphysik - Physik des Holzes und der Holzwerkstoffe (Physics of wood and wood materials). Fachbuchverlag Leipzig im Carl Hanser Verlag, Leipzig

Niemz P, Sonderegger W (2021) Holzphysik - Physik des Holzes und der Holzwerkstoffe (Physics of wood and wood materials). 2nd edn, Hanser Verlag, München

Niemz P, Mieczyslaw M, Poblete H (1999) Dickenquellung und Gleichgewichtsfeuchtigkeit von Span-, Faser- und mitteldichten (MDF) Platten (Thickness swelling and equilibrium moisture content of chipboard, fiberboard and medium-density (MDF) boards). Annals of Warsaw Agricultural University (SGGW), Forestry Wood Technol 49:209–212

Rowell RM (1998) The state of art and future development of bio-based composite science and technology towards the 21st century. In: Proceedings of the fourth Pacific Rim bio-based composites symposium, November 2–5, Indonesia

SNV (2003) Schweizer Norm SIA 265/1 Timber structures - Supplementary specifications. Schweizerischer Ingenieur- und Architekenverein, Zürich

Sonderegger W, Alter P, Niemz P (2008) Untersuchungen zu ausgewählten Eigenschaften von Fichtenklangholz aus Graubünden (Investigations on selected properties of spruce tonewood from Graubünden). Holz Roh- Werkstoff 66(5):345–354

Sonderegger W, Kránitz K, Bues C-T et al (2015) Aging effects on physical and mechanical properties of spruce, fir and oak wood. J Cult Herit 16(6):883–889

Stöckl F, Konnerth J, Gindl-Altmutter W (2013) Mechanical properties of adhesives for bondig wood-A review. Int J Adhes Adhes 41:32–41

Wagenführ A (ed) (2022) Holzbasierte Werkstoffe im Maschinenbau (HOMABA) -Berechnungskonzepte, Kennwerte Anforderungen, Kennwertermittlungen, Vol. 35 (Wood-based materials in mechanical engineering (HOMABA) - calculation concepts, characteristic values requirements, characteristic value determinations, Vol. 35). TU Dresden

Wimmer R, Kläusler O, Niemz P (2013) Water sorption mechanisms of commercial wood adhesive films. Wood Sci Technol 47(4):763–775

Yan L, Chouw N, Jayaraman K (2014) Flax fibre–A review. Composites Part B 56:296–317

Prof.Dr.-Ing.habil. Dr.h.c. Peter Niemz studied at the Faculty of Mechanical Engineering of TU Dresden. From 1972 onwards, he worked at the Institute of Wood Technology Dresden (IHD) and at TU Dresden. From 1993 to 1996, he served as a Full Professor at the Universidad Austral de Chile in Valdivia before joining the Swiss Federal Institute of Technology (ETH Zurich) in 1997. He was Professor of Wood Physics from 2002 until his retirement in 2015.

Prof.Dr. Dick Sandberg received his Doctorate in Mechanical Engineering, Wood Technology and Processing from the Royal Institute of Technology (KTH), Stockholm, Sweden, in 1998. He is a Chaired Professor of Design and Fabrication of Timber Construction at the Norwegian University of Science and Technology (NTNU). He specialises in using new measurement technology, advanced information processing, and wood communication from felling to finished product. He is an Elected Fellow of the International Academy of Wood Science.

Dr. Rico Emmler (Sect. 5.5) after an apprenticeship as a cabinetmaker, Dr. Rico Emmler studied Process Engineering at the Technische Universität Dresden, Germany, specialising in Wood Technology and Fibre Materials Technology. Since 1993, he has worked in surface testing and surface technology at the Institut für Holztechnologie Dresden gemeinnützige GmbH (IHD). In 2005, he was appointed head of the Surface Department, and in 2016, he became head of the Entwicklungs- und Prüflabor Holztechnologie GmbH (EPH), a subsidiary of IHD. He received his doctorate from the Technical University of Dresden, Germany 2001. His tasks include standardising furniture surfaces and MMF/Laminate floor coverings as a Convenor of five ISO/CEN Standardization working groups.

6 Selected Important Process Parameters

Peter Niemz, Dirk Grunwald, Sabine Heinemann and Dick Sandberg

6.1 Introduction

Peter Niemz, Dick Sandberg

For technological calculations, values such as energy consumption, material consumption (adhesive, fibres, wood etc.) and mechanical properties are required for products such as cross-laminated timber, fibreboard, glued-laminated timber, plywood, and particleboard. These data are usually difficult to find and rarely published. In addition, these values show a large variation between products in the same category and can be significantly different than the average values. In the present chapter, essential production data are listed as far as they are available and accessible. With the assistance of specialists with many years of industrial experience with wood-particle materials as well as pulp and paper, detailed data sets are presented for these materials. Values have also been taken from other technical books. Still, unfortunately, these are often older works and more recent reference books with a technological focus are rarely available, as wood technology is not in the focus of current research and is often being carried out in industry or at universities of applied science or technologically oriented research institutes. The classification of materials is described in several chapters in the Springer Handbook of Wood Science and Technology (Niemz et al. 2023).

6.1.1 Comparison Between Wood-Material Groups

See Table 6.1

Table 6.1 Volume yield and wood consumption for different wood-refinement processes. EWPs—engineered wood products

Production process	Volume yield (% of log top-diameter cylinder volume)	Wood consumption (m^3/m^3)	Reference
Softwood sawn timber	57–85	1.76–1.18	Kehr (1975)
Hardwood sawn timber	60–78	1.67–1.25	Kehr (1975)
Sleepers	74–83	1.35–1.20	Kehr (1975)
Veneer	55–70	1.80–1.43	Kehr (1975)
EWPs:			
High-density fibreboards (HDF), wet process	35–48	2.90–2.10	Kehr (1975)
Medium-density fibreboards (MDF)	55–67	1.80–1.50	Kehr (1975)
Low-density fibreboards (LDF)	110–200	0.90–0.50	Kehr (1975)
Particleboard (PB)	55–67	1.80–1.50	Kehr (1975)
Plywood	35–50	1.80–1.43	Kehr (1975)
Plywood	50	2.00	Gong (2021)

(continued)

P. Niemz (✉)
Department Wood Physics (today, Wood Materials Science), Institute of Building Materials, ETH Zürich, Zurich, Switzerland
e-mail: niemzp@retired.ethz.ch

D. Grunwald
Follmann GmbH & Co. KG, Minden, Germany
e-mail: dirk.grunwald@follmann.com

S. Heinemann
Wood Science and Fiber Material Science, Paper Technology Group, Technische Universität Dresden (TU Dresden), Dresden, Germany
e-mail: sabine.heinemann@tu-dresden.de

D. Sandberg
Department of Manufacturing and Civil Engineering, Norwegian University of Science and Technology (NTNU), Gjøvik, Norway
e-mail: dick.sandberg@ntnu.no

P. Niemz et al. (eds.), *Wood Material and Processing Data*, https://doi.org/10.1007/978-3-031-38329-8_6

Table 6.1 (continued)

Production process	Volume yield (% of log top-diameter cylinder volume)	Wood consumption (m³/m³)	Reference
Plywoood of European white or silver birch	37	2.70	
Plywood of Norway spruce	48	2.10	
Laminated veneer lumber (LVL)	52	1.92	Gong (2021)
Parallel strand lumber (PSL)	64	1.56	Gong (2021)
Oriented strand board (OSB)	88	1.14	Gong (2021)
Laminated strand lumber (LSL)	76	1.32	Gong (2021)
Scrimber/scrim lumber	83	1.20	Gong (2021)
Blockboard	37–50	2.70–2.00	Kehr (1975)

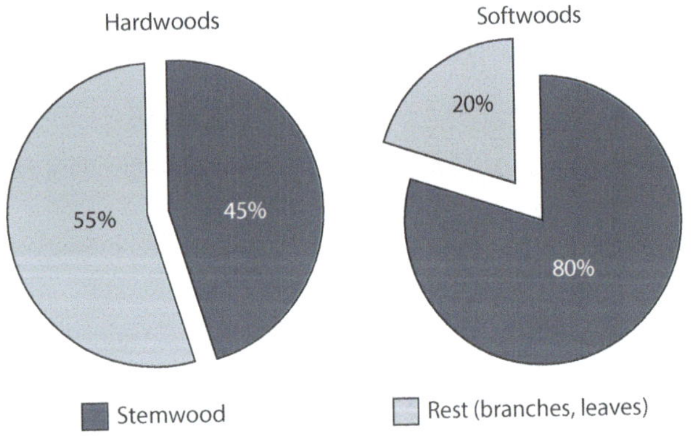

Fig. 6.1 Stem part of a softwood and hardwood trees (Krackler et al. 2010, 2011)

6.2 Sawmilling

Peter Niemz, Dick Sandberg

6.2.1 Outcome of the Sawing Process

Hardwood has a lower percentage of usable stem wood than softwood (Figs. 6.1, 6.2, 6.3 and 6.4, Tables 6.2 and 6.3).

6.2.2 Volume Yield of Sawn Timber

See Tables 6.4, 6.5, 6.6 and 6.7

6.2.3 Debarking

See Figs. 6.5 and 6.6 also Table 6.8

6.2.4 Energy Consumption in Different Sawmill Processes

See Tables 6.9, 6.10, 6.11, 6.12

6.2.5 Economic Numbers in Sawmilling

See Figs. 6.7 and 6.8

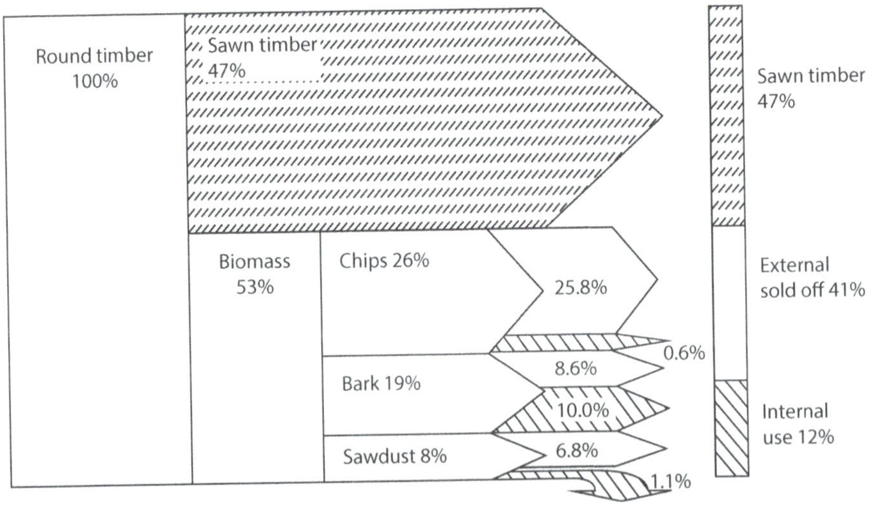

Fig. 6.2 Sawn-timber and biomass interchange (Anon 1975; Li et al. 2006). For Scandinavian relationships see Staland et al. (2000)

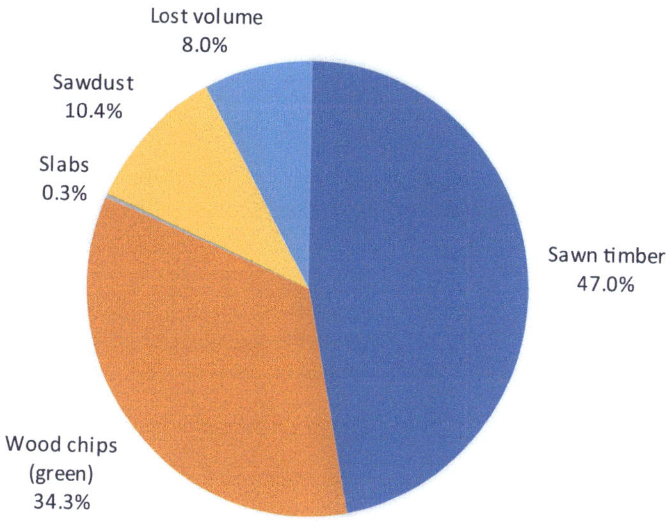

Fig. 6.3 Annual distribution of the outcome from the sawmill process as a weighted mean value of all Swedish sawmills with an annual output of more than 5000 m^3 based on a top-to-root yield (Staland et al. 2000). The main species are Norway spruce and Scots pine. Still, smaller volumes of contorta pine are also included. The mean log volume is approx. 0.175 m^3. Lost volume is due to shrinkage, dimension allowance, etc

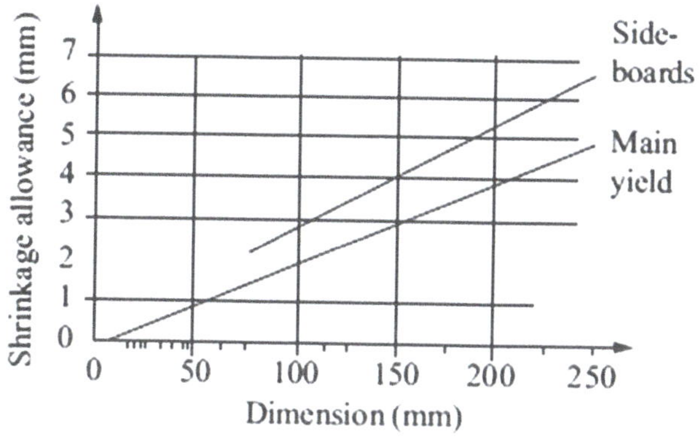

Fig. 6.4 Relationship between the shrinkage allowance and the nominal dimensions at a moisture content of 20% (Bråkenhielm et al. 1969)

Table 6.2 Average surface roughness values (R_a and R_{max}) for different sawmill machines (Sandberg and Teischinger 2023)

Machine type	Centre (main) yield			Sideboards		
	R_{max} (μm)	R_a (μm)	2Δ (mm)	R_{max} (μm)	R_a (μm)	2Δ (mm)
Framesaw	650	120	0.4	720	130	0.5
Bandsaw	350	60	0.2	300	50	0.2
Circular saw	360	70	0.2	360	60	0.6
Chipper canter	400	70	0.2	400	80	0.2

Table 6.3 Conversion figures for sawmill by-products

Wood material	Measure	Equivalent solid-wood volume
Chips	1 m^3 bulk (loose) volume	0.36
Sawdust	1 m^3 bulk (loose) volume	0.32
Slabs	1 m^3 stacked volume	0.45

Table 6.4 Volume yield of sawn timber in through-and-through sawing (live-sawing) with a framesaw. Sawn-timber thickness 40 mm (Anon 1975). The volume yield is based on log volume measured as top-diameter-based cylinder volume

Top diameter	(mm)	190–220	230–250	260–290	300–330	340–370
Mean volume yield	(%)	78.1	79.2	83.4	82.6	85.7

Table 6.5 Volume yield of sawn timber in block-sawing with a framesaw. Sawn-timber thickness 40 mm and 140 mm in width (Anon 1975). The volume yield is based on log volume measured as top-diameter-based cylinder volume

Top diameter	(mm)	180–200	210–230	240–260	270–290	300–320	330–350	360–380
Mean volume yield	(%)	63.0	66.3	68.0	69.8	70.9	71.3	72.5

Table 6.6 Volume yield in sawing of European beech of different stem middle-diameter with a framesaw. Sawn-timber thickness 30 mm (Kohler 1954)

Middle diameter	(mm)	200–290	300–390	400–490	500–590	600–690
Mean volume yield	(%)	73.0	76.5	78.5	79.5	80.0

Table 6.7 Volume yield in sawing of European beech in a framesaw as a function of stem middle-diameter and the thickness of the sawn-timber (Fehrendsen 1956)

Middle diameter	Mean volume yield (%) for board thickness (mm)		
(mm)	18	30	40 and 26
150–190	57.6		
200–240	62.1	65.0	
250–290	65.0	67.7	69.6
300–340		68.7	71.2
350–390		71.5	73.0
400–490		73.2	75.2
500–590			75.8

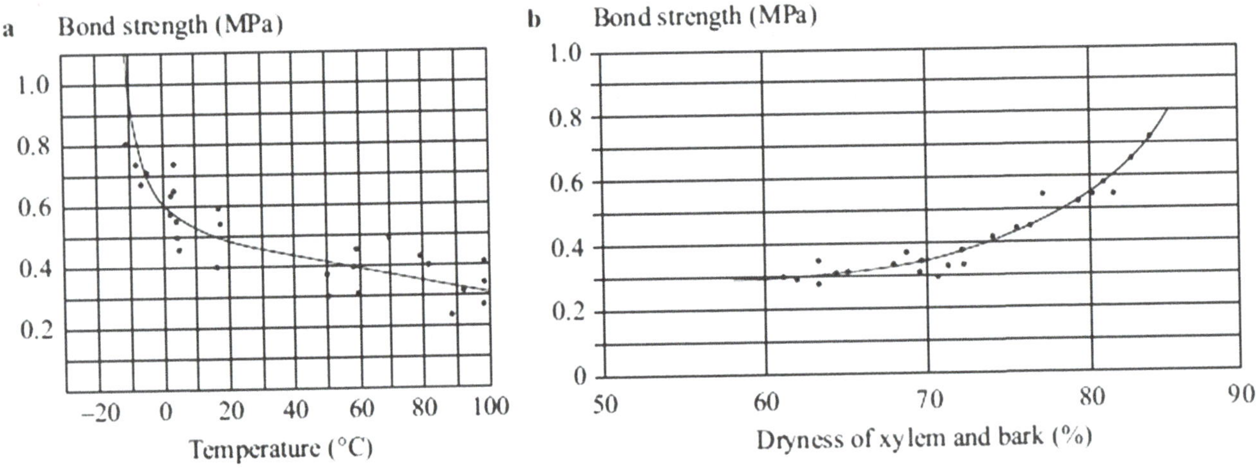

Fig. 6.5 The bond strength between bark and wood as a function of (a) temperature, and (b) dryness for Norway spruce (Ebner 1977). Dryness–the mass of absolutely dry material divided by the wet weight of the same material

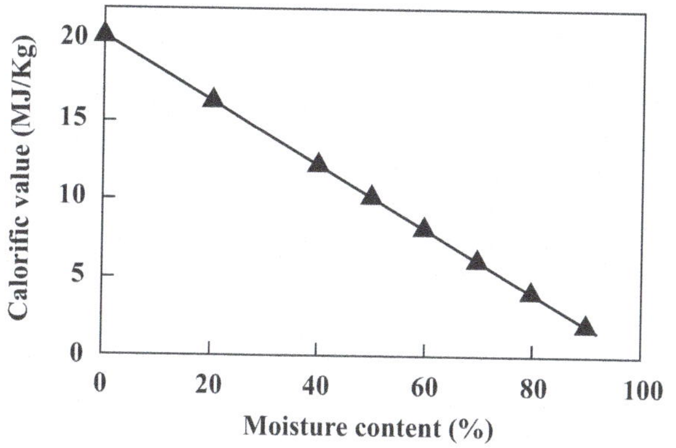

Fig. 6.6 Calorific value of bark as a function of moisture content (Fardim 2011)

Table 6.8 Example of work needed for debarking different species (Fardim 2011)

Ranking	Bark-to-wood adhesion (MPa)	Species
Almost impossible to debark	2.0	American basswood, hickory (*Carya* spp.)
Very difficult to debark	1.2–1.4	black poplar, "ironwood"
Difficult to debark	0.8–1.0	silver birch, elm (*Ulmus* sp.)
Normal to debark	0.5–0.6	Norway spruce, European beech, hemlock
Easy to debark	0.0–0.4	Scots pine, southern yellow pine, Norway maple, European oak, most tropical hardwoods

Table 6.9 Volume-weighted electricity consumption (kWh/m^3 output) of different processes (coniferous)

Study	Debarking	Milling	Total (debarking and milling)	Kiln drying	Planing
Diederichs (2014)	2.2–2.7	15.3–17.7	19.4–21.8	21.6–22.0	22.8
Bergman and Bowe (2008a)			52.6	24.5	19.5
Puettmann et al. (2010)	1.8	19.0	20.8	16.7	34.1
Milota et al. (2005): West USA			45.4	26.7	16.8
Milota et al. (2005): South USA			34.4	28.7	13.9
Werner et al. (2007)	4.5	28.2	32.7	25	31.0
Frühwald et al. (2000)			23.2	20.3	27.2
Average of all studies:	3.2	23.6	34.9	23.7	23.8

Table 6.10 Electricity consumption, according to Diederichs (2014) and Bergman and Bowe (2008b)

Process	Output unit	Diederichs	Bergmann and Bowe
Debarking	(kWh/m^3)	6.0	
Milling	(kWh/m^3)	49.7	
Debarking and milling	(kWh/m^3)	55.7	62.3
Kiln drying	(kWh/m^3)	57.8	31.1

Table 6.11 Primary energy consumption (Diederichs 2014)

	Energy consumption (MJ)				
	Coniferous sawn timber	Coniferous sawn timber	Coniferous sawn timber	Broad-leaved sawn timber	Share renewables
	green	kiln dried	kiln dried, planed	kiln dried	
Forestry transport					
Sawlogs (material)	8672	8672	8762	13 104	100%
Forest operations	204	244	253	336	1%
Transport to mill	71	93	97	186	2%
Total (non-material)	275	337	350	522	
Sawmill					
Diesel	41	66	88	79	0%
Electricity	227	514	806	1515	13%
Wood fuel (on site)		1677	806	4913	100%
Wood fuel (of site)		290	242	1285	100%
Fossil fuel		2	5		0%
Others (material)	7	48	58	39	0%
Total (non-material)	268	2549	3173	7791	
Total (non-material)	544	2886	3522	8313	
Total (all)	9223	11 605	12 253	21 645	

Table 6.12 Heat and electricity consumption in sawn-timber production processes (Anderson and Westerlund 2014)

	Electricity	Heat	Process temperature
	(kWh/m^3 sawn timber)	(kWh/m^3 sawn timber)	(°C)
Debarking	4		
Sawing	23	10	30
Sorting	2	5	30
Drying	31	299	75
Handling around drying	4	5	30
Grinding	13	5	30
Office		5	30
Total	77	15	30

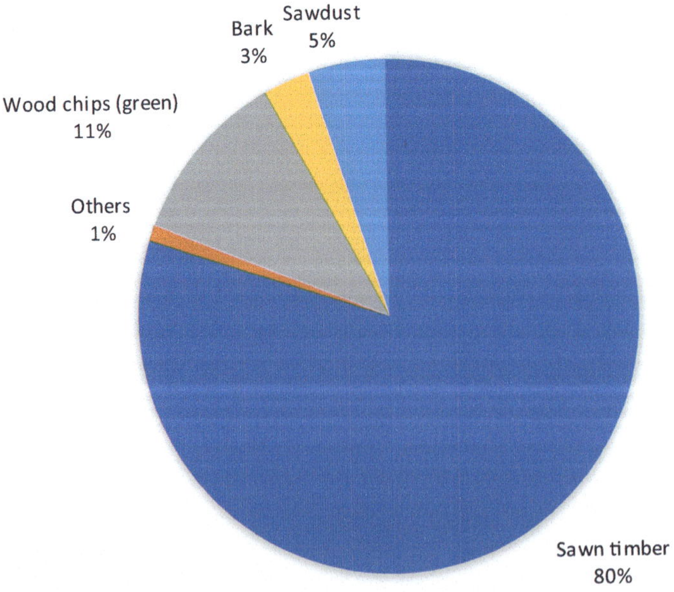

Fig. 6.7 The turnover shares of different products in a typical large-scale softwood sawmill in Northern Europe (Varis 2018a)

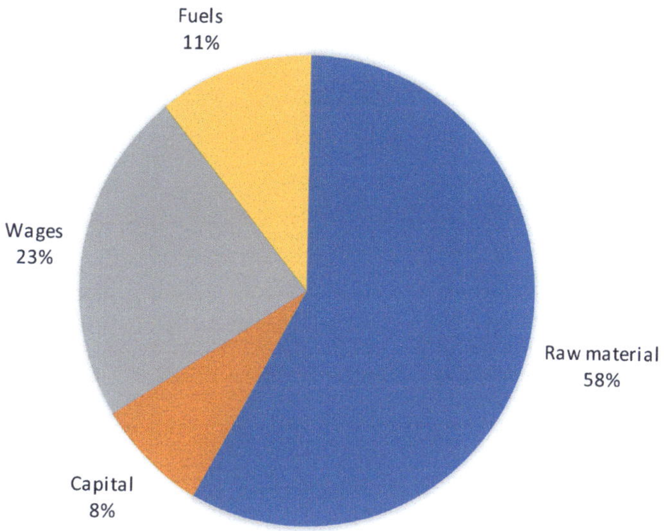

Fig. 6.8 The costs of different features (share of total cost) of a typical large-scale softwood sawmill in Northern Europe (Varis 2018a)

6.3 Veneer Production and Veneer-Based Products

Peter Niemz, Dick Sandberg

6.3.1 General

See Table 6.13

6.3.2 Veneer Production

See Tables 6.14 and 6.15

6.3.3 Manufacture of Veneer-Based Products

See Tables 6.16, 6.17, 6.18, 6.19, 6.20 and 6.21

Table 6.13 Utilisation of veneers in the furniture industry (Mette and Korell 1989)

Intended use	Degree of utilisation (%)
Barrier (cross-bonded) veneer	67
Face veneer	58
Interior veneer	58
Edge veneer	44

Table 6.14 Volume yield in veneer production (Mette and Korell 1989)

Species	Veneer type	Volume yield (%)
Norway maple, Norway spruce	sawn	41
European beech	peeled	65
European beech	sliced	75
European oak	sliced	70
Limba	peeled	67
Abachi	peeled	67

Table 6.15 Lathe capacities for different species and block dimensions (Varis 2018b)

Peeling machine data	Unit	Length = 130/160 cm Block diam. = 220 mm 1.5 mm birch veneer	Length = 190 cm Block diam. = 240 mm 2.1 mm softwood veneer	Length = 260 cm Block diam. = 340 mm 3.2 mm softwood veneer
Capacity	(m^3/h)	12	21	45
Spindle rotation	(rpm)	750–1000	750–1000	750–1000
Diameter of the inner spindle (smallest diameter)	(mm)	50	60	75

Table 6.16 Mean volume yield in the production of laminated veneer lumber (LVL) or plywood of furniture quality from European beech (Anon 1975a)

Type of plywood	Unit	LVL and plywood for furniture	Plywood for furniture
Volume yield	(%)	33	35–38

Table 6.17 Mean volume yield in the production of plywood of furniture quality from different species (Kollmann and Schulte-Brader 1954; Anon 1975)

	Species:	European beech	Limba	Gaboon
Volume yield	(%)	35.0–37.5–39.8	42.0–44.0–46.0	47.0–49.3–56.0

Table 6.18 Influence of round-timber diameter of European beech wood on volume yield in plywood production (Fehrendsen 1956)

Diameter:	(mm)	200–290	300–340	350–390	400–490	500–590	>600
Volume yield	(%)	24.4	33.2	37.0	40.3	42.8	43.5

Table 6.19 Mean values for different species (Dofinè 1962)

	Species:	Gaboon	Poplar spp.	European beech	Birch spp.
Volume yield	(%)	40–50–54	35–40–45	33–37–42	22–25–28

Table 6.20 Production data for plywood from Norway spruce and birch spp. (European white birch and silver birch). Data from the Raute Company

Process	Unit	Value
Raw material consumption including bark		
Birch	(m^3/m^3)	2.7
Spruce	(m^3/m^3)	2.1
Speed of rotary peeling		
Birch	(m/min)	250
Spruce	(m/min)	280–300–320
Time through which the production takes place		
A modern factory producing standard spruce plywood	(days)	<1
A modern factory producing made-to-order birch plywood	(days)	4–7
A modern LVL production	(days)	2–3
Working hours per produced volume		
A modern birch plywood factory producing high-quality grades "white panels"	(h/m^3)	4
A modern conifer plywood factory	(h/m^3)	1

Table 6.21 The average volume yield in plywood mills for birch (European white birch and silver birch) and Norway spruce and Scots pine (Varis 2018b). Volume of logs including the bark

Operation	Birch	Pine/spruce
	(%)	(%)
Round timber (veneer logs)	100	100
Debarked logs for cross-cutting	89	90
Blocks to peeling	85	89
Green veneer to drying	57	66
Veneer to lay-up	43	54
Pressed plywood	41	51
Formatted plywood	35	45
Total volume yield	35	45
Raw material consumption including bark for the manufacture of 1 m^3 plywood (m^3)	2.86	2.22

6.4 Fibre- and Particle-Based Processes for Manufacture of EWPs

Peter Niemz, Dick Sandberg

6.4.1 Adhesive Consumption

See Table 6.22 also Fig. 6.9

6.4.2 Fibre-Based Panel Material

See Tables 6.23, 6.24, 6.25 and 6.26

6.4.3 Organic-Bonded Particleboard

See Tables 6.27, 6.28, 6.29, 6.30, 6.31, 6.32 and Fig. 6.10

Table 6.22 Adhesive content in different wood-based materials

Material	Wood (%)	Adhesive content (% solid resin)
Glued-laminated timber (GLT)	95–97	3–5
Cross-laminated timber (CLT)	95–97	3–5 (0% when connected with dowels, nails or mechanical connectors)
Particleboard	86–93	7–14
Fibreboard	86–100	0–16 (for HDF up to 16%, for MDF with low density, the adhesive content may be higher)
Veneer-based materials	20–95	5–(80) (high contents for impregnated materials)

HDF—high-density fibreboard, MDF—medium-density fibreboard

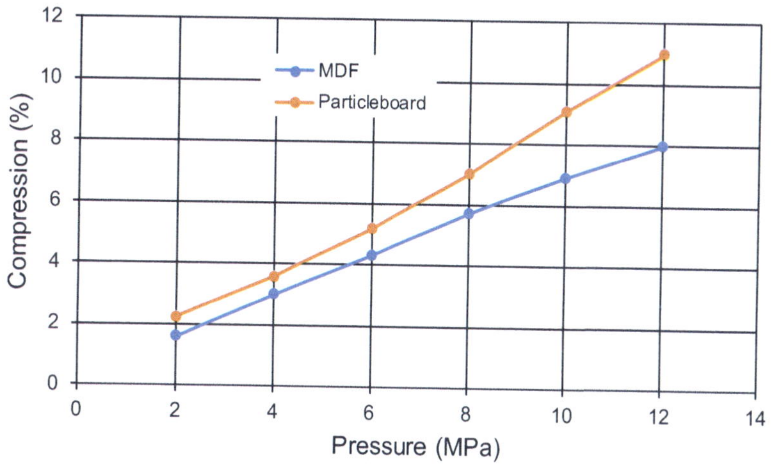

Fig. 6.9 Changes in thickness of particleboard and medium-density fibreboard (MDF) under pressure perpendicular to the plane of the board (Niemz 2002)

Table 6.23 Index values of energy and raw-material demand for medium-density fibreboard (MDF) production plant, depending on the board density (Anon 1975)

MDF grade:	Unit	Ultra-light	Light	Regular	Thin
Electricity	(kW/m^3)	211	257	314	358
Heat	(kW/m^3)	684	822	998	1108
Logs	(m^3/m^3)	1.13	1.62	2.01	2.27
Urea–formaldehyde (UF) resin solid	(kg/m^3)	80	81	79	65
Ratio of electricity to heat		0.32	0.32	0.32	0.32

Produced from radiata pine
Ratio of bark on logs is 5% on weight
Ratio of solid UF resin based on weight applied to dry fibres ranges from 8% to 17% depending on MDF grade
Ratio of fines in chip screenings is 2% based on weight
Ratio of loss in chip washing is 4% based on weight
Ratio of panel trim-off is 4% based on weight
Ratio rejected is 3% based on weight

Table 6.24 Influence of fibreboard density (board thickness 16–19 mm) on wood consumption and volume yield. Raw material: Scots pine with basic density 400 kg/m^3 (Kehr 1975)

Board density:	(kg/m^3)	600	650	700	750
Wood consumption	(m^3/m^3)	1.52	1.64	1.77	1.99
Volume yield	(%)	66	61	57	53

Table 6.25 Wood consumption in dry (3.2 mm thick HDF) and wet (MDF) processes for fibreboard (Swiderski 1963; Miels 1965; Gran 1968; Anon 1975)

Consumption of wood with bark in oven-dry weight	Unit	Process	
		dry process	wet process
Wood consumption	(m^3/t fibreboard)	1.16	1.15
Wood consumption	(m^3/t fibreboard)	1.16	2.30–2.90

Table 6.26 Wood consumption in fibreboard production as a function of wood density (Miels 1965)

Wood density	(kg/m^3)	400	500	600	700
Wood consumption	(m^3/ton fibreboard)	2.75–2.90	2.20–2.30	1.80–1.90	1.55–1.65

Table 6.27 Influence of species in particleboard production (Stegmann and von Bismarck 1968; Anon 1975)

	Unit	Norway spruce	Scots pine	European oak	European beech
Wood density	(kg/m^3)	440	490	650	680
Board density	(kg/m^3)	600–700	600–700	700	700
Wood consumption	(m^3/m^3)	2.4–2.8	2.1–2.4	2.0	1.8

Table 6.28 Influence of species and the density of the species, board density 600–640 kg/m^3 (Kehr 1975)

	Unit	Black poplar	Norway spruce	Scots pine	Common birch	European beech
Basic density	(kg/m^3)	330	380	400	520	560
Wood consumption	(m^3/m^3)	2.0–2.2	1.6–1.8	1.5–1.7	1.1–1.4	1.1–1.3
Volume yield	(%)	50–45	65–62	67–59	83–72	91–77

Table 6.29 Bulk and mat density at 5–10% moisture content for particles used in particleboard production (Niemz and Sandberg 2022)

Particle type	Bulk density	Mat density
	(kg/m^3)	(kg/m^3)
Surface-layer particles (normal particles)	50–60	70
Surface-layer particles (fine particles)	120–130	150
Core-layer particles	110–120	145
Framesaw sawdust	150–160	190
Sanding dust	160–200	200
Milled particles for particleboard	250–300	295

Table 6.30 Influence of board thickness on board density for Scots pine with basic density above 400 kg/m^3 (Kehr 1975)

	Unit	Particleboard thickness, sanded (mm)			
		10	16	19	22
Density	(kg/m$^{3)}$	700	675	650	620
Wood consumption	(m^3/m^3)	1.81	1.72	1.64	1.57
Volume yield	(%)	55	58	61	64

Table 6.31 Particle thickness for core and surface layers in particleboard production (from Deppe and Ernst 2000 in Niemz and Sandberg 2022)

Type of particle	Layer	Thickness
		(mm)
Coarse particles	core layer	>0.8
	surface layer	0.4–0.8
Optimal particles	core layer	0.2–0.8
	surface layer	0.1–0.3
Dust[1)]	not used	<0.3

[1)] removed out and used for energy production

Table 6.32 Characteristics of particles for particleboard manufacture (from Watzke 1969; Anon 1990; Wagenführ and Scholz 2012 in Niemz and Sandberg 2022)

Particle type	Particle size (mm)			Slenderness	Mat density
	length	thickness	width	L/T (-)	(kg/m³)
Surface-layer particles (normal particles)	5–10	0.20–0.30		20–50	60–120
Surface-layer particles (fine particles)	3–6	0.10–0.25		15–40	120–180
Grinding dust	0.4–0.6				160–200
Core-layer flakes	8–15	0.50–2.00	1.50–3.50	5–50	100–180
Core-layer particles from flaking (second shredding)	8–15	0.40–0.60	2.00–4.00	20–40	48–180
Molding and planing shavings	5–15	0.25–0.80	2.50–5.00	5–60	50–130
Framesaw dust	2–5	0.40–1.00	1.00–2.00	2–10	120–180

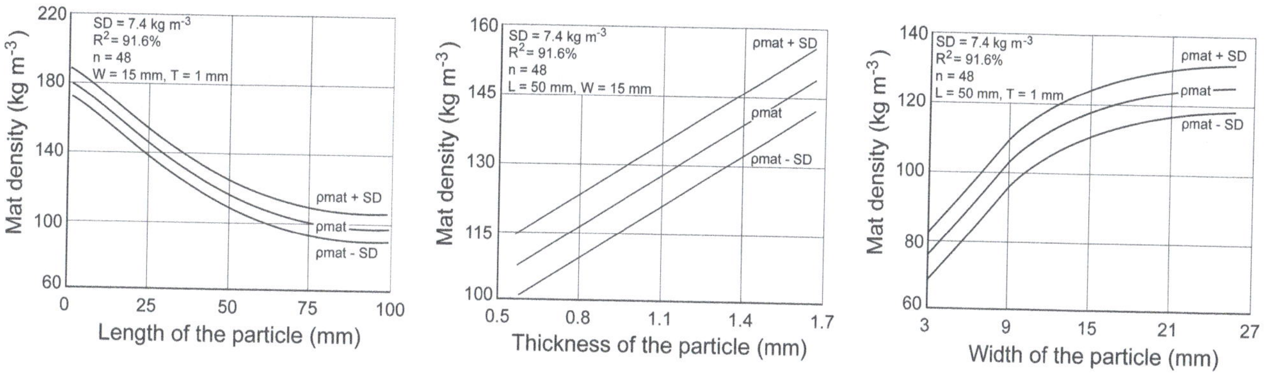

Fig. 6.10 Influence of particle geometry on mat density: a) particle length, b) particle thickness, and c) particle width. n—No. of tested specimens, SD—standard deviation, R^2—coefficient of determination (Niemz 1982; Niemz and Sandberg 2022)

6.5 Material Consumption for Wood-Based Materials

Dirk Grunwald

6.5.1 Particleboard P2

The following terminology is used for resin-bonded particleboard (chipboard) (CEN 2010):

P1: general purpose boards for use in dry conditions.
P2: boards for interior fitments (including furniture) for use in dry conditions.
P3: non load-bearing boards for use in humid conditions.
P4: load-bearing boards for use in dry conditions.
P5: load-bearing boards for use in humid conditions.
P6: heavy duty load-bearing boards for use in dry conditions.
P7: heavy duty load-bearing boards for use in humid conditions.
E1 Grade: It is a mandatory standard for the engineered wood and its products industry, and the formaldehyde emission is stipulated to be ≤ 0.124 mg/m³. It can be understood that it is a mandatory "safety standard line", and the panels that meet this limit value standard can be used.
CARB 2: The CARB phase 2 certification signifies that the product releases a low enough amount of formaldehyde to be considered safe. Any formaldehyde released by products with a CARB phase 2 compliant seal is small enough to pose absolutely no risk to your health.

See Tables 6.33, 6.34 and 6.35

6.5.2 Particleboard P3

See Tables 6.36 and 6.37

6.5.3 Particleboard P5

See Tables 6.38 and 6.39

6.5.4 Particleboard—WFI Rigid

See Table 6.40

Table 6.33 Material and energy consumption of particleboard. Data: P2 quality, density 610–670 kg/m³, 16–22 mm thickness, E1+CARB 2. MUF—melamine urea formaldehyde, ODW—oven-dry weight, UF—urea–formaldehyde

Material consumption	Minimum (kg/m³)	Maximum (kg/m³)	Average (kg/m³)	(%)
Wood (net, no losses)	515	560	530	83.8
Wood (gross, with losses), 100% green wood (pine)	567	616	583	
Wood (gross, with losses), 100% recycling wood (mix)	670	728	689	
Adhesive UF / MUF (solid ODW)	40	80	60	9.4
Adhesive UF / MUF (liquid ODW, 65% solid)	62	123	92	
Hardener (ammonium-salts, solid ODW)	1.5	2.2	1.7	0.3
Urea (solid ODW)	0.5	6.9	2.0	0.3
Water (board before storage)	25	40	35	5.5
Emulsion / paraffin (solid ODW)	2.0	4.0	3.0	0.5
pMDI / eMDI (accelerator)	0	2.2	1.0	0.2
Total:			**633**	**100**
Energy consumption	**Minimum (KWh/m³)**	**Maximum (KWh/m³)**	The values and wood loss depend on each other, due to the set-up of the energy plant. Wood moisture in the log yard and climate also play a role	
Gas / oil / Biomass	600	800		
Electricity	80	180		

Table 6.34 Material and energy consumption of particleboard. Data: P2 quality, density 630–700 kg/m³, 12–15 mm thickness, E1+CARB 2. eMDI—emulsified isocyanate, MUF—melamine urea formaldehyde, ODW—oven-dry weight, pMDI—polymetric 4,4-diphenylmethane diisocyanate

Material consumption	Minimum (kg/m³)	Maximum (kg/m³)	Average (kg/m³)	(%)
Wood (net, no losses)	525	560	540	83.0
Wood (gross, with losses), 100% green wood (pine)	578	616	594	
Wood (gross, with losses), 100% recycling wood (mix)	683	728	702	
Adhesive UF / MUF (solid ODW)	60	70	68	10.5
Adhesive UF / MUF (liquid ODW. 65% solid)	92	108	105	
Hardener (ammonium-salts. solid ODW)	1.6	2.2	1.8	0.3
Urea (solid ODW)	2.5	6.5	4.5	0.7
Water (board before storage)	27	41	33	5.1
Emulsion / paraffin (solid ODW)	1.8	2.3	2.0	0.2
pMDI / eMDI (accelerator)	0	2.2	1.0	0.2
Total:			**650**	**100**
Energy consumption	**Minimum (KWh/m³)**	**Maximum (KWh/m³)**	The values and the wood loss depend on each other, due to the set-up of the energy plant. Wood moisture in the log yard and climate also play a role	
Gas / oil / biomass	600	800		
Electricity	100	200		

6.5.5 Oriented Strand Board—OSB 2

See Tables 6.41 and 6.42

6.5.6 Oriented Strand Board—OSB 3

The following terminology is used for OSB, i.e. oriented strand boards (CEN 2006):

OSB/1: General purpose boards, and boards for interior fitments (including furniture) for use in dry conditions.

OSB/2: Load-bearing boards for use in dry conditions.

OSB/3: Load-bearing boards for use in humid conditions.

OSB/4: Heavy-duty load-bearing boards for use in humid conditions.

See Tables 6.43 and 6.44

6.5.7 Oriented Strand Board—OSB 4

See Tables 6.45 and 6.46

Table 6.35 Material and energy consumption particleboard. Data: P2 quality, density 590–640 kg/m³, 23–32 mm thickness, E1+CARB 2. eMDI—emulsified isocyanate, MUF—melamine urea formaldehyde, ODW—oven-dry weight, pMDI—polymetric 4,4-diphenylmethane diisocyanate

Material consumption	Minimum (kg/m³)	Maximum (kg/m³)	Average (kg/m³)	(%)
Wood (net, no losses)	495	525	515	84.3
Wood (gross, with losses), 100% green wood (pine)	545	578	567	
Wood (gross, with losses), 100% recycling wood (mix)	708	751	736	
Adhesive UF / MUF (solid ODW)	49	65	57	9.3
Adhesive UF / MUF (liquid ODW. 65% solid)	75	100	88	
Hardener (ammonium-salts. solid ODW)	1.5	2.1	1.9	0.3
Urea (solid ODW)	1.2	5.0	4.5	0.7
Water (board before storage)	26	35	30	4.9
Emulsion / paraffin (solid ODW)	1.5	2.2	1.8	0.3
pMDI / eMDI (accelerator)	0	2.2	1.0	0.2
Total:			**611**	**100**
Energy consumption	**Minimum (KWh/m³)**	**Maximum (KWh/m³)**	The values and the wood loss depend on each other, due to the set-up of the energy plant. Wood moisture in the log yard and climate also play a role	
Gas / oil / biomass	600	800		
Electricity	80	150		

Table 6.36 Material and energy consumption of particleboard. Data: P3 quality, density 640–680 kg/m³, 16–22 mm thickness, E1. eMDI—emulsified isocyanate, MUF—melamine urea formaldehyde, ODW—oven-dry weight, pMDI—polymetric 4,4-diphenylmethane diisocyanate

Material consumption	Minimum (kg/m³)	Maximum (kg/m³)	Average (kg/m³)	(%)
Wood (net, no losses)	550	570	550	83.5
Wood (gross, with losses), 100% green wood (pine)	589	610	589	
Wood (gross, with losses), 100% recycling wood (mix)	688	713	688	
Adhesive MUF (solid ODW)	55	70	65	9.9
Adhesive MUF (liquid ODW, 65% solid)	85	108	100	
Hardener (ammonium-salts, solid ODW)	1.9	2.1	2.0	0.3
Urea (solid ODW)	1.2	5.0	3.0	0.5
Water (board before storage)	33	35	34	5.2
Emulsion / paraffin (solid ODW)	2.0	4.8	2.5	0.4
pMDI / eMDI (accelerator)	0	3.0	2.0	0.2
Total:			**659**	**100**
Energy consumption	**Minimum (KWh/m³)**	**Maximum (KWh/m³)**	The values and the wood loss depend on each other, due to the set-up of the energy plant. Wood moisture in the log yard and climate also play a role	
Gas / oil / biomass	600	800		
Electricity	100	180		

Table 6.37 Material and energy consumption of particleboard. Data: P3 quality, density 640–680 kg/m³, 16–22 mm thickness, NAF certified (NAF=no-added formaldehyde, i.e. formaldehyde-free adhesive is used in the production of NAF products. ODW—oven-dry weight, pMDI—polymeric 4,4-diphenylmethane diisocyanate, PUR—polyurethane

Material consumption	Minimum (kg/m³)	Maximum (kg/m³)	Average (kg/m³)	(%)
Wood (net, no losses)	580	615	590	91.2
Wood (gross, with losses), 100% green wood (pine)	621	658	631	
Wood (gross, with losses), 100% recycling wood (mix)	725	769	738	
Adhesive pMDI / PUR (solid ODW)	24	31	25	3.9
Water (board before storage)	30	31	30	4.6
Emulsion / paraffin (solid ODW)	2.0	2.2	2.1	0.3
Total:			**647**	**100**
Energy consumption	**Minimum (KWh/m³)**	**Maximum (KWh/m³)**	The values and the wood loss depend on each other, due to the set-up of the energy plant. Wood moisture in the log yard and climate also play a role	
Gas / oil / biomass	600	800		
Electricity	100	160		

Table 6.38 Material and energy consumption of particleboard. Data: P5 quality, density 670–720 kg/m³, 16–22 mm thickness, E1. eMDI—emulsified isocyanate, MUF—melamine urea formaldehyde, ODW—oven-dry weight, pMDI—polymetric 4,4-diphenylmethane diisocyanate

Material consumption	Minimum (kg/m³)	Maximum (kg/m³)	Average (kg/m³)	(%)
Wood (net, no losses)	580	620	600	84.1
Wood (gross, with losses), 100% green wood (pine)	621	663	642	
Wood (gross, with losses), 100% recycling wood (mix)	737	787	762	
Adhesive MUF (solid ODW)	58	80	70	9.8
Adhesive MUF (liquid ODW, 65% solid)	89	123	108	
Hardener (ammonium-salts, solid ODW)	1.9	3.0	2.8	0.4
Urea (solid ODW)	1.2	5.0	1.4	0.2
Water (board before storage)	33	35	34	4.8
Emulsion / paraffin (solid ODW)	2.0	4.8	3.0	0.4
pMDI / eMDI (accelerator)	0	3.0	2.0	0.3
Total:			**713**	**100**
Energy consumption	**Minimum (KWh/m³)**	**Maximum (KWh/m³)**	The values and the wood loss depend on each other, due to the set-up of the energy plant. Wood moisture in the log yard and climate also play a role	
Gas / oil / biomass	600	800		
Electricity	100	180		

Table 6.39 Material and energy consumption of particleboard. Data: P5 quality, density 650–710 kg/m³, 16–22 mm thickness, NAF certified (NAF = no-added formaldehyde, i.e. formaldehyde-free adhesive is used in the production of NAF products. ODW—oven-dry weight, pMDI—polymeric 4,4-diphenylmethane diisocyanate, PUR—polyurethane

Material consumption	Minimum (kg/m³)	Maximum (kg/m³)	Average (kg/m³)	(%)
Wood (net, no losses)	640	660	650	90.4
Wood (gross, with losses), 100% green wood (pine)	685	706	696	
Wood (gross, with losses), 100% recycling wood (mix)	813	838	826	
Adhesive pMDI / PUR (solid ODW)	35	45	37	5.1
Water (board before storage)	30	31	30	4.2
Emulsion / paraffin (solid ODW)	2.0	2.2	2.1	0.3
Total:			**719**	**100**
Energy consumption	**Minimum (KWh/m³)**	**Maximum (KWh/m³)**	The values and the wood loss depend on each other, due to the set-up of the energy plant. Wood moisture in the log yard and climate also play a role	
Gas / oil / biomass	600	800		
Electricity	100	160		

Table 6.40 Material and energy consumption of particleboard. Data: WFI Rigid, density 120–220 kg/m³, 40–300 mm thickness, NAF certified (NAF = no-added formaldehyde, i.e. formaldehyde-free adhesive is used in the production of NAF products. ODW—oven-dry weight, PUR—polyurethane

Material consumption	Minimum (kg/m³)	Maximum (kg/m³)	Average (kg/m³)	(%)
Wood (net, no losses)	100	190	150	82.0
Wood (gross, with losses), in case of 100% green wood (pine)	110	209	165	
Adhesive PUR (solid ODW)	4	15	9	4.9
Water (board before storage)	10	25	21	11.5
Emulsion / paraffin (solid ODW)	0.6	4.0	3.0	1.6
Total:			**183**	**100**
Energy consumption	**Minimum (KWh/m³)**	**Maximum (KWh/m³)**	The values and the wood loss depend on each other, due to the set-up of the energy plant. Wood moisture in the log yard and climate also play a role	
Gas / oil / biomass	150	400		
Electricity	125	200		

Table 6.41 Material and energy consumption of OSB. Data: OSB 2 quality according to the EN 300 standard (CEN 2006), density 590–670 kg/m^3, 10–14 mm thickness, NAF certified (NAF = no-added formaldehyde, i.e. formaldehyde-free adhesive is used in the production of NAF products. ODW—oven-dry weight, pMDI—polymeric 4,4-diphenylmethane diisocyanate

Material consumption	Minimum (kg/m^3)	Maximum (kg/m^3)	Average (kg/m^3)	(%)
Wood (net, no losses)	550	580	570	92
Wood (gross, with losses), in case of 100% green wood (pine)	627	661	650	
Adhesive pMDI (solid ODW)	15	20	16	2.6
Release agent	0.8	1.2	0.9	0.1
Water (board before storage)	24	43	30	4.8
Emulsion / paraffin (solid ODW)	2.0	7.0	3.0	0.5
Total:			**620**	**100**
Energy consumption	**Minimum (KWh/m^3)**	**Maximum (KWh/m^3)**	The values and the wood loss depend on each other, due to the set-up of the energy plant. Wood moisture in the log yard and climate also play a role	
Gas / oil / biomass	600	800		
Electricity	100	160		

Table 6.42 Material and energy consumption of OSB. Data for OSB 2 quality according to the EN 300 standard (CEN 2006), density 570–640 kg/m^3, 15–25 mm thickness, NAF certified (NAF = no-added formaldehyde, i.e. formaldehyde-free adhesive is used in the production of NAF products. ODW—oven-dry weight, pMDI—polymeric 4,4-diphenylmethane diisocyanate

Material consumption	Minimum (kg/m^3)	Maximum (kg/m^3)	Average (kg/m^3)	(%)
Wood (net, no losses)	525	570	550	91.5
Wood (gross, with losses), in case of 100% green wood (pine)	599	650	627	
Adhesive pMDI (solid ODW)	14	23	17	2.8
Release agent	0.8	1.2	0.9	0.1
Water (board before storage)	25	40	30	5.0
Emulsion / paraffin (solid ODW)	2.0	7.0	3.0	0.5
Total:			**601**	**100**
Energy consumption	**Minimum (KWh/m^3)**	**Maximum (KWh/m^3)**	The values and the wood loss depend on each other, due to the set-up of the energy plant. Wood moisture in the log yard and climate also play a role	
Gas / oil / biomass	600	800		
Electricity	100	200		

Table 6.43 Material and energy consumption of OSB. Data: OSB 3 quality according to the EN 300 standard (CEN 2006), density 600–680 kg/m^3, 10–14 mm thickness, NAF certified (NAF = no-added formaldehyde, i.e. formaldehyde-free adhesive is used in the production of NAF products. ODW—oven-dry weight, pMDI—polymeric 4,4-diphenylmethane diisocyanate, PUR—polyurethane

Material consumption	Minimum (kg/m^3)	Maximum (kg/m^3)	Average (kg/m^3)	(%)
Wood (net, no losses)	550	650	575	91.4
Wood (gross, with losses), in case of 100% green wood (pine)	627	741	656	
Adhesive pMDI / PUR (solid ODW)	17	25	20	3.2
Release agent	0.8	1.2	0.9	0.1
Water (board before storage)	24	43	30	4.8
Emulsion / paraffin (solid ODW)	2.0	7.0	3.0	0.5
Total:			**629**	**100**
Energy consumption	**Minimum (KWh/m^3)**	**Maximum (KWh/m^3)**	The values and the wood loss depend on each other, due to the set-up of the energy plant. Wood moisture in the log yard and climate also play a role	
Gas / oil / biomass	600	800		
Electricity	100	160		

Table 6.44 Material and energy consumption of OSB. Data: OSB 3 quality according to the EN 300 standard (CEN 2006), density 580–640 kg/m^3, 15–25 mm thickness, NAF certified (NAF = no-added formaldehyde, i.e. formaldehyde-free adhesive is used in the production of NAF products. ODW—oven-dry weight, pMDI—polymeric 4,4-diphenylmethane diisocyanate, PUR—polyurethane

Material consumption	Minimum	Maximum	Average	
	(kg/m^3)	(kg/m^3)	(kg/m^3)	(%)
Wood (net, no losses)	525	570	560	90.9
Wood (gross, with losses), in case of 100% green wood (pine)	599	650	638	
Adhesive pMDI / PUR (solid ODW)	16	23	20	3.2
Release agent	0.8	1.2	0.9	0.1
Water (board before storage)	25	40	32	5.3
Emulsion / paraffin (solid ODW)	2	7	3	0.5
Total:			**616**	**100**
Energy consumption	**Minimum**	**Maximum**	The values and the wood loss depend on each other, due to the set-up of the energy plant. Wood moisture in the log yard and climate also play a role	
	(KWh/m^3)	**(KWh/m^3)**		
Gas / oil / biomass	600	800		
Electricity	100	200		

Table 6.45 Material and energy consumption of OSB. Data: OSB 4 quality according to the EN 300 standard (CEN 2006), density 680–690 kg/m^3, 10–14 mm thickness, NAF certified (NAF = no-added formaldehyde, i.e. formaldehyde-free adhesive is used in the production of NAF products. ODW—oven-dry weight, pMDI—polymeric 4,4-diphenylmethane diisocyanate, PUR—polyurethane

Material consumption	Minimum	Maximum	Average	
	(kg/m^3)	(kg/m^3)	(kg/m^3)	(%)
Wood (net, no losses)	600	620	610	90.5
Wood (gross, with losses), in case of 100% green wood (pine)	684	707	695	
Adhesive pMDI / PUR (solid ODW)	25	35	30	4.5
Release agent	0.8	1.2	0.9	0.1
Water (board before storage)	24	43	30	4.5
Emulsion / paraffin (solid ODW)	2	4	3	0.4
Total:			**674**	**100**
Energy consumption	**Minimum**	**Maximum**	The values and the wood loss depend on each other, due to the set-up of the energy plant. Wood moisture in the log yard and climate also play a role	
	(KWh/m^3)	**(KWh/m^3)**		
Gas / oil / biomass	600	800		
Electricity	100	160		

Table 6.46 Material and energy consumption of OSB. Data: OSB 4 quality according to the EN 300 standard (CEN 2006), density 580–620 kg/m^3, 15–22 mm thickness, NAF certified (NAF = no-added formaldehyde, i.e. formaldehyde-free adhesive is used in the production of NAF products. ODW—oven-dry weight, pMDI—polymeric 4,4-diphenylmethane diisocyanate, PUR—polyurethane

Material consumption	Minimum	Maximum	Average	
	(kg/m^3)	(kg/m^3)	(kg/m^3)	(%)
Wood (net, no losses)	570	590	580	90.2
Wood (gross, with losses), in case of 100% green wood (pine)	650	673	661	
Adhesive pMDI / PUR (solid ODW)	20	35	30	4.7
Release agent	0.8	1.2	0.9	0.1
Water (board before storage)	23	42	29	4.5
Emulsion / paraffin (solid ODW)	2	4	3	0.5
Total:			**643**	**100**
Energy consumption	**Minimum**	**Maximum**	The values and the wood loss depend on each other, due to the set-up of the energy plant. Wood moisture in the log yard and climate also play a role	
	(KWh/m^3)	**(KWh/m^3)**		
Gas / oil / biomass	600	800		
Electricity	100	200		

6.5.8 High-Density Fibreboard—HDF

The following terminology is used for fibreboard:

E1 Grade: It is a mandatory standard for the engineered wood and its products industry, and the formaldehyde emission is stipulated to be ≤ 0.124 mg/m^3. It can be understood that it is a mandatory "safety standard line", and the panels that meet this limit value standard can be used.

CARB 2: The CARB phase 2 certification signifies that the product releases a low enough amount of formaldehyde to be considered safe. Any formaldehyde released by products with a CARB phase 2 compliant seal is small enough to pose absolutely no risk to your health.

See Tables 6.47 and 6.48

6.5.9 Medium-Density Fibreboard—MDF

See Table 6.49

6.5.10 Low-Density Fibreboard—LDF

See Table 6.50

Table 6.47 Material and energy consumption of HDF. Data: HDF, density 850–920 kg/m^3, 6–9 mm thickness flooring, E1 + CARB 2. MUF—melamine urea formaldehyde, ODW—oven-dry weight, UF—urea formaldehyde

Material consumption	Minimum (kg/m^3)	Maximum (kg/m^3)	Average (kg/m^3)	(%)
Wood (net, no losses)	680	750	690	78.5
Wood (gross, with losses), 100% green wood (pine)	830	915	842	
Wood gross, with losses), 20% sawdust or board waste	864	953	876	
Adhesive UF/ MUF (solid ODW)	90	140	102	11.6
Adhesive UF / MUF (liquid ODW, 65% solid)	138	215	157	
Hardener (ammonium-salts, solid ODW)	2.5	10.0	7.0	0.8
Urea (solid ODW)	6	50	28	3.2
Water (board before storage)	25	50	43	4.9
Emulsion / paraffin (solid ODW)	6	12	9	1.0
Total:			**879**	**100**
Energy consumption	**Minimum (KWh/m^3)**	**Maximum (KWh/m^3)**	The values and the wood loss depend on each other, due to the set-up of the energy plant. Wood moisture in the log yard and climate also play a role	
Gas / oil / biomass	700	1200		
Electricity	250	330		

Table 6.48 Material and energy consumption of HDF. Data: HDF, density 820–920 kg/m^3, ≤ 3 mm thickness, back panels, E1 + CARB 2. MUF—melamine urea formaldehyde, ODW—oven-dry weight, UF—urea formaldehyde

Material consumption	Minimum (kg/m^3)	Maximum (kg/m^3)	Average (kg/m^3)	(%)
Wood (net, no losses)	710	770	740	80.7
Wood (gross, with losses), 100% green wood (pine)	781	847	814	
Wood gross, with losses), 20% sawdust or board waste	817	886	851	
Adhesive UF/ MUF (solid ODW)	93	130	110	12
Adhesive UF / MUF (liquid ODW. 65% solid)	143	200	169	
Hardener (ammonium-salts. solid ODW)	2.5	5.0	3.0	0.3
Urea (solid ODW)	0	15	9	1.0
Water (board before storage)	35	54	46	5.0
Emulsion / paraffin (solid ODW)	5	13	9	1.0
Total:			**917**	**100**
Energy consumption	**Minimum (KWh/m^3)**	**Maximum (KWh/m^3)**	The values and the wood loss depend on each other, due to the set-up of the energy plant. Wood moisture in the log yard and climate also play a role	
Gas / oil / biomass	800	1300		
Electricity	250	330		

Table 6.49 Material and energy consumption of MDF. Data: MDF, density 630–850 kg/m³, 16–19 mm thickness, E1+CARB 2. MUF—melamine urea formaldehyde, ODW—oven-dry weight, UF—urea formaldehyde

Material consumption	Minimum	Maximum	Average	
	(kg/m³)	(kg/m³)	(kg/m³)	(%)
Wood (net, no losses)	580	650	600	77.7
Wood (gross, with losses), 100% green wood (pine)	708	793	732	
Wood gross, with losses), 20% sawdust or board waste	737	826	762	
Adhesive UF/ MUF (solid ODW)	80	156	90	11.7
Adhesive UF / MUF (liquid ODW. 65% solid)	123	240	138	
Hardener (ammonium-salts. solid ODW)	2.5	13.0	10.0	1.3
Urea (solid ODW)	16	50	30	3.9
Water (board before storage)	26	40	32	4.1
Emulsion / paraffin (solid ODW)	3.5	15.0	10.0	1.3
Total:			**772**	**100**
Energy consumption	**Minimum (KWh/m³)**	**Maximum (KWh/m³)**	The values and the wood loss depend on each other, due to the set-up of the energy plant. Wood moisture in the log yard and climate also play a role	
Gas / oil / biomass	700	1200		
Electricity	250	300		

Table 6.50 Material and energy consumption of LDF. Data: LDF, density 550–650 kg/m³, 12–19 mm thickness, flooring, E1+CARB 2. MUF—melamine urea formaldehyde, ODW—oven-dry weight, UF—urea formaldehyde

Material consumption	Minimum	Maximum	Average	
	(kg/m³)	(kg/m³)	(kg/m³)	(%)
Wood (net, no losses)	460	570	500	79.6
Wood (gross, with losses), 100% green wood (pine)	561	695	610	
Wood gross, with losses), 20% sawdust or board waste	584	724	635	
Adhesive UF/ MUF (solid ODW)	65	80	72	11.5
Adhesive UF / MUF (liquid ODW. 65% solid)	100	123	111	
Hardener (ammonium-salts. solid ODW)	1.0	2.0	1.3	0.2
Urea (solid ODW)	16	30	22	3.5
Water (board before storage)	20	34	30	4.8
Emulsion / paraffin (solid ODW)	2.0	3.1	2.5	0.4
Total:			**628**	**100**
Energy consumption	**Minimum (KWh/m³)**	**Maximum (KWh/m³)**	The values and the wood loss depend on each other, due to the set-up of the energy plant. Wood moisture in the log yard and climate also play a role	
Gas / oil / biomass	600	1100		
Electricity	250	300		

6.6 Material and Energy Consumption for Pulp and Paper and Selected Paper Properties

Sabine Heinemann

6.6.1 General

See Tables 6.51 and 6.52

6.6.2 Properties of Various Papers

See Table 6.53

6.7 Transparent Wood

Dick Sandberg

See Tables 6.54, 6.55 and 6.56

Table 6.51 Examples of specific power and heat consumption in industrial plants, summarized from the Best Available Techniques (BAT) References Document for the Production of Pulp, Paper and Board (Suhr et al. 2015; Godin 2019). CTMP—Chemi-thermomechanical pulping, MTP—thermomechanical pulping

Process	Product	Power consumption (external and internal supplies)	Heat consumption (external and internal supplies)
Chemical pulp production	Sulphate pulp (including drying)	700 kWh/t pulp	3000 kWh/t pulp
	Sulphate pulp (integrated pulp and paper production plants)	550 kWh/t pulp and 650 kWh/t paper	2400 kWh/t pulp and 1700 kWh/t paper
	Bleached sulphite pulp (without drying)	550–750 kWh/ADt	2050–2300 kWh/ADt
Mechanical pulp production	Graphic paper (integrated TMP and paper production plants)	3000 kWh/t	1500 kWh/t
	CTMP (market pulp, including drying)	1800 kWh/t	800 kWh/t
Preparation of recycled pulp	Packaging paper (without deinking and dispersing)	150–250 kWh/t	0 kWh/t
	Newsprint paper	300–420 kWh/t DIP	125–250 kWh/t (=0.2–0.4 t_{steam}/t)
	Graphic paper	400–600 kWh/t DIP	180–350 kWh/t (=0.3–0.5 t_{steam}/t)
	Sanitary tissue and market pulp	400–500 kWh/t DIP	180–300 kWh/t (=0.3–0.5 t_{steam}/t)
Paper production (not integrated)	Wood-free fine paper (coated and uncoated)	700 kWh/t *)	900 kWh/t *)
	Sanitary tissue	900–3100 kWh/t	1270–5800 kWh/t
	Specialty paper	200 kWh/t	1700 kWh/t

*) There are differences in power consumption for wood-free paper grades due to differences pulp refining. The different basis weight of the finished paper, the applied coating technology and the coat weight have considerable impacts on the energy consumption

Table 6.52 Paper grades for use in Table 6.53

Paper grade	Product ID	Product
Graphic paper	1002	Newsprint (100% pulp from paper for recycling)
	1005	SC gravure paper
	1010	Art paper
	2001	Wood-free copying paper
	2002	Copying paper (100% pulp from paper for recycling)
Packaging paper	3003	Kraft liner
	3004	Test liner
	3007	Bleached lined folding box board
	3016	Vegetable parchment paper
Tissue	4001	Toilet tissue (100% pulp from paper for recycling)
	4004	Household tissue
	4005	Handkerchiefs (from virgin pulps)
	4006	Non-woven for wet wipes
Specialty paper	5004	Filter paper for analytical purposes
	5011	Capacitor paper
	5021	Duplex wall base paper, uncoated
	5026	Photographic base paper for PE-coating
	5027	Photographic paper, PE-coated
	5063	Cigarette paper
	5064	Tea-bag paper, heat-sealable

Table 6.53 Properties of different papers (Baumgarten 1996)

		Graphic papers					Packaging papers				Tissue				Specialty papers						
Property	**Unit**	**1002**	**1005**	**1010**	**2001**	**2002**	**3003**	**3004**	**3007**	**3016**	**4001**	**4004**	**4005**	**4006**	**5004**	**5011**	**5021**	**5026**	**5027**	**5063**	**5064**
Basis weight	(g/m^2)	44.4	57.1	135.1	80.6	80.7	202.0	177.0	203.0	49.4	52.3	21.4	16.4	51.5	66.8	16.7	111.0	178.0	237.0	24.7	16.7
Thickness	(μm)	68.0	49.1	107.0	98.9	111.0	270.0	275.0	326.0	61.6	189.6	93.8	99.9	137.0	141.0	25.9	132.0	166.0	224.0	39.9	79.1
Apparent density	(g/cm^3)	0.653	1.163	1.262	0.815	0.727	0.748	0.644	0.623	0.802	0.276	0.228	0.164	0.376	0.474	0.645	0.841	1.072	1.058	0.618	0.211
Specific volume	(cm^3/g)	1.532	0.860	0.793	1.228	1.375	1.337	1.554	1.606	1.247	3.625	4.383	6.091	2.660	2.111	1.551	1.189	0.933	0.945	1.617	4.737
Tensile strength MD	(kN/m)	2.33	2.14	6.80	4.97	4.03	22.47	12.07	0.09	3.31	0	0	0	0	2.39	1.83	5.54	13.07	12.40	0.89	0.80
Tensile strength CD	(kN/m)	0.66	0.85	3.81	2.17	2.01	7.87	4.67	5.71	1.58	0	0	0	0	1.23	0.87	1.79	7.53	6.93	0.36	0.27
Breaking length MD	(km)	5.36	3.82	5.13	6.29	5.09	11.34	6.95	0.04	6.84					3.64	11.15	5.09	7.48	5.33	3.66	4.88
Breaking length CD	(km)	1.52	1.51	2.87	2.74	2.54	3.97	2.69	2.87	3.26					1.88	5.33	1.65	4.31	2.98	1.50	1.62
Tensile Index MD	(Nm/g)	52.55	37.48	50.37	61.70	49.98	111.22	68.17	0.43	67.07					35.73	109.38	49.91	73.41	52.32	35.90	47.90
Tensile Index CD	(Nm/g)	14.86	14.83	28.20	26.88	24.93	38.94	26.40	28.14	31.98					18.46	52.30	16.16	42.32	29.25	14.76	15.93
Breaking elongation MD	(%)	1.0	1.1	3.0	1.7	1.3	2.3	2.0	1.6	1.3					1.5	1.8	1.8	3.0	2.4	1.4	2.0
Breaking elongation CD	(%)	1.9	1.7	8.6	3.4	2.2	5.6	3.2	5.2	2.8					2.8	3.1	3.9	7.0	7.0	5.9	2.6
Tear Index Elmendorf MD	(mNm^2/g)	5.01	3.35	5.01	6.66	5.00	12.86	8.44	9.60	5.08				16.76	9.49	3.57	7.73	6.31	5.88	4.70	13.77
Tear Index Elmendorf CD	(mNm^2/g)	7.67	4.72	5.58	7.26	5.99	16.52	10.50	10.78	5.43	40.39	5.54	6.03	16.76	10.12	4.64	9.46	6.54	5.98	5.66	21.05
Tear Index Brecht-Imset MD	(mNm^2/g)	8.17	5.78	9.80	11.81	8.91				10.75					14.82	7.41	10.68	11.18		6.17	
Tear Index Brecht-Imset CD	(mNm^2/g)	10.19	7.50	10.91	13.09	10.75				11.11					16.33		15.23	12.15		6.52	
Bekk smoothness (mean value of top side and bottom side)	(s)	42.08	1693.60	1939.20	23.24	22.20	3.06	1.94	18.24	18.40	0.52	1.34	0.52	1.09	1.97	52.15	31.36	164.00	180.10	59.36	1.75

(continued)

Table 6.53 (continued)

		Graphic papers					Packaging papers				Tissue				Specialty papers						
Bekk smoothness (difference between top side and bottom side)	(s)	9.93	106.40	177.60	3.20	7.76	0.60	0.68	28.97	14.68	0.05	0.30	0.08	0.11	0.98	37.70	37.52	26.64		28.56	1.14
Bendtsen roughness (mean value of top side and bottom side)	(ml/min)	180	15.05	<10	225	290	1282			320											
Bendtsen roughness (difference between top side and bottom side)	(ml/min)	0	0.3		150	20				200											
Roughness Parker Print Surf (0.5 MPa) (mean value of top side and bottom side)	(µm)	5.72	1.62	1.04	6.97	7.93	12.89	13.28	7.79	8.92	11.13	11.34	13.79	11.92	12.05	4.26	9.57	3	2.94	4.82	11.34
Roughness Parker Print Surf (0.5 MPa) (difference between top side and bottom side)	(µm)	0.09	0.02	0.05	0.48	0.64	0.47	0.63	6.8	1.84	0.64	1.1	0.7	0.09	1.33	0.17	5.25	0.32	2.62	0.07	1.65
Roughness Parker Print Surf (1.0 MPa) (mean value of top side and bottom side)	(µm)	4.53	1.24	0.77	6.08	6.84	11.95	12.14	6.28	8	8.91	8.8	11.6	10.34	10.47	3.57	8.16	2.5	1.67	3.8	9.36
Roughness Parker Print Surf (1.0 MPa) (difference between top side and bottom side)	(µm)	0	0.03	0.07	0.28	0.72	0.58	0.79	7.06	1.74	0.42	0.89	0.73	0.11	1.32	0.25	5.08	0.25	2.92	0.11	1.78

(continued)

Table 6.53 (continued)

		Graphic papers					Packaging papers				Tissue				Specialty papers						
Roughness Parker Print Surf (2.0 MPa) (mean value of top side and bottom side)	(µm)	3.45	0.96	0.57	5.31	5.68	10.55	10.43	5.02	6.97	6.78	6.47	8.87	8.52	8.64	2.92	6.6	2.15	2.46	2.96	7.16
Roughness Parker Print Surf (2.0 MPa) (difference between top side and bottom side)	(µm)	0.22	0.05	0.04	0.3	0.84	0.79	1.03	6.29	1.81	0.41	0.63	0.49	0	0.87	0.21	3.71	0.23	2.46	0.13	1.73
KAM Fogra (0.98 MPa) (mean value of top side and bottom side)	(%)	7.7	14.3	27.2	4.2	7.6	4.7	5.1	7.1	5.4	10	7.9	8.8	9.8	5.8	2.8	8.1	13.3	52.1	5.9	7.1
KAM Fogra (0.98 MPa) (difference between top side and bottom side)	(%)	0.6	0.6	0.4	0.4	0.4	1.8	1.4	3	0.4	10	0.6	2	0.8	0.4	0.4	0.6	1	59.4	1	2.2
KAM Fogra (2.45 MPa) (mean value of top side and bottom side)	(%)	13.7	26.5	49.2	7.7	13.3	8.6	9.5	12.6	9.8	16.4	12.8	12.9	18.1	10.1	5.2	12.7	21.8	38.2	10.8	11.7
KAM Fogra (2.45 MPa) (difference between top side and bottom side)	(%)	0.2	1.4	0.4	0.2	0.2	2	1.4	5.6	0.4	16.4	0.8	3	0.6	0.2	0	0.6	2	38.2	1.2	3.4
KAM Fogra (4.90 MPa) (mean value of top side and bottom side)	(%)	20.9	39.2	68.8	11.9	19.2	13.1	14.4	19.1	14.2	23.2	18.3	18.1	27.4	15.1	8.9	18	30.1	54.8	16.1	16.6

(continued)

Table 6.53 (continued)

		Graphic papers					Packaging papers				Tissue				Specialty papers						
KAM Fogra (4.90 MPa) (difference between top side and bottom side)	(%)	1	1.6	0.4	1	0	3.4	2	8.6	0.8	23.2	1	3.4	0.8	0.2	0.2	0.4	1.4	54.8	1.4	5.2
Bending stiffness (two-point method) MD	(Nmm)	0.704	0.004	0.3608		0.242	1.612	2.773	28.174	0.03			8.708	7.894	0.283	7.352	0.113	9.16	0.214	0.001	96.525
Bending stiffness (two-point method) CD	(Nmm)	0.458	0.002	0.176		0.106	0.799	0.722	12.528	0.017			3.22	3.273	0.148	2.998	0.058	4.272	0.126	0	54.577
Bending angle	(°)	5.1	7.5	7.5		7.5	6.4	6.4	4.8	7.5			7.5	6.4	7.5	7.5	5.1	7.5	7.5	7.5	4.8
Young's modulus MD	(MPa)	1918	993	4469		3807	2574	3732	3667	7743			643	472	8103	9479	2369	5628	4865	121	2177
Young's modulus CD	(MPa)	1249	513	2182		1676	1275	972	1631	4309			238	196	4240	3866	1215	2625	2865	33	1231
Opacity (mean value of top side and bottom side)	(%)	93.52	93.38	95.73	92.14	98.24			99.26	59.75	84.33	51.47	37.65	54.98	76.94	41.27	97.79	97.52	98.2	72.43	35.07
Opacity (difference between top side and bottom side)	(%)	0.23	0.04	0.06	0.4	0.02			0.16	0.69	0.86	0.85	0.18	0.18	0.04	0.46	0.14	0.04	0.04	0.5	0.46
Transparency (mean value of top side and bottom side)	(%)	19.76	18.61	10.68	18.05	7.95			4.87	48.68	29.31	57.53	66.28	50.49	29.25	75.47	5.48	7.93	6.04	31.12	70.54
Transparency (difference between top side and bottom side)	(%)	0.36	0.18	0.27	0.58	0.13			0.5	0.51	0.86	0.7	0.15	0.14	0.02	0.35	0.32	0.06	1.51	0.64	0.36

(continued)

Table 6.53 (continued)

		Graphic papers					Packaging papers				Tissue				Specialty papers						
Reflectance factor R 457 (mean value of top side and bottom side)	(%)	57.71	63.82	86.88	80.57	59.25	15.75	20.94	78.48	74.26	66.91	71.43	81.73	83.47	86.58	15.13	75.48	91.99	91.78	90.38	71.4
Reflectance factor R 457 (difference between top side and bottom side)	(%)	0.06	0.96	0.76	0.06	0.1	4.82	8.16	10.86	0.16	0.42	0.14	0.1	0.1	0.12	0.06	19.92	0.26	2.64	0.2	0.03
specific light scattering coefficient (mean value of top side and bottom side)	(m^2/kg)	62.96	53.02	50.97	45.14	58.46			48.1	19.9	38.01	32.88	30.85	18.91	36.16	9.12	122.93	48.8	42.63	88.21	24.2
specific light scattering coefficient (difference between top side and bottom side)	(m^2/kg)	1.2	1.09	2.11	1.4	0.03			3.08	0.32	1.1	0.68	0.23	0.17	0.07	0.09	9.3	0.08	0.46	2.62	0.44
specific light absorption coefficient (mean value of top side and bottom side)	(m^2/kg)	5.89	3.06	0.45	1.2	4.67			0.89	0.33	1.86	1	0.2	0.1	0.2	8.9	0.23	0.4	0.33	0.1	0.5
specific light absorption coefficient (difference between top side and bottom side)	(m^2/kg)	0.06	0.09	0.1	0.04	0.02			0.02	0.02	0.11	0	0	0	0	0.16	0.06	0	0.06	0	0

(continued)

Table 6.53 (continued)

		Graphic papers					Packaging papers				Tissue				Specialty papers						
Lehmann gloss 45° (mean value of top side and bottom side)	(%)	1.23	3.76	18.52	1.56	1.56	1.12	1.09	1.81	1.49	1.05	1	0.95	1.46	1.59	0.9	1.38	1.92	40.34	1	1.01
Lehmann gloss 45° (difference between top side and bottom side)	(%)	0.05	0.08	4.03	0.08	0.02	0.11	0.02	0.05	0.02	0.04	0	0.1	0.05	0.02	0	0.01	0.01	78.25	0	0.02
Ash content (575°C)	(%)	12.53	32.08	36.41	28.08	18.72	7.47	10.59	8.09	4.6	1.67	1.03	0.49	0.61	0.19	0.15	26.64	3.51	6.42	34.6	0.26
Ash content (900°C)	(%)	11.78	31.86	28.98	16.44	18.18	3.12	9.30	5.24	4.44	1.40	0.90	0.41	0.54	0.17	0.13	17.37	3.36	4.41	19.84	0.26

Table 6.54 Methods and properties of large-scale transparent wood (Wachter et al. 2023)

Species	Method	Lignin retained	Processing time	Transmittance	Haze	Tensile strength	Reflectivity	Thickness
		(%)	(hours)	(%)	(%)	(MPa)	(%)	(mm)
Balsa	Immersion (10% NaOH + 30% H_2O_2) + 20 W UV irradiation	≈ 80	1.0–6.5			≈ 20 (wet)	90–96	9.0
Balsa	Brushed H_2O_2 30 $_{wt.\%}$ 0 NaOH 10wt.% + solar irradiation	≈ 85	1 (UV index: 7 to 8)	>90	60–80	31.4–46.2	>88	3.3
Balsa	H_2O_2:DI water:ammonia = 5:5:1 + UV	83	2–8	82	75			1.5

Table 6.55 Properties of transparent wood and procedure conditions (Wachter et al. 2023)

Species	Method	Lignin retained	Transmittance	Haze	Tensile strength	Toughness	Power conversion efficiency	Thickness
		(%)	(%)	(%)	(MPa)		(%)	(mm)
Balsa	1 wt.% $NaClO_2$ solution	≈ 80	86	70	3.2 $MPa^{1/2}$	0.67 $MPa^{1/2}$	16.8	1.0
Basswood	5 wt.% NaClO solution	≈ 85	90	90			14	0.8
Sugarcane bagasse fibre	H_2O_2 + UV light	83	89	90	32			0.045
Southern yellow pine pulp	TEMPO-oxidized system		96	60	105	1.88 $J \cdot m^{-3}$	5.34–5.88	0.050

Table 6.56 Preparation of transparent bamboo (Wachter et al. 2023)

Species	Method	Resin	Dimensions (mm)			Transmittance	Haze
		(%)	length	width	thickness	(%)	(%)
Moso bamboo	$NaClO_2$ treatment	Expoxy, type E51	40	20	0.3	92.4	43.5
Moso bamboo	$NaClO_2$ treatment	Expoxy, type E51	40	20	1.2	78.6	70.0
Moso bamboo	Lignin modification	Expoxy, type E51	50	20	1.8	87.0	90.0
Moso bamboo	$NaClO_2$ + microwave processing	Expoxy, type E51	3	8	1.8	15.0	
Moso bamboo	1% NaOH precondition + $NaClO_2$ treatment	Expoxy, Clearcast 7000	165	13	1.0	80.0	81.0
Moso bamboo	Alkali pretreatment, cross-linking, delignification	Expoxy	135	135	1.5	76.9	79.7

6.8 Special Processes

Dick Sandberg

6.8.1 Thermal Modification

See Tables 6.57 and 6.58 also Fig. 6.11

Table 6.57 Process temperature and duration of thermal modification of various wood and plant species (Sandberg et al. 2021)

Species	Temperature (°C)	Duration (hours)
Acacia hybrid (sapwood)	210–230	2–6
Balkan beech	170, 190, 210	4
Bamboo	130, 180	2 or 5
	130, 220	2 or 5
Bamboo	140–200	0.5 – 2
Black pine	130, 180, 230	2 or 8
Black spruce	190, 200, 210	
Calabrian pine	130, 180, 230	2 or 8
Common aspen	195	
	160–170	1 or 3
	170	1
Common oak	160	
	160, 180, 210, 240	
European alder	150, 180, 200	2, 6 or 10
European ash	180, 200	3
	190	2.5
European beech	180, 200, 220	4
	160, 190	
	170, 180, 190, 212	2
	125–130	6.5
	170	3
European white birch	160, 190	
Fir	170, 180, 190, 212	2
	160, 220	
	100, 150, 200, 220, 240, 260, 280	1, 3 or 5
Flooded gum	180, 200, 220, 240	4 or 8
Grey alder	160–170	1 or 3
Gympie messmate	180, 200, 220, 240	4
Mantsurian poplar	160, 180, 200, 220, 240	4
	180–220	4
Maritime pine	170–200	2–24
Meranti	160, 180, 210	about 3
Merbau	160, 180, 210	about 3
Narrow-leaved ash	160, 180	2 or 4
	140, 180, 200, 220	2, 4 or 6
Norway spruce	160–260	2 – 8
	195	
	200	2, 4, 8, 10 or 24
	100, 150	24
	200	0.08, 0.5 or 1
	180, 200, 220	4
	160, 220	
	113, 134, 158, 187, 221, 237, 253, 271	1.5
	180	2
Oil palm mesocarp fiber	190–230	1, 2 or 3
Oleaster-leafed pear	160, 180, 200	3, 5 or 7
Oriental beech	130, 150, 180, 200	2, 6, or 10
	150, 160, 170	1, 3, 5, or 7
Paulownia	160, 180, 200	3, 5 or 7
	150, 170	0.75

(continued)

Table 6.57 (continued)

Species	Temperature (°C)	Duration (hours)
Rubberwood	210–240	1–8
Red-bud maple	120, 150, 180	2, 6 or 10
Scots pine	160	5
	160–260	2–8
	120, 150, 180	2, 6 or 10
	180, 200, 240	
	160, 180, 210, 240	
	125, 140, 155, 170	3
	180	2
	225	3
Sessile oak	130, 180, 230	2 or 8
Silver oak	210–240	1–8
	185–230	2–3
Small-leaved lime	140	up to 504
Sweet chestnut	130, 180, 230	2 or 8
	160, 180	2 or 4
Teak	160, 180, 210	approx. 3
Turkish hazel	120, 150, 180	2, 6 or 10
Warty birch	195	
	140, 160, 180	1
	160–170	1 or 3
	160, 180, 210, 240	
Western red cedar	220	1 or 2

Table 6.58 Example of thermal-modification processes (Sandberg et al. 2021). Treatment temperatures for different stages of the process are separated by /. MC—moisture content, FWD—Feuchte-Wärme-Druck, OHT—oil heat treatment. X/Y/Z = no values available

Process	Patent[1] (year, country code)	First ind.[2] (year)	Trademarks™	System[3] (Open/closed)	MC[4] (%)	Temperature[5] (°C)	Process duration[6] (hours)	Heat transfer[7]	Heating medium[8]	Atmosphere[9]	Pressure[10]0 (MPa)	Reactor type[11]
Bicos (RU)				Open	X/ wet_dry/Z	180–240	32–54	Fluid contact	Organic oil		0.3	Autoclave
FirmoLin (NL)	2006 (NL)	2008–2009	Firmolin	Closed	12/moist/8	150–175	up to 24/x	Convective	Steam	Pressurised	0.45–0.80	Autoclave
FWD (DE)	1979 (DE)	~1975, 2001	FWD-Holz, Baldur-Antikholz (CH)	Closed	8–12 (30)/ wet/Z	170–200	16–30/5–7	Convective	Steam	Pressurised	0.5–0.7	Autoclave
Le Bois Perdure (FR)		~ 1990	Perdure		green/Y/Z	200–230 (240)	12–36	Convective	Steam	Atmospheric	0.1	Autoclave
Lignius (NL)	Yes	2002	SmartHeat	Open	X/Y/Z	200–240	X/Y	Contact heating	Metal plate	Vacuum	0.1	
Stellac (FI)	1997 (FI)		StellacWood	Open	<15/OD/5	up to 250	X/Y	Convective	Dry-steam	Atmospheric	0.1	Kiln
OHT (DE)	1979 (DE)	2000	OHT, Menzholz, pannaq	Open	6/OD/Z	180–200 (240)	18–30/2–4	Fluid contact	Vegetable oils	Atmospheric	0.1	Immersion tank
Plato (NL)	1989 (EU)	2001	PlatoWood	Hybrid	14–18/ OD/4–6	150–180/150–190 (165–185)	70–120 up to 2 weeks/4–5/	Convective & condensing	Saturated steam	Pressurised	0.6–0.8	Auotclave/ kiln
Retification (FR)	1996 (FR)	1986 (FR)	Retibois, NOW—New Option Wood	Open	~12/OD/Z	210–240 (260)	8–24/Y	Convective	Nitrogen	Atmospheric	0.1	Autoclave
SilvaProduct (SI)	2004 (SI)	2012	SilvaproWood	Open	X/Y/Z		X/Y			Vacuum	0.09	Autoclave
Staybwood (US)	1942 (US)		Staybwood	Open	~10/0/0	~260–315	0.2–2/Y	Fluid contact	Molten metal, fused salts etc.	Atmospheric	0.1	Immersion tank
Tanwood				Closed	X/wet/Z	160–180	X/Y	Condensing	Steam		0.6–1.0	Autoclave
TERMOVUOTO (IT)	2011 (IT)	~ 2010	VacWood	Open	X/OD/Z	160–220	up to 25/Y	Convective	Steam	Vacuum	0.015–0.035	Auotclave
ThermoWood (FI)	1993 (FI)	1998–1999	ThermoWood	Open	10(G)/ OD/4–7	130/185–215/80–90*	30–70/Y	Convective	Dry-steam	Atmospheric	0.1	Kiln
Vacu3 (DE)		2008	Vacu3 (Timura)	Open	14–18/ OD/Z	200–240 (175–230)	11–28/1–4	Solid contact	Metal plate	Vacuum	0.015–0.060	Kiln with press

(continued)

Table 6.58 (continued)

Process	Patent[1] (year, country code)	First ind.[2] (year)	Trademarks™	System[3] (Open/ closed)	MC[4] (%)	Temperature[5] (°C)	Process duration[6] (hours)	Heat transfer[7]	Heating medium[8]	Atmosphere[9]	Pressure[10]0 (MPa)	Reactor type[11]
VAP Holzsysteme (BR)			VAP HolzSysteme	Closed	X/wet/Z	160–180	ca. 16/0.75	Condensing	Steam	Pressurised	0.3–1.0	Autoclave
WDE-Maspell (DE)					X/Y/Z		X/Y	Convective	Dry steam	Sub-athmospheric		
WTT (DK)		2005	WTT, THERMO VAPO (Corbat AG, CH)	Closed	X/Y/Z	160–210 (180)	up to 12/Y	Fluid contact	Vegetable oils	Pressurised	0.7–0.8	Autoclave
WTT 2.0 (DK)	2016 (DK)		WTT 2.0	Closed	X/wet/Z	170	up to 12/1.5-2.5	Convective	Nitrogen	Pressurised	0.7–1.4	Autoclave

[1]Patent: Priority year and country (abbreviations for country names) for the first patent application for the method
[2]First industrial plant: When an industrial production line has been built and the product starts to be available on the market
[3]System: If the thermal-modification process is in a open/closed hybrid system
[4]MC: Moisture content (MC) when the process start (X), during the thermal-modification regime (Y), and after the thermal-modification process (Z), i.e. X/Y/Z% (G—green wood; OD—oven-dry wood); N.B. for several of the processes information about process data is missing
[5]Temperature: Temperature at which the thermal-modification regime operate
[6]Process duration: Process time for the whole process (X), and only for the thermal modification regime (Y), i.e. X/Y hours
[7]Heat transfer: Convective heating transefer (dry/wet steam, inertgas), fluid contact (water, condensing water, hot oil, melted metal), solid contact to a e.g metal plate
[8]Heating medium:
[9]Athmosphere: Vacuum, sub-atmospheric, atmospheric or pressurised
[10]Pressure: Absolute pressure
[11]Reactor type: Kiln, immersion tank, autoclave

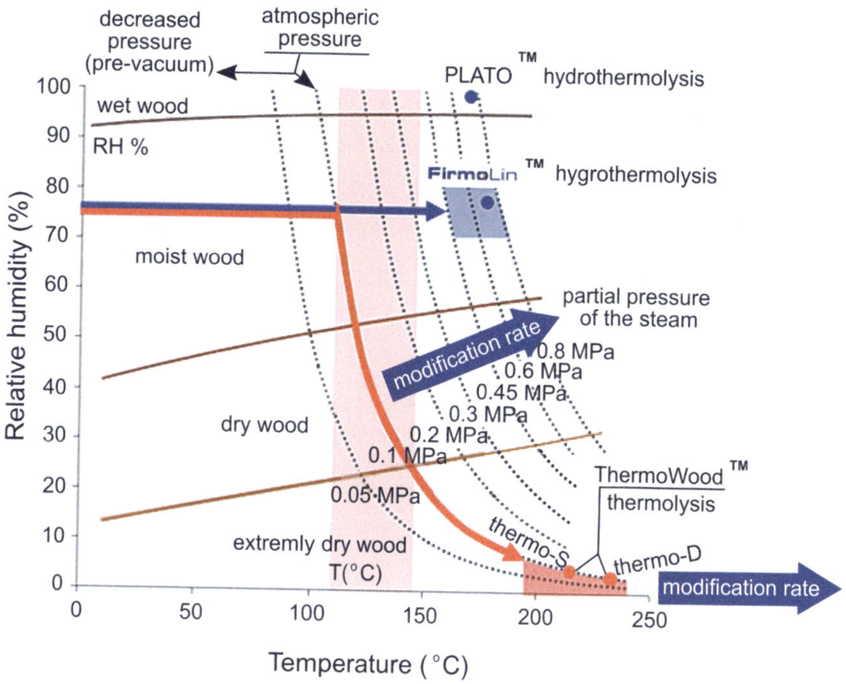

Fig. 6.11 Diagram of RH versus temperature showing regions relevant for different thermal-modification processes (from Sandberg et al. 2021)

6.8.2 Self-Bonding of Wood

See Table 6.59

Table 6.59 Process parameters of self-bonding of European beech veneer (approx. a thickness of 2 mm and a density of 600 kg/m³) and resulting water resistance of the bond-line (Cristescu 2015)

Parameter	Unit	Process intensity			
		mild	medium	harsh	harsh
Temperature of press plates	(°C)	200	225	225	250
Pressure	(MPa)	4	5	6	4–6
Pressing time	(s)	240	300	300	360
Delamination of bond-line under water immersion	(-)	Yes	Yes	No	No

6.8.3 Friction Welding of Wood

See Fig. 6.12 also Tables 6.60, 6.61 and 6.62

6.8.4 Wood Densification

See Fig. 6.13 and Table 6.63

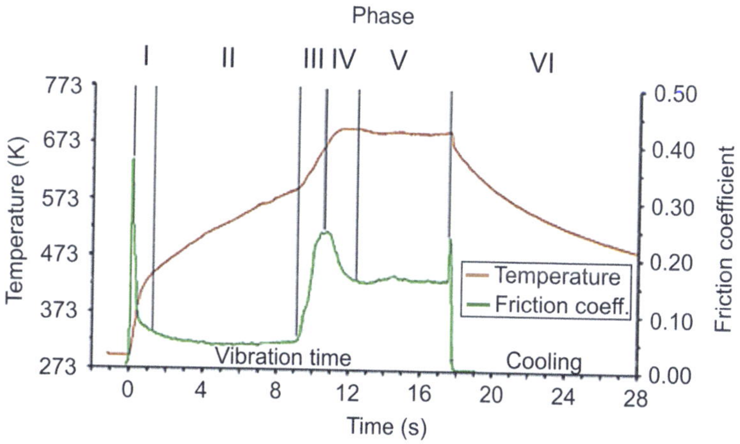

Fig. 6.12 The division of the welding of Norway spruce into six different stages is based on the coefficient of friction and the bond-line temperature (Stamm 2005)

Table 6.60 Parameters and values of normal pressure for the welding of Norway spruce (Stamm 2005)

Parameters	Unit	Pressure
Amplitude	(mm)	0.95
Normal pressure	(MPa)	0.31 0.47 0.63 0.79 0.94
Frequency	(Hz)	130
Thickness of specimen	(mm)	10
Welding displacement	(mm)	2
Cooling time	(s)	20
Cooling pressure	(MPa)	1.57

Table 6.61 Parameters and values of frequency for the welding of Norway spruce (Stamm 2005)

Parameters	Unit	Frequency
Amplitude	(mm)	0.95
Normal pressure	(MPa)	0.63
Frequency	(Hz)	80, 95, 110, 125, 150
Thickness of specimen	(mm)	10
Welding displacement	(mm)	2
Cooling time	(s)	20
Cooling pressure	(MPa)	1.57

Table 6.62 Some typical linear-welding parameters for different wood species (Sandberg et al. 2021). MC – moisture content, pres. – pressure, W. – welding

Parameter	Unit	European beech	European beech	Silver birch	Scots pine	Scots pine	Scots pine	Scots pine
MC	(%)	12	12	12	12	12	12	12
W. area	(mm²)	3000	4600	4000	4000	4000	48,000	4600
W. frequency	(Hz)	100	240	240	150	150	115	240
W. amplitude	(mm)	3	2	2	1+1.5+2+0	2	4	2
Welding time	(s)	3	1+1.8	2.5+3	1+3+3.5	2+2.8	2+10.2	2+3
W. pres.	(MPa)	1.3	1.6+1.6	1.3+1.8	0.5+0.75+1.25	0.75+1.3	1.5+1.5	1.3+1.7
Holding pres.	(MPa)	2.0	2.5	2	2.75	1.85	1.87	2.7
Holding time	(s)	5	10	10	50	70	60	10
Reference		Gfeller et al. 2003	Vaziri et al. 2020	Zor et al. 2019	Mansouri et al. 2011	Vaziri et al. 2012	Vaziri et al. 2014	Vaziri et al. 2021
Machine type		Branson 2700	Branson M-624	Branson M-624	KLN Ultraschall LVW-2261	Mecasonic LVW-2361	Branson M-DT24L	Branson M-624

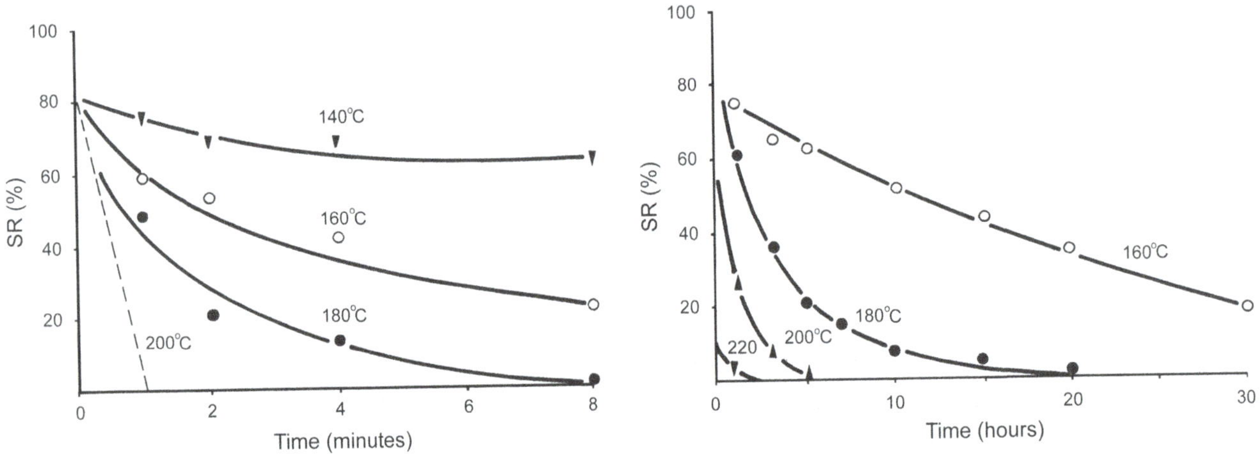

Fig. 6.13 Set-recovery (SR) of sugi specimens, 20 × 20 x 30 mm in size (LxRxT) after compressive deformation (CR = 51%): a) densified in saturated vapour as a function of treatment time at different temperatures, and b) densified in dry air in an oven as a function of treatment time at different temperatures (Inoue et al. 1993)

Table 6.63 Wood-densification processes including impregnation (Sandberg et al. 2021)

Product name (patent)	Short description	Type of wood
Compreg™ (EP3470189A1); A similar product is produced in Europe under the trade names Delignit™ – Panzerholz™	Green or dry veneer impregnated with 50% aqueous solution of phenol–formaldehyde resin prior to compression at 150 – 170 °C for 15 to 30 min and laminated. High density (up to 1,400 kg/m^3) product with very low volumetric swelling (< 4%) when soaked in water	Applied to veneer and produces a laminated wood composite The Delignit™ – Panzerholz™ produced only from European beech veneer
Dehonit™ – Deutsche Holzveredelung, Germany	Beech veneer impregnated with phenolic resin, assembled into a pack, and compressed under high pressure and temperature. Density of 850 – 1,250 kg/m^3 used for high voltage power and distribution transformers	Applied only to European beech veneer
Permali™	Selected beech veneers impregnated under vacuum with synthetic resin and densified under heat and pressure. Water-resistant general-purpose densified wood with applications in cryogenic, electrical and general engineering	Applied only to European beech veneer
Permatred™	Beech veneers impregnated with phenolic resin and consolidated under high pressure and temperature	Applied only to European beech veneer
RANPREX™	Beech veneer impregnated with phenol resin, pressed at high temperature and pressure Homogeneous flooring sheets of high mechanical strength	Applied only to European beech veneer
Insulcul Services Ltd	Densified wood manufactured from beech veneers, impregnated with a synthetic adhesive, and densified under high pressure or a temperature of 90 °C	Applied only to European beech veneer
Wood densification method developed by Haygreen and Daniels (1969)	Densifying and drying green sapwood, including 72 h of impregnation of green wood, compression for 20 h and drying to 0% while in the press for 75–125 min. The set-recovery is 1–2%	Applied only to European beech veneer
Mywood2™ KK (Japan, JPH11151703A	Coniferous wood impregnated with aqueous boric acid solution and then THM treated. Woods like cedar or larch are impregnated with a proprietary boric-acid solution to provide resistance to water. The increase in density is about 50%	Applied only to European beech veneer
Process for densifying low-density woods (US4606388A)	Green solid wood impregnated with anhydrous ammonia to plasticize the wood pressed in a cyclic press with perforated platens, while keeping the temperature of the wood below 100 °C. Up to 50% reduction in thickness, density increase up to 1000 kg/m^3	Wood species like poplar, alder, oak, maple and birch can be densified. Set-recovery after water soaking can reach 30%
Hartzel method of compressing wood (US2101542A)	Steam heating of wood veneer and solid wood of thickness to 25 mm at 150 °C and then compressed and laminated. Set recovery prevented by coating with Bakelite, varnish or other moisture-resisting material after compression	Primarily applied to veneer, but also to thin solid wood. Set recovery prevented by coating finished compressed wood preventing any further processing of material
Compressed wood product and manufacture (WO2004056542A1; developed from NZ52329502A)	Two-step compression of conifer and broad-leaved woods including coating and impregnation with fatty acid at 60–120 °C as an intermediate step. Temperature between 145–185 °C for softwoods and up to 200 °C for hardwoods. Process completed with 12 h keeping the densified wood under pressure in a cold press. The densified wood should be formed into laminated products	Intended to be used only for laminated products. Clear information about compression-related properties not provided

References

Anderson J, Westerlund L (2014) Improved energy efficiency in sawmill drying system. Appl Energy 113:891–901

Anon (1975) Werkstoffe aus Holz und andere Werkstoffe der Holzindustrie (Materials made of wood and other materials used in the wood industry). Fachbuchverlag, Leipzig

Anon (1990) Lexikon der Holztechnik (Encyclopaedia of wood technology). 4th edn. Fachbuchverlag, Leipzig

Baumgarten HL (ed) (1996) TUD-Paper samples—Paper-Board-Pulp. Selbstverlag für Papiertechnik, Dresden

Bergman RD, Bowe SA (2008a) Environmental impact of producing hardwood lumber using life-cycle inventory. Wood Fibre Sci 40(3):448–458

Bergman RD, Bowe SA (2008b) Life-cycle inventory of hardwood lumber manufacturing in the northeast and north central United States, CORRIM Phase II: Final Report Module C. University of Washington, Seattle WA

Bråkenhielm C, Jonsson T, Sundqvuist I (1969) Förslag till tärningsmått för lättmetalltärningar och trätärningar vi försågning till metriska mått (Suggestions for dice measurements for light-metal dice and wooden dice we saw for metric measurements). STFI Meddelande serie B No. 2, Swedish Forest Products Laboratory, Stockholm

CEN (2006) EN 300, Oriented strand boards (OSB)—Definitions, classification and specifications. CEN–European Committee for Standardization, Brussels

CEN (2010) EN 312, Particleboards. Specifications. CEN–European Committee for Standardization, Brussels

Cristescu C (2015) Self-bonding of beech veneer. Dissertation, Luleå University of Technology

Deppe H-J, Ernst K (2000) Taschenbuch der Spanplattentechnik (Encyclopaedia of wood particleboard technology 4th edn. DRW-Verlag, Leinfelden-Echterdingen

Diederichs SK (2014) Status quo for life-cycle inventory and environmental Impact assessment of the core sawmill products in Germany. Wood Fib Sci 46(1):65–84

Dofinè G (1962) Standortfragen der Lagenholzindustrie (Location issues in the laminated wood industry). In: Kollmann F (1962) Furniere, Lagenhölzer und Tischlerplatten. (Veneers, laminated wood and blockboard) 4th edn. Springer Verlag Berlin

Ebner B (1977) Entrindungswiderstand (Debarking resistance). Wochenblatt für Papierfabrikation 105(15):595–596

Fardim P (ed) (2011) Papermaking science and technology. Book 6: Chemical pulping Part 1. Finnish Paper Engineers' Association and TAPPI, Helsinki

Fehrendsen W (1956) Untersuchungen zum Verhältnis Holzpreis und Holzdurchmesser (Studies on the relationship between the price of wood and the diameter of the wood). Dissertation, Universität Göttingen

Frühwald A, Speckels L, Scharai-Rad M et al (2000) Oekologische Bilanzierung von Bauschnittholz und veredelten Vollholzprodukten (Ecological assessment of sawn timber and processed solid wood products). Ordinariat für Holztechnologie der Universität Hamburg, Hamburg

Gfeller B, Zanetti M, Properzi M et al (2003) Wood bonding by vibrational welding. J Adh Sci Techn 17(11):1573–1589

Godin H (2019) Energiewende in der Industrie. Potenziale und Wechselwirkungen mit dem Energiesektor, Branchensteckbrief der Papierindustrie (Energy transition in industry. Potential and interactions with the energy sector, profile of the paper industry. Processing of deciduous wood for chipboard production). Navigant Energy Germany GmbH, Vorhaben I C 4—80 14 38/42, Projekt 42/17, Projektnr. SISDE17915, Köln

Gong M (ed) (2021) Engineered wood products for construction. IntechOpen, London

Gran G (1968) Comparison between the dry and wet process of fibreboard manufacture. FAO Committee on wood-based panel products, Rom

Haygreen JG, Daniels DH (1969) The simultaneous drying and densification of sapwood. Wood Fib Sci 1(1):38–53

Inoue M, Norimoto M, Tanahashi M et al (1993) Steam or heat fixation of compressed wood. Wood Fib Sci 25(3):224–235

Kehr E (1975) Spanplatten (Particleboard). In: Anon (1975) Werkstoffe aus Holz und andere Werkstoffe der Holzindustrie (Materials made of wood and other materials used in the wood industry). Fachbuchverlag, Leipzig

Kohler H (1954) Das Problem der richtigen Messzahlen (The problem of the right metrics). Holz-Zent.bl 80(73):883–885

Kollmann F, Schulte-Brader H-J (1954) Studien über die Ausbeute der deutschen Sperrholzindustrie (Studies on the yield of the German plywood industry). Holz Roh- Werkstoff 12(12):465-471

Krackler V, Keunecke D, Niemz P (2010) Verarbeitung und Verwendungsmöglichkeiten von Laubholz (Processing and uses of hardwood). ETH Zürich

Krackler V, Keunecke D, Niemz P et al (2011) Possible fields of hardwood applications. Wood Res 56(1):125–136

Li J, Mc Curdy M, Pang S (2006b) Energy demand in wood processing plants. N Z J For Sci 51(2):13–18

Mansouri HR, Pizzi A, Leban JM et al (2011) Causes for the improved water resistance in pine wood linear welded joints. J Adhes Sci Technol 25:1987–1995

Mette H, Korell U (1989) Richtzahlen und Tabellen für die Forstwirtschaft (Indicative numbers and tables for forestry). Deutscher Landwirtschaftsverlag, Berlin

Miels G (1965) Stand und Tendenzen der technologischen Entwicklung der Faser-und Spanplattenindustrie und die Auswirkungen der Kapazität auf den ökonomischen Wirkungsgrad (State and trends of technological development in the fiber and chipboard industry and the impact of capacity on economic efficiency). Holztechnologie 1:4–9

Milota M, West CD, Hartley ID (2005) Gate-to-gate life-cycle inventory of softwood lumber production. Wood Fiber Sci 37:47–57

Niemz P (1982) Untersuchungen zum Kriechverhalten von Spanplatten (Investigations on the creep behaviour of particleboards). Dissertation, TU Dresden

Niemz P (2002) Changes in thickness of chipboard and MDF under pressure perpendicular to the plane of the board. HOB 49(12):23–24

Niemz P, Sandberg D (2022) Critical wood-particle properties in the production of particleboard. Wood Mater Sci Eng 17(5):386–387

Niemz P, Teischinger A, Sandberg D (eds) (2023) Springer Handbook of Wood Science and Technology. Springer Nature, Cham

Puettmann ME, Wagner FG, Johnson L (2010) Life cycle inventory of softwood lumber from the inland northwest US. Wood Fiber Sci 42:52–66

Sandberg D, Teischinger A (2023) Sawmilling. In: Niemz P, Teischinger A, Sandberg D (eds) Springer handbook of wood science and technology. Springer Nature, Cham

Sandberg D, Kutnar A, Karlsson O et al (2021) Wood modification technologies. Principles, sustainability, and the need for innovation. CRC Press, Boca Raton

Staland J, Navrén M, Nylinder M (2000) Såg 2000—Resultat från sågverkinventeringen 2000 (Sawmill 2000—Results from The 2000 Sawmill Inventory). Report No. 3, The Swedish University of Agricultural Sciences

Stamm B (2005) Development of friction welding of wood—physical, mechanical and chemical studies. Dissertation, Swiss Federal Institute of Technology in Lausanne (EPFL)

Stegmann G, Bismarck CV (1968) Verarbeitung von Laubhölzern für die Spanplattenherstellung (Processing of deciduous wood for chipboard production). Holzforsch Holzverwert 20:1–11

Suhr M, Klein G, Kourti I et al (2015) Best available techniques (BAT) references document for the production of pulp, paper and board. EUR—Scientific and Technical Research Reports, Publications Office of the European Union, Brussels

Swiderski J (1963) Vergleich der Verfahren zur Herstellung von Hartfaserplatten: Nass-, Halbtrocken-und Trockenverfahren (Comparison of hardboard manufacturing processes: wet, semi-dry and dry processes). Holz Roh- Werkstoff 6:217–225

Varis R (ed) (2018a) The sawmill industry handbook. Finnish Woodworking Engineers Association, Helsinki

Varis R (ed) (2018b) Wood-based panels industry. Finnish Woodworking Engineers Association, Helsinki

Vaziri M, Lindgren O, Pizzi A (2012) Optimization of tensile-shear strength for linear welded Scots pine. J Adhes Sci Technol 26:109–119

Vaziri M, Berg S, Sandberg D et al (2014) Three-dimensional finite element modelling of heat transfer for linear friction welding of Scots pine. Wood Mater Sci Eng 9(2):102–109

Vaziri M, Abrahamsson L, Hagman O et al (2020) Welding wood in the presence of wollastonite. BioResources 15(1):1617–1628

Vaziri M, Karlsson O, Abrahamsson L et al (2021) Wettability of welded wood-joints investigated by the Wilhelmy method: Part 1. Determination of apparent contact angles, swelling and water sorption. Holzforschung 75(1):65–74

Wachter I, Rantuch P, Štefko T (2023) Transparent wood materials: Properties, applications, and fire behaviour. Springer Nature, Zug

Wagenführ A, Scholz F (eds) (2012) Taschenbuch der Holztechnik (Encyclopaedia of wood technology). 2nd edn. Fachbuchverlag Leipzig im Carl Hanser Verlag, München

Watzke H (ed) (1969) Taschenbuch der Holztechnik (Encyclopaedia of wood technology). 2nd edn. Fachbuchverlag, Leipzig
Werner F, Althaus H-J, Kuenniger T et al (2007) Life cycle inventories of wood as fuel and construction material. Ecoinvent report No. 9. Swiss Centre for Life Cycle Inventories, Dübendorf
Zor M, Vaziri M, Sandberg D (2019) Shear strength of welded birch wood produced by linear friction. In: Dindaroğlu T (ed) 3th International Mediterranean Forest and Environment Symposium (IMFES 2019)—Future of Mediterranean Forests: Sustainable Society and Environment, Kahramanmaraş, Turkey

Prof.Dr.-Ing.habil. Dr.h.c. Peter Niemz (Sects. 6.1–6.4) studied at the Faculty of Mechanical Engineering of TU Dresden. From 1972 onwards, he worked at the Institute of Wood Technology Dresden (IHD) and at TU Dresden. From 1993 to 1996, he served as a Full Professor at the Universidad Austral de Chile in Valdivia before joining the Swiss Federal Institute of Technology (ETH Zurich) in 1997. He was Professor of Wood Physics from 2002 until his retirement in 2015.

Dr. Dirk Grunwald (Sect. 6.5) received his PhD on the analysis of wood adhesives at TU Braunschweig and at Fraunhofer WKI (2002), meanwhile and afterwards research assistant at Fraunhofer WKI, 2004–2010 Technologist (Chemistry) at Swiss Krono Group in Germany and Poland, 2010–2013 Chief Technologist at IKEA Industry (Swedspan) in Slovakia, 2014–2019 Business Development Manager and Platform Manager Binders at Huntsman in Belgium, since 2020 Head of R&D Adhesives at Follmann in Germany.

Dr. Sabine Heinemann (Sect. 6.6) received her doctoral degree in Paper Technology from Dresden University of Technology (TU Dresden) in 1985. She has worked in pulp and paper research at TU Dresden, OY Keskuslaboratorio (KCL), Finland, and VTT Ltd., Finland. Today, she is a freelancer, training students at TU Dresden, DHBW Karlsruhe, and University of Applied Forest Sciences Rottenburg mainly in the field of physical and microscopic characterisation of fibrous materials.

Prof.Dr. Dick Sandberg (Sects. 6.1–6.4 and 6.7–6.8) received his Doctorate in Mechanical Engineering, Wood Technology and Processing from the Royal Institute of Technology (KTH), Stockholm, Sweden, in 1998. He is a Chaired Professor of Design and Fabrication of Timber Construction at the Norwegian University of Science and Technology (NTNU). He specialises in using new measurement technology, advanced information processing, and wood communication from felling to finished product. He is an Elected Fellow of the International Academy of Wood Science.

7 Books in the field of wood science and technology

Peter Niemz, Alfred Teischinger and Dick Sandberg

7.1 Introduction

This chapter contains a list of useful books in the field of wood science and technology. The books are divided into six groups with a special focus ion the following fields:

- wood science,
- wood-based materials,
- wood mechanical processing,
- timber engineering,
- chemical engineering, pulp and paper, and
- energy from biomass.

The selection includes both old books of an essentially historical value, and contemporary books for current use in daily work. The suggested books are mainly in English and German, although there are many interesting and high-quality books in other languages.

7.2 List of selected books

P. Niemz (✉)
Department Wood Physics (today, Wood Materials Science), Institute of Building Materials, ETH Zürich, Zurich, Switzerland
e-mail: niemzp@retired.ethz.ch

A. Teischinger
Institute of Wood Science Technology and Renewable Materials, University of Natural Resources and Life Sciences (BOKU), Vienna, Tulln, Austria
e-mail: alfred.teischinger@boku.ac.at

D. Sandberg
Department of Manufacturing and Civil Engineering, Norwegian University of Science and Technology (NTNU), Gjøvik, Norway
e-mail: dick.sandberg@ntnu.no

P. Niemz et al. (eds.), *Wood Material and Processing Data*, https://doi.org/10.1007/978-3-031-38329-8_7

Literature	Wood science	Wood-based materials	Wood mechanical processing	Timber engineering	Chemical engineering, pulp and paper	Energy from biomass
Aicher, S., Reinhardt, H.-W. & Garrecht, H. (2014). Materials and Joints in Timber Structures. Recent Developments of Technology. Dordrecht Heidelberg: Springer.				•		
Aguilera, A. & Davim, J. P. (Eds.) (2014). Research Developments in Wood Engineering and Technology. Hershey (PA): IGI Global.		•	•			
Aguilera, A. & Davim, J. P. (Eds.) (2017). Wood Composites. Materials, Manufacturing and Engineering. Berlin, Boston: De Gruyter.		•	•			
Anderson, D., Wu, Q. & Han, G. (2013). Introduction to Wood and Natural Fiber Composites. Wiley Series in Renewable Resources, West Sussex: John Wiley & Sons.		•				
Ansell, M. P. (Ed.) (2015). Wood Composites. Cambrige. Woodhead Publisching Series in Composites Science and Engineering, Elsevier.		•				
Atkins, T. (2009). The Science and Engineering of Cutting - The Mechanics and Processes of Separating, Scratching and Puncturing Biomaterials, Metals and Non-Metals. Oxford: Elsevier.		•	•			
Author collective (1975). Werkstoffe aus Holz und andere Werkstoffe der Holzindustrie. Leipzig: Fachbuchverlag.	•	•				
Author collective (1985). Wissensspeicher Holztechnik. Leipzig: Fachbuchverlag.			•			
Author collective (1990). Lexikon der Holztechnik (4th ed.). Leipzig: Fachbuchverlag.	•	•	•		•	
Author collective (2010). BSP Handbuch, Holz-Massivbauweise in Brettsperrholz. Nachweise auf Basis des neuen europäischen Normenkonzepts. Graz: Verlag der Technischen Universität Graz.				•		
Bajpai, P. (2013). Biorefinery in the Pulp and Paper Industry. London: Elsevier Academic Press.					•	
Bajpai, P. (2015). Green Chemistry and Sustainability in Pulp and Paper Industry. Heidelberg: Springer.					•	

Literature	Wood science	Wood-based materials	Wood mechanical processing	Timber engineering	Chemical engineering, pulp and paper	Energy from biomass
Baldwin, R. (1981). Plywood Manufacturing Practices. New York: Miller Freeman Publications Inc.		•	•			
Baldwin, R. (1984). Operations Management in Forest Products Industry. New York: Miller Freeman Publications Inc.			•			
Baldwin, R. F. (1995). Plywood and Veneer-based Products. Madison (WI): Forest Products Society.		•	•			
Baldwin, R. F. (2016). Plywood and Veneer-based Products. Montgomery (AL): The Donnel Group.		•	•			
Barkas, W. W. (1950). The Swelling of Wood under Stress. Stockholm: STFI.	•					
Barnett, J. R. & Jeronimidis, G. (Eds.) (2003). Wood Quality and its Biological Basis. Oxford: Blackwell Publishing Ltd.	•					
Begemann, H. F. (1981). Das grosse Lexikon der Nutzhölzer Band 1-12. Berlin: DBV, Deutsche Betriebswirte-Verlag.	•					
Belgacem, M. & Pizzi, A. (Eds.) (2016). Lignocellulosic Fibers and Wood Handbook: Renewable Materials for Today's Environment. Salem (MA): Scrivener Publishing.	•		•		•	
Bhat, R., Kumar, A., Nguyen, T.A. & Sharma, S. (2022). Nanotechnology in Paper and Wood Engineering: Fundamentals, Challenges and Applications. Amsterdam: Elsevier.		•	•	•	•	
Bianconi, F. & Filippucci, M. (Eds.) (2019). Digital Wood Design - Innovative Techniques of Representation in Architectural Design, Volume 1 and 2. Cham: Springer.		•	•	•		
Blaß, H.-J. & Sandhaas, C. (2016). Ingenieurholzbau - Grundlagen der Bemessung. Karlsruhe: KIT Scientific Publishing.				•		
Blechschmidt, J. (Ed.) (2013). Taschenbuch der Papiertechnik. München: Fachbuchverlag Leipzig im Carl Hanser Verlag.					•	
Blunck, A. (1901). Die Konstruktionslehre für Bau- und Möbeltischler. Berlin: Bruno Hessling.		•	•			
Bodig, J. & Jayne, B. A. (1993). Mechanics of Wood and Wood Composites (2nd ed.). Malabar (FL): Krieger Publishing Company.	•	•				

Literature	Wood science	Wood-based materials	Wood mechanical processing	Timber engineering	Chemical engineering, pulp and paper	Energy from biomass
Boerhave Beekman Door, W. (Ed.) (1949). Hout in alle Tijden, Deel 1-6. N. V. Uitgevers-Maatschappij Ae. E. Kluwer.	•					
Borgström, E. (Ed.). (2016). Design of Timber Structures. 1. Structural Aspects of Timber Construction. Stockholm: Swedish Wood.				•		
Borgström, E. (Ed.) (2016). Design of Timber Structures. 2. Rules and Formulas According to Eurocode 5. Stockholm: Swedish Wood.				•		
Borgström, E. (Ed.) (2016). Design of Timber Structures. 3. Rules and Formulas According to Eurocode 5. Stockholm: Swedish Wood.				•		
Bosshard, H. H. (1982-1984). Holzkunde I-III (2nd ed.). Basel: Birkhäuser.	•	•	•			
Boulton, B. C. (1920). The Manufacture and use of Plywood & Glue. Melbourne, Toronto, New York: Sir Isaac Pitman & Sons, Ltd.		•	•			
Boutelje, J. B. (1980). Encyclopedia of World Timbers - Names and Technical Literature. Stockholm: Swedish Forest Products Research Laboratory (STFI).	•					
Breyer, D., Fridley, K., Cobeen, K. & Pollock, D. (2006). Design of Wood Structures - ASD/LRFD. New York: McGraw-Hill.				•		
Brown, T. D. (Ed.) (1982). Quality Control in Lumber Manufacturing. Madison (WI): Forest Products Society.			•			
Buchanan, A. & Östmann, B. (2022). Fire Safe Use of Wood in Buildings. Boca Raton: CRC Press, Taylor & Francis.				•		
Bucur, V. (2003). Nondestructive Characterization and Wood Imaging. Heidelberg: Springer.	•					
Bucur, V. (2006). Acoustics of Wood (2nd ed.). Berlin: Springer.	•					
Bucur, V. (2011). Delamination in Wood, Wood Products and Wood-based Composites. Dordrecht: Springer Verlag.		•	•			
Bulian, F. & Graystone, J. A. (2009). Wood Coatings. Theory and Practice. Amsterdam: Elsevier.					•	

Literature	Wood science	Wood-based materials	Wood mechanical processing	Timber engineering	Chemical engineering, pulp and paper	Energy from biomass
Butterfield, B. (1997). Proceedings of IAWA/IUFRO International Workshop of Significance of the Microfibril Angle to Wood Quality. Westport, New Zealand.	•					
Böhme, P. (1980). Industrielle Oberflächenbehandlung von plattenförmigen Werkstoffen aus Holz. Leipzig: Fachbuchverlag.		•		•		
Böhme, P. (1986). Dekorfolien. Leipzig: Fachbuchverlag.		•				
Capareda, S. C. (2013). Introduction to Biomass Energy Conversions. Boca Raton (FI): CRC Press.						•
Carlquist, S. (1988). Comperative Wood Anatomy. Berlin, Heidelberg: Springer.	•					
Chen, H. (2015). Lignocellulose Biorefinery Engineering: Principles and Applications. Cambridge: Woodhead Publishing.					•	
Chugg, W. A. (1964). GLULAM – The Theory and Practice to the Manufacture of Glued Laminated Timber Structures. London: Ernest Benn.		•	•	•		
Christopher, L. P. (Ed.) (2012). Integrated Forest Biorefineries. London: Royal Society of Chemistry Publishing.					•	
Core, H. A., Côté, W. A. & Day, A. C. (1976). Wood Structure and Identification. Ayeacuse: Syracuse University Press.	•					
Coronel, E. (1994, 1995). Fundamentos de las Propiedades Fisicas y Mechanicas de la Maderas 1-2. [Basics of Physical and Mechanical Properties of Wood, Vol. 1 and 2.] Santiago del Estero, Argentina: Instituto de Technologia de Madera, Universidad Nacional de Santiago del Estero.	•					
Coutts, M. P. & Grace, J. (1995). Wind and Trees. Cambridge: Cambridge University Press.	•					
Csanady, E. & Magoss, E. (2013). Mechanics of Wood Machining. Berlin, Heidelberg: Springer.			•			
Csanady, E., Magoss, E. & Tolvaj, L. (2015). Quality of Machined Wood Surfaces. Berlin, Heidelberg: Springer.			•			
Csanady, E., Kovacs, Z., Magoss, E. & Ratnasingam, J. (2019). Optimum Design and Manufacture of Wood Products. Berlin, Heidelberg: Springer.			•			

Literature	Wood science	Wood-based materials	Wood mechanical processing	Timber engineering	Chemical engineering, pulp and paper	Energy from biomass
Denig, J. (1993). Small Sawmill Handbook. Doing it Right and Making Money. San Francisco: Miller Freeman Publications.			•			
Denig, J. & Wengert, G. (2005). Sawing, Edging and Trimming Hardwood Lumber. Putting Theory into Practise. Madison (WI): Forest Products Society.			•			
Deppe, H.-J. & Ernst, K. (1996). MDF - Mitteldichte Faserplatten. Leinfelden-Echterdingen: DRW-Verlag.		•				
Deppe, H.-J. & Ernst, K. (2000). Taschenbuch der Spanplattentechnik (4th ed.). Leinfelden-Echterdingen: DRW-Verlag.		•				
Desch, H. E. (1968). Timber - Its Structure and Properties. (4th ed.) New York: Macmillan Press Ltd.	•		•			
Desch, H. E. & Dinwoodie J. M. (1996). Timber - Structure, Properties, Conversion and Use. New York: Macmillan Press Ltd.	•	•	•			
Dietsch, P., Kreuzinger, H. & Winter, S. (2012). Holzbau. Werkstoffe und Bemessung. In: Zilch, K., Diederichs, C.J., Katzenbach, R. & Beckmann, K.J. (Eds.) Handbuch für Bauingenieure. Heidelberg: Springer.				•		
Dinwooodie, J. M. (2000). Timber: Its Nature and Behaviour (2nd ed.). London, New York: E&FN Spon.	•	•	•			
Dufresne, A. (2017). Nanocellulose. From Nature to High Perfomance Tailored Materials (2nd ed.) Berlin, Boston: Walter de Gryter.	•					
Duhamel du Monceau, H. (1758). La Physique des Arbres. Saint Thomas dé Aquin: Chez H.L Guerin & L.F. Delatour.	•					
Dunky, M. & Niemz, P. (2002). Holzwerkstoffe und Leime: Technologie und Einflussfaktoren. Berlin: Springer.		•			•	
Dunky, M. & Mittal, K. (Eds.) (2024). Biobased Adhesives. Wiley-Scrivener.		•			•	
Eckelman, C. A. (2003). Textbook of Product Engineering and Strength Design of Furniture. West Lafayette (IN): Purdue University Press.			•			

Literature	Wood science	Wood-based materials	Wood mechanical processing	Timber engineering	Chemical engineering, pulp and paper	Energy from biomass
Ek, M., Gellerstedt, G. & Henriksson, G. (2009). Pulp and Paper Chemistry and Technology, Vol. 1-4. Berlin: Walter de Gryter GmbH & Co. KG.				•	•	
Eriksson, K-E., Blanchette, A. & Ander, P. (1990). Microbiological and Enzymatic Degradation of Wood and Wood Based Components. Berlin, Heidelberg: Springer.	•					
Exner, W. F. (1875). Studien über das Rotbuchenholz. (2nd ed.) Wien: Mödlinger Lehrmittel-Ausstellung.	•					
Exner, W. F. (1878). Werkzeuge und Maschinen zur Holz-Bearbeitung, deren Construction, Behandlung und Leistungsfähigkeit. Erster Band: Handsägen und Sägemaschinen Desciptiver Theil. Weimar: Bernhard Friedrich Voigt.			•			
Exner, W. F. (1880). Das Biegen des Holzes, für Möbelfabrikanten, Wagen- und Schiffbauer wichtiges Verfahren. Weimar: Bernhard Friedrich Voigt.			•			
Fan, M., Ohlmeyer, M., Irle, M., Haelvoet, W., Athanassiadou, E. & Rochester, I. (Eds.) (2009). Performance in Use and New Products of Wood-based Composites. London: Brunel University Press.		•				
Farmer, R. H. (1972). Handbook of Hardwoods. (2nd ed.) London: Her Majesty´s Stationery Office.	•					
Fengel, D. & Wegener, G. (1984). Wood: Chemistry, Ultrastructure, Reactions. Berlin, New York: De Gruyter.	•					
Flynn Jr., J. H. & Holder, C. D. (Eds.) (2001). A Guide to Useful Woods of the World. Madison: Forst Products Society.	•					
Flynn Jr., J. H. (Ed.) (2007). A Guide to more Useful Woods of the World. Madison: Forst Products Society.	•					
Forest Products Laboratory (2021). Wood Handbook. Wood as an Engineering Material. General Technical Report FPL-GTR-282. Madison (WI): Dept. of Agriculture, Forest Service.	•	•	•	•	•	•
Frey-Wyssling, A. (1984). Lehre und Forschung. Autobiographische Erinnerungen. Stuttgart: Wissenschaftliche Verlagsgesellschaft.	•					

Literature	Wood science	Wood-based materials	Wood mechanical processing	Timber engineering	Chemical engineering, pulp and paper	Energy from biomass
Gardiner, B., Barnett, J., Saranpää, P. & Grill, J. (Eds.) (2014). The Biology of Reaction Wood. Berlin, Heidelberg: Springer.	•					
Gibson, L. J. & Ashby, M. F. (1988). Cellular Solids: Structure & Properties. Oxford: Pergamon Press	•					
Glavonjić, B., Kitek Kuzman, M. & Sandberg, D. (2022). Engineered Wood Products in Contemporary Architecture – New ways to use timber in future. [Kompozitni proizvodi od drveta – Drvo kao materijal u građevinarstvu i arhitekturi], Luleå: Luleå University of Technology.		•	•			
Gong, M. (Ed.) (2022). Engineered Wood Products for Construction. London: IntechOpen Ltd.		•		•		
Gottlöber, C. (2015). Zerspanung von Holz und Holzwerkstoffen. Leipzig: Fachbuchverlag Leipzig im Carl Hanser Verlag.			•			
Grabner, M. (2017). WerkHolz. Remagen-Oberwinter: Verlag Dr. Kessel.		•	•	•		
Grabner, M., Weber, A., Nemestothy, S. & Reschreiter, H. (2021). Holzverwendung. Altes Wissen für eine nachhaltige Zukunft. Graz: Stocker.		•	•	•		
Gulichsen, J. & Paulauro, H. (Serie Eds.) (2009). Papermaking Science and Technology, 20 volumes. Helsinki: Papet Oy.					•	
Harlow, W. M. (1970). Inside Wood - Masterpiece of Nature. Washington D.C.: The American Forstry Association.	•					
Harris, A. (1989). Spiral Grain and Wave Phenomena in Wood Formation. Berlin, Heidelberg: Springer-Verlag.	•					
Hartig, G. (1894). Textbook of the Diseases of Trees. (Translation from Lehrbuch der Baumkrankheiten, 2. verb. und vermehrte Auflage, Berlin 1889). London: Macmillan. (On-line version).	•					
Hartig, R. (1890). Das Holz der deutschen Nadelwaldbäume. Stuttgart: J. G. Cotta Verlag. (On-line version)	•					
Haygreen, J. & Bowyer, J. (1982). Forest Products and Wood Science. Ames (IA): Iowa State University Press.	•	•	•			

Literature	Wood science	Wood-based materials	Wood mechanical processing	Timber engineering	Chemical engineering, pulp and paper	Energy from biomass
Herzog, T., Natterer, J., Schweitzer, R., Volz, M. & Winter, W. (2008). Timber Construction Manual. Basel: Birkhäuser Verlag.				•		
Higuchi, T. (1997). Biochemistry and Molecular Biology of Wood. Berlin, Heidelberg: Springer-Verlag.	•					
Hill, C. A. (2006). Wood Modification: Chemical, Thermal and other Processes. Chichester: Wiley.			•		•	
Harris, A. (1987). Heartwood and Tree Exudates. Berlin, Heidelberg: Springer-Verlag.	•					
Hoadley, R. B. (1990). Identifying Wood - Accurate Results with Simple Tools. Newtown: The Taunton Press.	•					
Hoadley, R. B. (2000). Understanding Wood – A Craftsman's Guide to Wood Technology. Newtown: The Taunton Press.	•					
Hon, D. N. S. (Ed.) (1996). Chemical Modification of Lignocellulose Materials. New York: Marcel Dekker.					•	
Hon, D. N. S. & Shiraishi, N. (Eds.) (1991). Wood and Cellulosic Chemistry. New York: Marcel Dekker.					•	
Hoyer, E. (1878). Lehrbuch der vergleichenden Mechanischen Technologie. Wiesbaden: C. W. Kreidel´s Verlag.			•			
Hoyer, E. (1888). Lehrbuch der vergleichenden Mechanischen Technologie. Zweite neu bearbeitete Auflage. Wiesbaden: C. W. Kreidel´s Verlag.			•			
Hänsel, A. & Prieto, J. (Eds.) (2018). Industrielle Beschichtung von Holz und Holzwerkstoffen im Möbelbau. München: Carl Hanser Verlag.		•	•			
Hänsel, A. (2012). Holz und Holzwerkstoffe. Prüfung - Struktur - Eigenschaften. Berlin: Logos Verlag.	•	•	•			
Ilvessalo-Pfäffli, M.-S. (1995). Fibre Atlas: Identification of Papermaking Fibres. Berlin, Heidelberg: Springer.	•				•	
Jawald, M., Boufi, S. & Abdul Khalil, H. (Eds.) (2017). Cellulose - Reinforced Nanofibre Composites. Cambridge: Woodhead Publishing.		•				
Jawaid, M., Khan, T. A., Nasir, M. & Asim, M. (Eds.) (2021). Eco-Friendly Adhesives for Wood and Natural Fiber Composites. Characterization, Fabrication and Applications. Singapore: Springer Nature.		•	•			

Literature	Wood science	Wood-based materials	Wood mechanical processing	Timber engineering	Chemical engineering, pulp and paper	Energy from biomass
Jayne, B. A. (1972). Theory and Design of Wood and Fiber Composite Materials. Syracuse: Syracuse University Press.		•				
Jones, D. & Brischke, C. (Eds.) (2017). Performance of Bio-based Building materials. Sawston, Cambridge: Woodhead Publishing	•			•		
Jones, D., Sandberg, D., Goli, G. & Todaro, L. (2019). Wood Modification in Europe: A State-of-the-art about Processes, Products, Applications. Florence: Florence University Press.	•		•			
Kaltschmitt, M., Hartmann, H. & Hofbauer, H. (Eds.) (2016). Energie aus Biomasse. Grundlagen, Techniken und Verfahren. Berlin Heidelberg: Springer-Verlag.					•	
Kaltschmitt, M. (Ed.) (2019). Energy from Organic Materials (Biomass). New York: Springer.					•	
Kamm, B., Gruber, P. & Kamm, M. (Eds.) (2010). Biorefineries-Industrial Processes and Products. Status Quo and Future Directions. Weinheim: Wiley-VCH Verlag GmBH & CoKG.					•	
Karacabeyli, E. & Douglas, B. (Eds.) (2013). CLT Handbook US Edition. Pointe-Claire (QC): FPInnovations and Binational Softwood Lumber Council.				•		
Karacabeyli, E. & Gagnon, S. (Eds.) (2019). Canadian CLT Handbook 2019 Edition. Quebec: FPInnovations.				•		
Karimi, K. (Ed.) (2015). Lignocellulose-based Bioproducts. Berlin: Springer.						•
Karmarsch, K. (1837). Grundriβ der mechanischen Technologie. Erster Band: Die Bearbeitung der Metalle. Hannover: Helwingsche Verlagsbuchhandlung.			•			
Karmarsch, K. (1839). Grundriβ der mechanischen Technologie. Zweiter Band: Die Bearbeitung des Holzes, Spinnerei und Weberei, Papierfabrisation, Bearbeitung der Thon und Glaswaren. Hannover: Helwingsche Verlagsbuchhandlung.			•			
Karmarsch, K. (1851). Handbuch der mechanischen Technologie. Hannover: Hellwingsche Hofbuchhandlung.			•			

Literature	Wood science	Wood-based materials	Wood mechanical processing	Timber engineering	Chemical engineering, pulp and paper	Energy from biomass
Kasal, B., Friebel, S., Gunschera, J., Schirp, A., Schwab, H. & Thole, V. (2000). Wood-based Materials. Ulmanns Encyclopedia of Industrial Chemistry. Weinheim: Wiley-VCH Verlag GmbH & Co. KGaA.		•				
Kaufmann, H., Krötsch, S. & Winter, S. (2021). Atlas Mehrgeschossiger Holzbau. (2nd ed.) Munich: Detail Business Information.				•		
Keey, R., Langrish, T. A. G. & Walker, J. F. C. (2000). Kiln-drying of Lumber. Berlin, Heidelberg: Springer-Verlag.			•			
Kehr, E. (Ed.) (1975). Werkstoffe aus Holz und andere Werkstoffe der Holzindustrie. Leipzig: VEB Fachbuchverlag.		•				
Kermani, A. (1999). Structural Timber Design. Oxford: Blackwell Science.				•		
Kim, J. & Kaushik, P. (2011). Recent Advances in Processing of Wood-plastic Composites. Heidelberg: Springer.		•				
Kitek Kuzman, M., Sandberg, D. & Moutou Pitti, R. (2018). Produits d'Ingénierie en Bois pour l'Architecture Contemporaine – Cas d´étude. Ljubljana: University of Ljubljana.		•	•			
Klyosov, A. A. (2007). Wood-plastic Composites. Hoboken: John Wiley & Sons.		•	•			
Knigge, W. & Schulz, H. (1966). Grundriβ der Forstbenutzung. Hamburg: Parey.	•					
Koch, P. (1972). Wood Machining Processes. New York: The Ronald Company.			•			
Koch, P. (1972). Utilization of the Southern Pines: I. The Raw Material, II. Processing. Washington DC: U. S. Department of Agricultural Forest Service.	•		•			
Koch, P. (1996). Lodgepole Pine in North America I-III. Madison (WI): Forest Products Society	•		•			
Kocsis, Z. & Csanády, C. (2019). Theory and Practice of Wood Pellet Production. Cham: Springer Nature.						•
Koehler, A. & Thelen R. (1926). The Kiln Drying of Lumber. New York: McGraw-Hill Book Company.			•			

Literature	Wood science	Wood-based materials	Wood mechanical processing	Timber engineering	Chemical engineering, pulp and paper	Energy from biomass
Kokocinski, W. (2004). Drewno. Poznan: Wydawnictwo-Drukarnia PRODRUK.	•	•	•		•	
Kolb, J. (2008). Systems in Timber Engineering. Basel: Birkhäuser Architecture.				•		
Kollmann, F. (1936). Technologie des Holzes. Berlin: Verlag von Julius Springer.	•	•	•		•	
Kollmann, F. F. P. (1962). Furniere, Lagenhölzer und Tischerplatten. Berlin, Heidelberg: Springer.		•	•			
Kollmann, F. (1951). Technologie des Holzes und der Holzwerkstoffe, Vol. 1. Berlin, Göttingen, Heidelberg: Springer.	•	•	•		•	
Kollmann, F. (1955). Technologie des Holzes und der Holzwerkstoffe, Vol. 2. Berlin, Göttingen, Heidelberg: Springer.			•			
Kollmann, F. & Côté Jr., W. A. (1968). Principles of Wood Science and Technology, Vol. 1. Berlin, Heidelberg: Springer.	•	•	•		•	
Kollmann, F., Kuenzi, E. & Stamm, A. (1975). Principles of Wood Science and Technology, Vol. 2. Berlin, Heidelberg: Springer.			•		•	
Koon-Yang, L. (Ed.) (2018). Nanocellulose and Sustainability. Production, Properties, Aplications and Case Studies. Boca Raton: CRC Press.	•				•	
Koppejan, J. & van Loo, S. (Eds.) (2008). The Handbook of Biomass Combustion and Co-firing. London: Earthscan.						•
Kozakievicz, P. (2012). Fizyka Drewna W Theorii I Zadaniach. Warschau: Wydawnictwo SGGW (Landwirtschaftliche Akademie Warschau).	•					
Krzysik, F. (1957). Nauka o Drevnie. Warschau: PNW.	•					
Kumari Thakur, V., Kumari Thakur, M. & Kessler, M. R. (Eds.) (2017). Handbook of Composites from Renewable Materials, Structure and Chemistry. Hoboken, New Jersey: John Wiley & Sons.		•				
Kutnar, A. & Muthu, S. S. (Eds.) (2016). Environmental Impacts of Traditional and Innovative Forest-based Bioproducts. Singapore: Springer Science+Buisness media.		•			•	
Lampert, H. (1966). Faserplatten. Leipzig: Fachbuchverlag.		•				

Literature	Wood science	Wood-based materials	Wood mechanical processing	Timber engineering	Chemical engineering, pulp and paper	Energy from biomass
Langendorf, G., Schuster, E. & Wagenführ, R. (1990). Rohholz (4th ed.). Leipzig: Fachbuchverlag.	•					
Larsen, H. & Enjily, V. (2009). Practical Design of Timber Structures to Eurocode 5. London: ICE Publishing.				•		
Ledebur, A. (1881). Die Verarbeitung des Holzes auf mechanischem Wege. Braunschweig, Verlag Friedrich Vieweg und Sohn.			•			
Liese, W. & Köhl, M. (Eds.) (2015). Bamboo - The Plants and its Uses. Cham: Springer.	•					
Lin, S. Y. & Dence, W. D. (Eds.) (1992). Methods in Lignin Chemistry. Berlin, Heidelberg: Springer-Verlag.					•	
Lohmann, U. (Ed.) (2003). Holz-Lexikon (4th ed.) Leinfelden-Echterdingen: DRW-Verlag.	•	•	•		•	
Lukowsky, D. (2015). Failure Analysis of Wood and Wood-based Products. New York: McGraw-Hill Education.	•	•				
Madsen, B. (1992). Structural Behaviour of Timber. Vancouver: Timber Engineering.				•		
Madsen, B. (2000). Behaviour of Timber Connections. Vancouver: Timber Engineering.				•		
Maloney, T. (1993). Modern Particleboard & Dry Process Fibre Board Manufacturing. San Francisco: Miller Freemann Inc.		•				
Mark, R. E. (1967). Cell Wall Mechanics of Tracheids. New Haven and London: Yale University Press.	•					
Marra, A. (1992). Technology of Wood Bondings Principles in Practice. New York: Van Norstrand Reinold.					•	
Matejak, M. & Niemz, P. (2011). Das Holz in deutschen Texten zwischen 1587 und 1922. Zürich: ETH Zürich, Institut für Baustoffe, Holzphysik (on-line at the e-collection at the ETH library).	•	•	•		•	•
Mattek, K. & Kubler, H. (1995). Wood - The Internal Optimization of Trees. Berlin, Heidelberg.	•					
McKenzie, W. M. (1961). Fundamental Analysis of the Wood-cutting Process. Ann Arbor (MI): The University of Michigan.			•			
McKenzie, W. & Zhang, B. (2007). Design of Structural Timber to Eurocode 5. London: Palgrave.				•		

Literature	Wood science	Wood-based materials	Wood mechanical processing	Timber engineering	Chemical engineering, pulp and paper	Energy from biomass
Mette, H. J. (1984). Holzkundliche Grundlagen der Forstnutzung. Berlin: Dt. Landwirtschaftsverlag.	•					
Molnar, S. (2004). Faanyagismeret. Budapest: Mezogasagi Szaaktudas Kiago.	•	•	•			
Natterer, J., Sandoz, J. & Rey, M. (1997). Construction en Bois. Lausanne: EPFL Press.				•		
Navi, P. (2006). Propriétés acoustiques des matériaux. Propagation des ondes planes harmoniques. Lausanne: Presses polytechniques et universitaires romandes.	•			•		
Navi, P. & Heger, F. (2005). Comportement thermo-hydro-mécanique du bois. Applications technologiques et dans les structures. Lausanne: Presses polytechniques et universitaires romandes.	•	•	•			
Navi, P. & Sandberg, D. (2012). Thermo-hydro-mechanical Processing of Wood. Lausanne: EPFL Press, CRC Press.	•	•	•			
Neroth, G. & Vollenschaar, H. (2011). Wendehorst Baustoffkunde (27th ed.) Wiesbaden: Vieweg+Teubner.		•		•		
Neuhaus, H. (2011). Ingenieurholzbau (3rd ed.). Wiesbaden: Vieweg +Teubner.		•		•		
Neuhaus, H. (2017). Ingenieurholzbau. Grundlagen, Bemessung, Nachweise, Beispiel. Wiesbaden: Springer-Vieweg.				•		
Niemz, P. (1993). Physik des Holzes und der Holzwerkstoffe. Leinfelden-Echterdingen: DRW-Verlag Weinbrenner GmbH & Co.	•	•				
Niemz, P. & Sander, D. (1989). Prozessmesstechnik in der Holzindustrie. Leipzig: Fachbuchverlag.			•		•	
Niemz, P. & Sonderegger, W. (2017/2021). Holzphysik. Physik des Holzes und der Holzwerkstoffe (2nd, 3rd eds.) Leipzig: Fachbuchverlag Leipzig im Carl Hanser Verlag.	•	•				
Niemz, P, Teischinger, A. & Sandberg, D. (Eds.) (2023). Springer Handbook of Wood Science and Technology. Springer Nature Switzerland AG, Cham, Switzerland, 2100 pp.	•	•	•	•	•	•
Niska, K. O. & Sain, P. (Eds.) (2008). Wood Polymer Composites. Boca Raton (FL): CRC Press, Woodhead Publishing Ltd.		•				

Literature	Wood science	Wood-based materials	Wood mechanical processing	Timber engineering	Chemical engineering, pulp and paper	Energy from biomass
Nördlinger, H. (1860). Die technischen Eigenschaften der Hölzer für Forst- und Baubeamte, Technologen und Gewerbetreibende. Stuttgart: J.G. Cotta'scher Verlag.	•			•		
Obućina, M., Kitek Kuzman, M. & Sandberg, D. (2017). Use of Sustainable Wood Buildings Materials in Bosnia and Herzegovina, Slovenia and Sweden. Sarajevo: University of Sarajevo.			•	•		
Oksman, K., Mathew, A. P., Bismarck, A., Rojas, O. & Sain, M. (Eds.) (2014). Handbook of Green Materials: Bionanomaterials. Singapore: World Scientific Pub.		•				
Magoss, E., Sitkei, G. & Kocsis, Z. (2022). Dust Extraction and Handling in the Wood Industry. Cham: Springer Nature.			•			
Ozelton, E. C. & Baird, J. A. (2002). Timber Designers' Manual. Oxford: Blackwell Science.				•		
Pandey, K. K., Ramakantha, V., Chauhan, S. S., & Arun Kumar, A. N (Eds.) (2017). Wood is Good: Current Trends and Future Prospects in Wood Utilization. Singapore: Springer Nature Singapore Pte Ltd.		•		•		
Panshin, A. J. & de Zeeuw, C. (1964). Textbook of Wood Technology. Part 1: Structure, Identification, Uses and Properties of the Commercial Woods of the United States. (2nd ed.) New York: McGraw-Hill Book Company.	•	•				
Paulitsch, M. & Barbu, M. C. (2015). Holzwerkstoffe der Moderne. Leinfelden-Echterdingen: DRW-Verlag.		•				
Peck, E. C. (1957). Bending Solid Wood to Form. Washington: U.S. Department of Agriculture, Forest Service, Agriculture.			•			
Pellérin, R. F. & Ross, R. J. (2002). Nondestructive Evaluation of Wood. Madison (WI): Forest Products Society.	•					
Perre, P. (Ed.) (2007). Fundamentals of Wood Drying. COST E15. Nancy: A.R.Bo: LOR ENGREF.			•			
Pizzi, A. & Mittal, K. L. (Eds.) (2018). Handbook of Adhesive Technology. (3rd ed.). Boca Raton (FL): CRC Press.					•	
Porteous, J. & Kermani, A. (2013). Structural Timber Design to Eurocode 5. Hoboken, New Jersey: Wiley-Blackwell.				•		

Literature	Wood science	Wood-based materials	Wood mechanical processing	Timber engineering	Chemical engineering, pulp and paper	Energy from biomass
Porteous, J. & Ross, P. (2013). Designers' Guide to Eurocode 5: Design of Timber Buildings: EN 1995-1-1. Eurocode Designers' Guide. London: ICE Publishing.				•		
Powis Bale, M. (1896). Saw-mills. Their Arrangement and Management; and the Economical Conversion of Timber, Being a Companion Volume to Woodworking Machinery. London: Crosby Lockwood and Son.			•			
Pozgaj, J., Chonavec, D., Kurjatko, S. & Babiak, M. (1997). Struktura a vlasnosti dreva. Bratislava: Priroda.	•					
Pretzsch, H. (2018). Grundlagen der Waldwachstumsforschung. 2. überarbeitete Auflage Hrsg. Berlin: Springer Verlag.	•		•			
Qureshi, N., Hodge, D. & Vertes, A. (Eds.) (2014). Biorefineries: Integrated Biochemical Processes for Liquid Biofuels. Boston (MA): Elsevier.					•	
Radkau, J. (2007). Holz - Wie ein Naturstoff Geschichte schreibt. München: oecom Verlag.	•					
Radkau, D. (2012). Wood: A History. Cambridge: Polity Press.	•					
Reinprecht, L. (2016). Wood Detoration, Protection and Maintance. Chester: John Wiley & Sons.	•					
Richardson, B. (1993). Wood Preservation. London: E&FN Spon, Chapman & Hall.	•					
Richter, C. (2010). Holzmerkmale (3rd ed.) Leinfelden-Echterdingen: DRW-Verlag.	•	•				
Richter, C. (2015). Wood Characteristics: Description, Causes, Prevention, Impact on use and Technological Adaption. Cham: Springer-Verlag.	•	•	•			
Rijsdijk, J. F. & Laming, P. B. (1994). Physical and Related Properties of 145 Timbers. Dordrecht, Boston, London: Kluwer Academic Publishers.	•					
Robinson, J. C. (2016). Spalted Wood. The History, Science, and Art of a Unique Material. Atglen: Schiffer Publishing	•					
Roffael, E. (2017). Formaldehyde in Nature in Wood and in Wood-based Panels. Leinfelden-Echterdingen: DRW-Verlag.				•		

Literature	Wood science	Wood-based materials	Wood mechanical processing	Timber engineering	Chemical engineer-ing, pulp and paper	Energy from biomass
Rosendahl, L. (Ed.) (2013). Biomass Combustion Science, Technology and Engineering. Cambrige: Woodhead Publishing.						•
Ross, R. J. & Wang, X. (Eds.) (2012). Nondestructive Testing and Evaluation of Wood - 50 Years of Research: International Nondestructive Testing and Evaluation of Wood Symposiums Series. Madison (WI): U.S. Department of Agriculture, Forest Service, Forest Product Laboratory.	•					
Ross, R. J. (Ed.) (2015); Nondestructive Evaluation of Wood (2nd ed.). General Technical Report FPL-GTR-238, Madison (WI), US. Department of Agriculture, Forest Service, Forest Products Laboratory	•	•				
Ross, R. J. (Ed.) (2021). Wood Handbook. Wood as an Engineering Material. General Technical Report FPL-GTR-282. Madison (WI): Dept. of Agriculture, Forest Service.	•	•	•	•	•	•
Rowe, J. W. (Ed.) (1989). Natural Products of Wood Plants. Berlin, Heidelberg: Springer.	•					
Rowell, R. M. (Ed.) (2013). Handbook of Wood Chemistry and Wood Composites (2nd ed.). Boca Raton (FL): CRC Press.	•	•				
Ruffinatto, F. & Alan Crivellaro, A. (2019). Atlas of Macroscopic Wood Identification: With a Special Focus on Timbers Used in Europe and CITES-listed Species. Cham: Springer Nature Switzerland AG.	•					
Ruiz, H. A., Thomsen, M. H. & Trajano, H. L. (Eds.) (2017). Hydrothermal Processing in Biorefineres: Production of Bioethanol and High Added-value Compounds of Second and Third Generation Biomass. Berlin, Heidelberg: Springer.					•	
Rug, W. & Mönck, W. (2021). Holzbau: Bemessung und Konstruktion. 16., vollständig überarbeitete Auflage Hrsg. Berlin, Wien: Beuth Verlag GmbH.				•		
Sandberg, D., Kitek Kuzman, M. & Gaff, M. (2018). Kompozitní výrobky na bázi dřeva - Dřevo jako kompozitní a konstrukční materiál. Engineered Wood Products - Wood as an Engineering and Architectural Material. Prague: Czech University of Life Sciences Prague.		•	•			

Literature	Wood science	Wood-based materials	Wood mechanical processing	Timber engineering	Chemical engineering, pulp and paper	Energy from biomass
Sandberg, D., Kutnar, A., Karlsson, O. & Jones, D. (2021). Wood Modification Technologies. Principles, Sustainability, and the need for Innovation. Boca Raton (FL): CRC Press.	•	•	•			
Sandhaas, H. J. & Blaß, C. (2020). Timber Engineering - Principles for Design. Sheridan (WY): Creative Media Partners, LLC.				•		
Scheiding, W., Grabes, P., Haustein, T., Haustein, V., Nieke, N., Urban, H. & Weiss, B. (2014). Holzschutz: Holzkunde, Pilze und Insekten, konstruktive und chemische Massnahmen, technische Regeln, Praxiswissen. Leipzig: Fachbuchverlag im Carl Hanser Verlag.	•					
Schickhofer, G., Bogensberger, T. & Moosbrugger, T. (Eds.) (2010). BSP Handbuch, Holz- Massivbauweise in Brettsperrholz. Nachweise auf Basis des neuen europäischen Normenkonzepts. [CLT Handbook, Structural Design According to the new European Design Code Concept.] Graz: Verlag der Technischen Universität Graz.				•		
Schniewind, A. P. (1989). Wood and Wood-based Materials. Oxford: Pergamon Press.	•	•				
Schweingruber, F. (1993). Trees in Wood Dendrochronlogy. Berlin, Heidelberg: Springer-Verlag.	•					
Schweingruber, F. H., Steiger, P. & Börner, A. (2019). Bark Anatomy of Trees and Shrubs in the Temperate Northern Hemisphere. Cham: Springer Nature	•					
Siau, J. F. (1984). Transport Processes in Wood. Berlin: Springer-Verlag.	•		•			
Sixta, H. E. (2006). Handbook of Pulp. Weinheim: WILEY-VC.					•	
Siau, J. F. (1995). Wood: Influence of Moisture on Physical Properties. Keene, New York: Department of Wood Science and Forest Products, Virginia Polytechnic Institute and State University.	•		•			
Sjöström, E. & Raimo, A. (Eds.) (1999). Analytical Methods in Wood Chemistry, Pulping, and Papermaking. Berlin, Heidelberg: Springer-Verlag.	•				•	

Literature	Wood science	Wood-based materials	Wood mechanical processing	Timber engineering	Chemical engineering, pulp and paper	Energy from biomass
Skaar, C. (1988). Wood-Water Relations. Berlin, Heidelberg: Springer-Verlag.	•		•			
Smith, I., Landis, E. & Gong, M. (2003). Fracture and Fatigue in Wood. Chichester: Wiley.	•					
Smook, G. (2015). Handbook for Pulp & Paper Technologist. (4th ed.) Peachtree Corners: Tappi Press.					•	
Smulisky, R. & Jones, P. (2019). Forest Products & Wood Science - An Introduction. Hoboken: Wiley-Blackwell.	•	•	•		•	
Spellmann, F. (2017). Forest-based Biomass Energy: Concepts and Applications. Boca Raton (FL): CRC Press.						•
Stalnaker, J. J. & Harris, E. C. (1997). Structural Design in Wood. Alphen aan den Rijn: Kluwer Academic Publishers.				•		
Stamm, A. J. (1964). Wood and Cellulose Science. New York: Ronald Press.	•				•	
Steinsiek, M. P. (2008). Forst- und Holzforschung im "Dritten Reich". Remhagen: Verlag Kessel.	•					
Stevanovic, T. (Ed.) (2017). Chemistry of Lignocellulosics: Current Trends. Boca Raton: CRC Press, Taylor & Francis.	•					
Stevens, W. C. & Turner, N. (1948). Solid and Laminated Wood Bending. London: Forest Products Research Laboratory.			•			
Stokke, D. D. & Groom, L. H. (Eds.) (2008). Characterization of the Cellulosic Cell Wall. Hoboken, New Jersey: Wiley-Blackwell.	•					
Suchsland, O. (2004). The Swelling and Shrinking of Wood. A Practical Technology Primer. Madison (WI): Forest Products Society.	•					
Szalai, J. (1994). A Fannyag és Faalapú Anyagok Anizotróp Rugalmasság- és Szilárdságtana. 1 Rész. A Mechanikai Tulajdonságok Anizotrópiája. [The anisotropic elasticity and strength of wood and wood-based materials. Part I: The anisotropy of mechanical properties.] Sopron: University of Sopron.	•					
Szymani R. (Ed.) (1999). Scanning Technology & Process Optimization. Advances in Wood Industry. San Francisco: Miller Freeman books.			•			

Literature	Wood science	Wood-based materials	Wood mechanical processing	Timber engineering	Chemical engineer-ing, pulp and paper	Energy from biomass
Taghiyari, H. R., J. Morrell, J. J. & Husen, A. (Eds.) (2023). Emerging Nanomaterials: Opportunities and Challenges in Forestry Sectors. Cham: Springer Nature.		•	•	•	•	
Thelandersson, S. & Larsen, H. J. (Eds.) (2003). Timber Engineering. West Sussex: John Wiley & Sons.				•		
Tiemann, H. D. (1942). Wood Technology - Constitution, Properties and Uses. New York, Chicago: Pitman Publishing.	•	•	•			
Timell, T. E. (1986). Compression Wood in Gymnosperms, Vol. I-III. Berlin, Heidelberg: Springer-Verlag.	•					
Torgovnikov, G. I. (1993). Dielectric Properties of Wood and Wood-based Materials. Berlin, Heidelberg: Springer-Verlag.	•					
Trendelenburg, R. (1939). Das Holz als Rohstoff. München: J. F. Lehmanns Verlag.	•					
Trendelenburg, R. & Mayer-Wegelin, H. (1955). Das Holz als Rohstoff (2nd ed.). München: Carl Hanser Verlag.	•					
Tsoumis, G. (1991). Science and Technology of Wood - Structure, Properties, Utilization. New York: Van Nostrand Reinhold.	•	•	•			
Ugolev, B. (1969). Internal Stresses in Wood caused by Drying. Washington: U.S. Department of Agriculture.	•		•			
Ugolev, B. N. (1986). Drevesina s osnovami lesnogo tovarove-denija. [Wood science and basics of wood products.] Moskau: Lesn. Prom. (in Russian).	•					
Ugolev, B. (2014). Istoricheskie vehi drevesinovedenija i vzgljad v budushhee [Historical milestones in wood research in Russia and a look into the future.] Moskau: Moskauer Forsttechnische Universität (in Russian).	•					
Unger, A. (1988). Holzkonservierung. Leipzig: Fachbuchverlag.	•					
Unger, A., Schniewind, A. P. & Unger, W. (2001). Conservation of Wood Artifacts. Berlin, Heidelberg: Springer-Verlag.	•					
Van Loo, S. & Koppejan, J. (Eds.) (2008). The Handbook of Biomass Combustion and Co-firing. London: Earthscan.			•			•
Varis, R. (Ed.) (2018). The Sawmill Industry Handbook. Helsinki: Association of Finnish Sawmill men and the Finnish Woodworking Engineers Association.		•	•			

Literature	Wood science	Wood-based materials	Wood mechanical processing	Timber engineering	Chemical engineering, pulp and paper	Energy from biomass
Varis, R. (Ed.) (2018). Wood-based Panels Industry. Helsinki: Association of Finnish Sawmill men and the Finnish Woodworking Engineers Association.		•	•			
Vernon, K. E., Wulpi, M. & Wegner-Cohn, L. M. (1930). Furniere und Sperrholz. Erster Band: Eigenschaften und Herstellung von Furnieren und Sperrholz. Berlin: M. Kryn, Technischer Verlag.		•	•			
Vorreiter, L. (1949-1963). Holztechnologisches Handbuch, Vol. 1-3. Wien: Fromme.	•	•	•		•	
Wachter, I., Rantuch, P. & Štefko, T. (2023). Transparent Wood Materials: Properties, Applications, and Fire Behaviour. Zug: Springer Nature Switzerland AG.		•		•		
Wagner, J. B. (1917). Seasoning of Wood. A Treatise on the Natural and Artificial Processes Employed in the Preparation of Lumber for Manufacture, With Detailed Explanations of Its Uses, Characteristics and Properties.New York: D. Van Nostrand.			•			
Wagenführ, R. (1989). Anatomie des Holzes (4th ed.). Leipzig: Fachbuchverlag.	•					
Wagenführ, R. (2007). Holzatlas (6th ed.). München: Fachbuchverlag Leipzig im Carl Hanser Verlag.	•					
Wagenführ, R. & Wagenführ, A. (2021). Holzatlas. (7th ed.) München: Carl Hanser Verlag.	•					
Wagenführ, A. & Scholz, F. (Eds.) (2012). Taschenbuch der Holztechnik (2nd ed.) München: Fachbuchverlag Leipzig im Carl Hanser Verlag.	•	•	•			
Walker, J. C. (2006). Primary Wood Processing: Principles and Practice (2nd ed.). Dordrecht: Springer.	•		•			
Wang, S. & Luo, Z. (2017). Pyrolysis of Biomass. Berlin: Walter de Gruyter, Beijing : Science Press.						•
Wangaard, F. F. (1950). The Mechanical Properties of Wood. New York: John Wiley & Sons.	•	•				
Weinand, Y. (Ed.) (2017). Advanced Timber Structures. Architectural Designs and Digital dimensioning. Basel: Birkhäuser.				•		

Literature	Wood science	Wood-based materials	Wood mechanical processing	Timber engineering	Chemical engineering, pulp and paper	Energy from biomass
Werner, G. & Zimmer, K.-H. (2009). Holzbau 1 (4th ed.) Berlin: Springer-Verlag.				•		
Wilcox, W., Botsai, E. & Kubler, H. (1991). Wood as a Building Material. New York, Chister, Brisbane, Toronto, Singapore: John Wiley & Sons.		•		•		
Williamson, T. G. (2002). APA Engineered Wood Handbook. New York: McGraw-Hill Handbooks.		•		•		
Williston, E. M. (1976). Lumber Manufacturing: The Design and Operation of Sawmill and Planer Mills. San Francisco: Miller Freeman Publications.			•			
Williston, E. M. (1981). Small Logs Sawmill, Profitable Product Selection, Process Design and Operation. San Francisco: Miller Freeman Publications.			•			
Williston, E. M. (1991). Value Added Products. Manufacturing and Marketing Strategies. Madison (WI): Forest Products Society.			•			
Winter, S. & Peter, M. (Eds.) (2021). Holzbau-Taschenbuch: Grundlagen. (10th ed.) Berlin: Ernst & Sohn.				•		
Xie, H. & Gathergood, N. (Eds.) (2012). The Role of Green Chemistry in Biomass Processing and Conversion. Hoboken, New Jersey: John Wiley & Sons.					•	
Zabel, R. & Morrel, J. (2020). Wood Microbiology. Decay and Its Prevention. (2nd ed.). Cambridge (MA): Academic Press Inc.	•					
Zehui, J. (2013). Wood Properties of Global of Important Tree Species. Bejing: Science Press.	•					
Zimmermann, M. H. (1983). Xylem Structure and the Ascent of Sap. Berlin, Heidelberg: Springer-Verlag.						
Zobel, B. J. & van Buijtenen, J. P. (1989). Wood Variation. Its Causes and Control. Berlin, Heidelberg: Springer-Verlag.	•					
Zobel, B. J. & Sprauge, J. R. (1998). Juvenile Wood in Forest Trees. Berlin, Heidelberg: Springer-Verlag.	•					
Zugenmaier, P. (2008). Crystalline Cellulose and Derivatives Characterization and Structures. Berlin, Heidelberg: Springer-Verlag.	•					

Prof.Dr.-Ing.habil. Dr.h.c. Peter Niemz studied at the Faculty of Mechanical Engineering of TU Dresden. From 1972 onwards, he worked at the Institute of Wood Technology Dresden (IHD) and at TU Dresden. From 1993 to 1996, he served as a Full Professor at the Universidad Austral de Chile in Valdivia before joining the Swiss Federal Institute of Technology (ETH Zurich) in 1997. He was Professor of Wood Physics from 2002 until his retirement in 2015.

Prof.Dr. Alfred Teischinger, a wood technologist, received his PhD from the University of Natural Resources and Life Sciences (BOKU), Vienna. He moved to the higher technical college Mödling as a professor of wood technology and head of an accredited testing laboratory. In 2000, he was appointed professor of wood technology at BOKU; from 2001 to 2015, he also acted as scientific director of the Competence Centre for Wood Composites and Wood Chemistry (Wood Kplus).

Prof.Dr. Dick Sandberg received his Doctorate in Mechanical Engineering, Wood Technology and Processing from the Royal Institute of Technology (KTH), Stockholm, Sweden, in 1998. He is a Chaired Professor of Design and Fabrication of Timber Construction at the Norwegian University of Science and Technology (NTNU). He specialises in using new measurement technology, advanced information processing, and wood communication from felling to finished product. He is an Elected Fellow of the International Academy of Wood Science.

Standards and Certification

8

Johannes Welling, Galina Gorbacheva, Jiali Jiang, Robert J. Ross and Christopher Adam Senalik

8.1 Introduction

Johannes Welling

Many technologies, material and product specifications are now largely subject to standardisation in order to obtain reliable material and product properties with comparable test methods. Acceptable levels of emissions from materials and many constructions are also standardised. Manufacturing is also carried out using a complex system of standards, and many terms are standardised. In general, standards are voluntary, but they also may be implemented as mandatory documents.

For example, in timber construction, characteristic values for materials, test methods for monitoring quality (e.g. moisture resistance, strength, long-term behaviour, fire resistance), for fire protection measures, fasteners, and adhesives are standardised. The construction is based mainly on standards. Some of these standards relate to property levels, but many relate to test procedures defining and standardising the measurement of these properties.

The procedures are similar worldwide, but there are clear differences between the regions. Based on a historical overview of standardisation in Europe, Russia, the USA and China, a short overview of standardisation in the individual regions is given, which enables an introduction to the standardisation system.

Numerous technical standards for test engineering, materials and technology are listed in detail in the Springer Handbook of Wood Science and Technology. This chapter is to be seen as a brief introduction to the international system of standardisation. Specialists from the respective regions, who are also involved in the national system, were consulted for processing.

8.2 Development and Status of Standardisation Related to Wood and Wood Products

Our modern world with its interlinked global markets and a division of work in processes and production sites would not exist without standardisation. Wood, being a natural product, exhibits many property variations and, due to its anisotropic behaviour, it is not engineers' and architects' best friend. Without standards describing the behaviour of wood, listing its properties, and giving guidance for engineering, the planning of structural frameworks for modern wooden buildings would not be possible.

Standardisation, and especially the harmonisation of standards at the international level, is hard work and time-consuming and, for those active in this field, sometimes frustrating and disappointing. Thousands of man-years have been invested to bring standardisation to its current level, and the current status is certainly not the end of the story. There is plenty of room for further development and improvement.

The original version of the chapter has been revised. A correction to this chapter can be found at https://doi.org/10.1007/978-3-031-38329-8_9

J. Welling (✉)
Sachverständigenbüro Holz-Expertise, Reinbek, Germany
e-mail: welling@holz-expertise.de.de

G. Gorbacheva
Bauman Moscow State Technical University, Moscow, Russian Federation
e-mail: gorbacheva@bmstu.ru

J. Jiang
Chinese Academy of Forestry, State Key Laboratory of Tree Genetics and Breeding, Research Institute of Wood Industry, Beijing, P. R. China
e-mail: Jialiwood@caf.ac.cn

R. J. Ross · C. A. Senalik
Forest Products Laboratory, USDA Forest Service, Madison, USA
e-mail: robert.j.ross@usda.gov

C. A. Senalik
e-mail: christopher.a.senalik@usda.gov

P. Niemz et al. (eds.), *Wood Material and Processing Data*, https://doi.org/10.1007/978-3-031-38329-8_8

8.2.1 China

Jiali Jiang

In 1949, the standardisation work was carried out by the Central Technical Authority, which was authorised by the Central Finance and Economic Commission of the People's Republic of China. In 1979, the Regulations on Standardisation Administration of the People's Republic of China were approved and promulgated by the State Council, and the Technical Committee for Professional Standardisation was created. The Standardisation Law of the People's Republic of China was adopted at the Fifth Meeting of the Standing Committee of the Seventh National People's Congress on December 29, 1988, and came into force on 1 April, 1989.

Nowadays, the Standardisation Administration of the People's Republic of China (SAC) is the standards organisation authorised by the State Council to undertake unified management, supervision and overall coordination of standardisation work in China. This standardisation work is carried out on the basis of the Standardisation Law of the People's Republic of China, which was revised and adopted at the 30th Meeting of the Standing Committee of the Twelfth National People's Congress on 4 November, 2017. It was promulgated and came into force on 1 January, 2018.

Standards include national standards, sector standards, local standards, association standards and enterprise standards. National standards are divided into mandatory standards and voluntary standards. Both sector standards and local standards are voluntary standards. Mandatory standards must be implemented. Searching is possible using the keywords or serial numbers of the standard designations on the website of the National Public Service Platform for Standards Information (http://std.samr.gov.cn).

8.2.2 Europe

Johannes Welling

In Europe, standardisation has a long history. It started in UK during the industrial revolution and helped to produce machine parts with standardised sizes and features. Without going into details, it can be said that standardisation in Europe dates back to the early twentieth century. The "Engineering Standards Committee" (now the British Standards Institution) was founded in 1901. In 1906 the "International Electrotechnical Commission" (IEC) started its work supported by the German "Verband Deutscher Elektrotechniker (VDE)". In the following years, national standardisation bodies were created in most European countries, e.g. AFNOR in France, Austrian Standard, DIN in Germany, and UNI in Italy. In the USA, the standardisation process started in 1898 with the foundation of the "American Society for Testing Materials (ASTM). The American National Standards Institute (ANSI) was founded in 1918. The International Organization for Standardization (ISO) started its work in 1947.

National and international standards helped the industry to produce mass products at competitive prices, and it facilitated the exchange of spare parts and the production of semi-finished or partially assembled parts in the automotive and other sectors.

Despite its many advantages, the large variety of national standards led to a situation in which small countries had problems in adhering to their national specifications. Cross-border trade and business were hindered by the many different rules, regulations and specifications. With the creation of the European Union the "European Committee for Standardization (CEN") was formed in 1961 to harmonise the many national standards which had been developed over decades in the European countries. This harmonisation process lasted for at least 20–30 years. Some people say it is still ongoing. Standards developed at the European level must be transformed into national standards. The development of national standards in European countries is not forbidden, but new national standards are only produced or maintained if very specific national conditions exist.

The harmonisation of European standards has facilitated the division of labour and the enhancement of trade within Europe. CEN standards and their related European and national regulations are so strong that imports into the European Union normally have to comply with CEN standards.

Affiliation with ISO and Other Standardisation Bodies Worldwide

Standardisation bodies interact in many ways. Many national and transnational standardisation bodies are members of ISO and support the development of ISO standards. In contrast to the situation in Europe, where European member countries in CEN are obliged to transform CEN standards into national standards, ISO standards need not be adopted by all the ISO member states.

Users of standards can easily determine whether or not a standard has been adopted from a transnational standardisation body. Standards are encoded according to the following system: Code of national standardisation body, code of transnational standardisation body from which the standard was adopted, code number of the standard followed by the year and month of its publication,

For example:

DIN EN 13183–1:2002–07 Feuchtegehalt eines Stückes Schnittholz - Teil 1: Bestimmung durch Darrverfahren. English title: Moisture content of a piece of sawn timber - Part 1: Determination by oven dry method.

This CEN standard has been adopted by DIN from CEN. It was developed in CEN/TC 175 Round and sawn timber by WG 1 General matters, definitions, measurement methods. The corresponding committee in DIN was NA 042–01-02 AA "Steering Committee Round Wood and Sawn Wood Products".

DIN EN ISO 14040:2009–11 Umweltmanagement - Ökobilanz - Grundsätze und Rahmenbedingungen (ISO 14040:2006); English Title: Environmental Management - Life Cycle Assessment - Principles and Framework.

This ISO standard has been adopted by CEN and as a consequence also by DIN. The mirror committee in DIN is NA 172–00-03 AA Ökobilanzen und umweltbezogene Kennzeichnung. It was elaborated by ISO/TC 207/SC 5 Life cycle assessment.

BS EN 14298:2017–11 Sawn Timber. Assessment of Drying Quality; Deutscher Titel: Schnittholz. Ermittlung der Trocknungsqualität.

This European standard was adopted by BS. It was edited by CEN TC 175 Round and sawn timber WG 2 Sawn timber. The mirror committee in DIN is NA 042–01-16 AA "Sawn timber".

DIN 52189–1:1981–12 Prüfung Von Holz; Schlagbiegeversuch; Bestimmung Der Bruchschlagarbeit; English Title: Testing of Wood; Impact Bending Test; Determination of Impact Bending Strength.

This German standard does not have an exact equivalent in international or transnational standards. It will be checked and reviewed at a five-year interval. It will be maintained until it is substituted by an international standard or until the DIN experts decide that it is not needed anymore.

Technical committees in any standardisation body are always dedicated to specific topics to ensure the well-focused cooperation of experts from industry, associations, interested parties, and science. To prevent overlapping or the development of contradictory rules or regulations, affiliations are established between technical committees or working/task groups within and between standardisation bodies.

8.2.3 Russia

Galina Gorbacheva

The standardisation work is carried out in Russia on the basis of the Federal law N184-FL of December 27, 2002, "On Technical Regulation". The Committee for Standardisation was created on 15 September 1925. The state standards (GOST) were compulsory documents for all enterprises and organisations regardless of their rank within the various branches of industry until 1992. Since 1992, the state standards are voluntary, but they nevertheless contain obligatory requirements. In accordance with the Federal Law «On Technical Regulation», the obligatory requirements for products, processes of production, operation, storage, transportation, marketing and utilisation are included in the Technical Regulations adopted by Federal Laws. The standards thus have a voluntary nature.

Nowadays, the Federal Agency on Technical Regulation and Metrology (a short title is Rosstandart) is the Federal Executive Body, implementing inter-industry coordination and functional regulation in the fields of standardisation, metrology and conformity assessment. It acts directly through the territorial bodies and organisations under its jurisdiction, and represents Russia in international (ISO, CEN, IEC/CEI, etc.) and regional organisations for standardisation.

In Russia, the Technical Regulations (TR) of the Eurasian Economic Union (EAEU) and Technical Regulations of the Russian Federation are in force (http://www.eurasiancommission.org/en/act/texnreg/deptexreg/tr/pages/TRVsily.aspx), (https://www.gost.ru/portal/gost//home/standarts/technicalregulationses).

These Technical Regulations are adopted by the Eurasian Economic Commission (EEC) and establish mandatory requirements for the application and enforcement on the territory of the EAEU to the objects of technical regulation. The Member-States of the EAEU are the Republic of Armenia, the Republic of Belarus, the Republic of Kazakhstan, the Kyrgyz Republic and the Russian Federation. Technical regulations of the Russian Federation are valid until the TR of the EAEU come into force. There is a list of products for which mandatory requirements are established within the EAEU and national legislation.

The following documents are used in the territory of the Russian Federation:

- national standards,
- rules of standardisation, norms and recommendations in the field of standardisation, and
- classifications, All-Russian Standards Classifiers of technical, economic and social information, used in accordance with the established procedure; factory standards.

Information about current Russian Interstate Standards (GOST) and State Standards of the Russian Federation (GOST R) can be found in the Standards catalogue https://www.gost.ru/portal/eng/home/standards/catalogue

Searching is possible by using standards' designations (numbers), titles (keywords) or OKS (classification groups of All-Russian Standards Classifier).

8.2.4 United States

Robert J. Ross, Christopher Adam Senalik

Many standards are utilised in the evaluation and assessment of wood and wood-based materials and it is beyond the scope of this document to summarise all the standards that are used in the United States. The following focuses on the fundamental standards used to evaluate the properties and performance of wood and wood-based materials.

The foundation standards used in research and industry in the United States are those maintained by ASTM International, formerly known as the American Society for Testing and Materials (ASTM). ASTM *Committee D-7: Wood* is responsible for the technical accuracy of the standards relating to fundamental test methods and properties, sawn timber and engineered wood products, pole and pile products, treatments for wood products and the fire performance of wood (Table 8.19). An excellent publication on the evolution of ASTM Committee D-7 published by Green et al. (2004) includes a discussion of the evolution of the standards widely used today and a list of current D-7 subcommittees, sections and standards. It was used extensively in the preparation of this chapter.

The current structure of Committee D-7: Wood was approved in 1984. It is divided into a number of subcommittees (Table 8.18) so that fundamental test methods and properties are under the purview of subcommittee D07.01, while structural product standards are consolidated in a single structural subcommittee (D07.02), with individual sections for each product line.

The development and expansion of the railroads in the United States in the mid to late 1800s spurred interest in standardisation with respect to wood utilisation. Prior to 1895, grading procedures, sawn-timber testing procedures, and methods for the calculation of allowable properties were proprietary procedures developed by consulting engineers, university professors, and a few government scientists active in the field of timber engineering.

In 1895, a committee of the American International Association of Railway Superintendents of Bridges and Buildings reported on the strength of bridge and trestle timbers (Berg et al. 1907) and their report presented fifteen recommendations for timber engineering, factors used for converting breaking strengths from laboratory tests to allowable properties, and average safe design stresses for the principal species used for bridge and trestle timbers. In response to the increased marketing opportunities provided by rail transportation, the wood industry had begun to develop standardised sizes of sawn timber surfaced on two sides—primarily to limit shipping costs.

Organised in 1898, ASTM provided an opportunity for further standardisation. In 1905, the ASTM "Committee Q on Standard Specifications for the Grading of Structural Lumber," was formed (Fulweiler 1955), and in 1907 the first standard established by this Committee D 9–07, the Standard Specification for Structural Timbers, was approved by a committee Society ballot. The standard contained a definition of structural timber, definitions of standard defects, a list of standard names for structural timbers, and specifications for bridge and trestle timbers. In 1910, the name of Committee D–7 was assigned, a Standard Specification for the Grading of Yellow-Pine Bridge and Trestle Timbers was approved, and work on similar standards for Douglas-fir and western hemlock were in progress. Committee activities were however changed to meet other challenges, especially with respect to the preservative treatment of railroad ties and structural timbers. Some of the enduring standards approved over the next few years included D 25–15, which provided a specification for the physical characteristics of round timber piles; D 38–15, standard methods of sampling and analysis of creosote oil; D 52–18, wood paving blocks for exposed pavements; and D 93–21, determination of the flashpoint of hazardous liquids under laboratory conditions.

For ASTM standards cited in this paper, the year the standard was first approved is indicated after the number, e.g. D 9–07 was established in 1907. The letter "T" after the year indicates a "tentative" standard; a designation no longer used.

8.3 Standards

8.3.1 China

Jiali Jiang

The work in the Standardisation Administration of the People's Republic of China (SAC) is divided into Technical Committees (TC). In total, the SAC standardisation body comprises approximately 579 Technical Committees, 7 of which are directly dedicated to wood/bamboo and wood/bamboo products (Table 8.1).

SAC is authorised by the State Council to undertake unified management, supervision and overall coordination of standardisation work in China. The main duties of SAC are:

1. to deliver national standards plans, approve and publish national standards, deliberate and release important documents such as standardisation policies, administrative rules and programs,
2. to inform the public of mandatory national standards,
3. to coordinate, guide and supervise standards work concerning industry, local regions, organisations and enterprises,

Table 8.1 Standards developed by SAC Technical committees dedicated to the forestry/wood/ bamboo and rattan chain

SAC Technical Committee	Topic / field of work	Number of published standards
SAC /TC 41	Wood, round and sawn timber, structural timber	203
SAC /TC 66	Woodworking machines	29
SAC /TC 198	Wood-based panels	114
SAC /TC 263	Bamboo and rattan	21
SAC /TC 362	Forest engineering	2
SAC /TC 416	Forest biomass materials	13
SAC /TC 480	Furniture	44
	Total number of published standards:	**426**

4. to represent China in ISO, IEC and other international and regional standardisation organisations,
5. to sign and execute international standards cooperation agreements, and.
6. to undertake the daily work of the standardisation coordination mechanism under the State Council.

Mandatory national standards shall be developed to meet the technical requirements to ensure people's health and the security of their lives and property, safeguarding national and eco-environmental security, and meeting the basic need of economic and social management. Administrative departments under the State Council shall, according to their duties and responsibilities, propose mandatory national standards and organise drafts, solicit opinions and conduct technical reviews thereof. The administrative department in charge of standardisation under the State Council shall be responsible for approving proposals, and for the numbering and notification of mandatory national standards. The administrative department in charge of standardisation under the State Council shall assess whether a proposed mandatory national standards conform with the provisions of the State Council and grant approval for proposals which conform with the provisions. Departments in charge of standardisation under people's governments of provinces, autonomous regions and municipalities directly under the central government may submit recommendations for mandatory national standards to the administrative department in charge of standardisation under the State Council, and this administrative department shall make decisions in conjunction with other relevant administrative departments. Social organizations, enterprises, public institutions and citizens may submit recommendations proposing mandatory national standards to the administrative department in charge of standardisation under the State Council. If the department in charge of standardisation believes a recommendation should be approved, it shall make a decision in conjunction with other relevant administrative departments. Mandatory national standards shall be authorised for approval and publication by the State Council.

Voluntary national standards may be developed relating to technical requirements that are needed to serve basic and generic purposes, to support mandatory national standards or to play a leading role in relevant industries. Voluntary national standards shall be developed by the administrative department in charge of standardisation under the State Council.

Where, in the absence of voluntary national standards, unified national technical requirements are needed within a sector, sector standards may be developed. They shall be developed by relevant administrative departments under the State Council and submitted to the administrative department in charge of standardisation under the State Council for registration (Tables 8.2, 8.3, 8.4, 8.5, 8.6, 8.7 and 8.8).

Table 8.2 SAC /TC 41 Wood, round and sawn timber, structural timber

Committee ID	Name	International Class Number
GB/T 36,870–2018	Code names of the main commercial timbers	79.040
GB/T 36,408–2018	Laminated veneer lumber for timber structures	91.080.20
GB/T 15,036.1–2018	Solid wood flooring—Part 1: Technical requirements	79.080
GB/T 15,036.2–2018	Solid wood flooring—Part 2: Test methods	79.080
GB/T 11,717–2018	Pulp logs	79.040
GB/T 36,407–2018	Machine stress-graded lumber	79.040
GB/T 36,202–2018	Sawn-timber inspection terminology	79.040
GB/T 15,779–2017	Logs for peeled veneer	79.040
GB/T 20,445–2017	Planed sawn timber	79.080
GB/T 143–2017	Ripping logs	79.040
GB/T 155–2017	Defects in logs	79.040
GB/T 35,475–2017	Specifications for wood used in Hongmu products	79.080
GB/T 18,107–2017	Hongmu	79.040
GB/T 33,041–2016	Decay and termite hazard zone for above-ground applications of wood in China	79.020
GB/T 18,513–2001	Names of main Chinese imported woods	79.040
GB/T 16,734–1997	Names of main Chinese woods	65.020.01

Table 8.3 SAC /TC 66 Woodworking machines

Committee ID	Name	International Class Number
GB/T 18,264–2018	General acceptance rules for particleboard production	79.120
GB/T 18,514–2018	Safety codes for wood-based panel machinery	79.120.01
GB/T 5051–2017	The particle forming machine	79.120.99
GB/T 18,002–2011	General acceptance rules for medium-density fibreboard production	79.120.99
GB/T 26,912–2011	General acceptance rules for bamboo-wooden compound floor production	79.120.99
GB/T 14,713–2009	General specifications for veneer lathes	65.060.80
GB/T 18,515–2001	The veneer lathe safety for construction	79.120.01
GB/T 18,262–2000	General specifications for wood-based panel machinery	79.120.10
GB/T 18,263–2000	Wood-based panel machinery–hot presses terminology	79.120.10
GB/T 6197–2000	Veneer roller dryer	79.120.10
GB/T 6199–2000	Veneer net-belt dryer	79.120.10
GB/T 18,004–1999	General rules for roller sanders	79.120.01

Table 8.4 SAC /TC 198 Wood-based panels

Committee ID	Name	International Class Number
GB/T 37,005–2018	Painting of wood-based panels	79.060
GB/T 36,872–2018	Glued-laminated timber production	79.060.99
GB/T 18,259–2018	Terminology for wood-based panels and their surface decoration	79.060.01
GB/T 17,656–2018	Plywood for concrete shuttering	79.060.10
GB/T 14,732–2017	Wood adhesives: urea–formaldehyde, phenol–formaldehyde and melamine–formaldehyde resins	83.180
GB/T 14,074–2017	Test methods for wood adhesives and resins	83.180
GB/T 15,102–2017	Surface decorated fibreboard and particleboard with paper impregnated with thermosetting resins	79.060
GB 18,580–2017	Indoor decorating and refurbishing materials—Limit of formaldehyde emission for wood-based panels and finished products	79.060.01
GB/T 5849–2016	Blockboards and battenboards	79.060.10
GB/T 31,765–2015	High-density fibreboard	79.060.20
GB/T 4897–2015	Particleboard	79.060.20
GB/T 9846–2015	Plywood for general use	79.060.10
GB/T 18,103–2013	Engineered wood flooring	79.080
GB/T 26,899–2011	Structural glued-laminated timber	79.060.99
GB/T 11,718–2009	Medium-density fibreboard	79.060.20
GB/T 24,508–2009	Wood-plastic composite flooring	79.080
GB/T 17,657–2013	Test methods for evaluating the properties of wood-based panels and surface decorated wood-based panels	79.060.01
GB/T 21,723–2008	Wheat/rice-straw particleboard	79.060.20
GB/T 18,102–2007	Laminate flooring	79.080

8.3.2 Europe

Johannes Welling

Technical Committees in CEN deal with wood and wood products. The standardisation work in CEN is divided among approximately 440 Technical Committees, six of which are directly dedicated to wood and wood products and three of which work on standards which have a direct impact on the forestry/wood sector. In total, approximately 500 CEN standards so far have been issued (Table 8.9).

All TCs in CEN are reflected in the national standardisation bodies of the EU member states. This means that topics for CEN standards have to be proposed to the cooresponding CEN Technical Committee by one or more national standardisation bodies. The CEN TC decides whether or not the proposed topic is to be adopted and registered as a new work item (WI), for which, after adoption, a work plan has to be developed. The WI is then allocated to a working group (WG) in the TC. The WG develops a first draft for the new standard. When the draft is accepted by the WG it is submitted to the secretariat of the TC which submits the draft to the national committees of the European CEN member countries. Experts

Table 8.5 SAC /TC 263 Bamboo and rattan

Committee ID	Name	International Class Number
GB/T 36,394–2018	Terminology of bamboo products	01.040.65
GB/T 35,242–2017	Commercial seedling quality grading of major bamboo species	65.020.40
GB/T 20,240–2017	Glued-laminated bamboo flooring	79.060.99
GB/T 32,770–2016	Bamboo terminology	01.040.65
GB/T 30,364–2013	Bamboo-scrimber flooring	79.080
GB/T 28,669–2012	Bamboo charcoal for fuel	75.160
GB/T 27,649–2011	Bamboo and wood laminated-composite flooring	79.060.10
GB/T 27,650–2011	Technical regulations for harvesting and processing of rattan fruits and seeds	65.020.20
GB/T 26,913–2011	Bamboo charcoal	65.020
GB/T 15,780–1995	Testing methods for physical and mechanical properties of bamboo	65.020.40
GB/T 13,123–2003	Bamboo-mat plywood	79.060.10

Table 8.6 SAC /TC 362 Forest engineering

Committee ID	Name	International Class Number
GB/T 33,890–2017	Silviculture–Engineering implementation guide	65.020
GB/T 33,892–2017	Symbols for planning and design of timber logistics	65.020

Table 8.7 SAC /TC 416 Forest biomass materials

Committee ID	Name	International Class Number
GB/T 23,471–2018	Laminated impregnated paper-strawboard-composite flooring	79.060.99
GB/T 36,055–2018	Analysis of forestry biomass—Determination of moisture content	65.020.01
GB/T 36,056–2018	Analysis of forestry biomass—Determination of soluble saccharides content	79.010
GB/T 36,057–2018	Analysis of forestry biomass—Determination of ash content	79.010
GB/T 36,058–2018	Analysis of forestry biomass—Determination of insoluble saccharides content	65.020.01
GB/T 35,820–2018	Analysis of forestry biomass — Sampling methods	79.010
GB/T 35,821–2018	Determination of biomass content in biomass/plastic composites	79.010
GB/T 35,905–2018	Testing of biomass—Determination of total solids in biomass	65.020.01
GB/T 35,809–2018	Analysis of forestry biomass—Determination of protein content	79.010
GB/T 35,808–2018	Analysis of forestry biomass—Determination of cellulase activity	79.010
GB/T 35,814–2018	Analysis of forestry biomass—Sample preparation	79.010
GB/T 35,812–2018	Analysis of forestry biomass—Determination of insoluble solids following pretreatment	79.010
GB/T 35,811–2018	Analysis of forestry biomass—Determination of starch content	79.010

Table 8.8 SAC /TC 480 Furniture

Committee ID	Name	International Class Number
GB/T 37,646–2019	Terms and definitions of Chinese traditional furniture	97.140
GB/T 37,648–2019	Assessment indicator system for cleaner production—Wood furniture manufacture	13.020.01
GB/T 36,022–2018	Test methods for ammonia emission of wooden furniture	97.140
GB/T 36,021–2018	Assessment of the heavy metals Sb, As, Ba, Se and Cr (VI) in furniture	97.140
GB/T 35,607–2017	Green product assessment—Furniture	13.020.10
GB/T 3324–2017	General technical requirements for wooden furniture	97.140
GB/T 3326–2016	Furniture—Main sizes of tables, chairs and stools	97.140
GB/T 3328–2016	Furniture—Main sizes of beds	97.140
GB/T 3327–2016	Furniture—Main sizes of storage units	97.140
GB/T 32,437–2015	Test methods for harmful substances in furniture—General guidelines	97.140
GB/T 32,442–2015	Assembly requirements for knock-down furniture	97.140
GB/T 32,445–2015	Classification of furniture materials	97.140
GB/T 31,106–2014	Determination of volatile organic compounds in furniture	97.140

all over Europe then have the chance to comment on the draft version using a harmonised template. All comments are collected, analysed and decided on by the working group. When a final draft has been produced, it is submitted to the technical committee which decides whether to submit the draft to the voting process at the European level. Each national standardisation body submits one vote in the final voting process, and the votes are weighted according to the size of the voting nation (number of citizens). In the case of a positive vote the primary version of the CEN standard (normally in English) is translated into at least French, German and Italian. All national standardisation bodies which are members to CEN are obliged to adopt the CEN standard as a national standard. Each CEN standard is reviewed every five years. The national committees are questioned as for whether the standard is in use or should be withdrawn. In addition, the national experts can decide whether a standard shall stay in its present form or whether it should be revised. If it is to be revised, the committees are asked to nominate experts who can support the revision process.

The following paragraphs show the structure of the important, wood-related CEN Technical Committees. By visiting the CEN webpage https://standards.cen.eu and by using the search function a list of all standards which are related to a Technical Committee can be seen. The same procedure can be followed using the web-pages of the national standardisation bodies, e.g. DIN, AFNOR, NBN and many others (Tables 8.10, 8.11, 8.12, 8.13, 8.14 and 8.15).

Table 8.9 Numbers of standards developed by CEN technical committees dedicated to the forestry/wood chain

CEN Technical Committee	Field of work	Number of published standards
CEN/TC 38	Durability of wood and wood-based products	59
CEN/TC 112	Wood-based panels	63
CEN/TC 124	Timber structures	40
CEN/TC 142	Woodworking machines - Safety	34
CEN/TC 172	Pulp, paper and board	79
CEN/TC 175	Round and sawn timber	73
CEN/TC 207	Furniture	67
CEN/TC 335	Solid biofuels	42
CEN/TC 351	Construction products–Assessment of release of dangerous substances	24
	Total number of CEN standards elaborated until 2019:	**481**

Table 8.10 CEN/TC 38 Durability of wood and wood products

Committee ID	Name	Secretariat	DIN Mirror committee
CEN/TC 38	Durability of wood and wood-based products	AFNOR	NA 042–03-06 AA
CEN/TC 38/WG 21	Durability - Classification (Use classes-natural durability)	AFNOR	NA 042–03-06 AA
CEN/TC 38/WG 22	Performance - Assessment and specifications (treated wood - Wood preservatives)	UNI	NA 042–03-06 AA
CEN/TC 38/WG 23	Fungal testing (basidiomycetes-microfungi)	NBN	NA 042–03-06 AA
CEN/TC 38/WG 24	Insect testing - (beetles - termites)	DIN	NA 042–03-06 AA
CEN/TC 38/WG 25	External factors and preconditioning	DS	NA 042–03-06 AA
CEN/TC 38/WG 26	Physical/chemical factors (analytical methods)	SIS	NA 042–03-06 AA
CEN/TC 38/WG 27	Exposure aspects	AFNOR	NA 042–03-06 AA
CEN/TC 38/WG 28	Performance classification	BSI	NA 042–03-06 AA

Table 8.11 CEN/TC 112 Wood-based panels

Committee ID	Name	Secretariat	DIN Mirror committee
CEN/TC 112	Wood-based panels	DIN	NA 042–02-15 AA
CEN/TC 112/WG 2	Plywood	AFNOR	NA 042–02-15 AA
CEN/TC 112/WG 4	Test methods	DIN	NA 042–02-15 AA
CEN/TC 112/WG 5	Regulated dangerous substances	DIN	NA 042–02-15 AA
CEN/TC 112/WG 7	Semi-finished and finished products	AFNOR	NA 042–02-15 AA
CEN/TC 112/WG 8	Oriented strand boards (OSB)	DIN	NA 042–02-15 AA
CEN/TC 112/WG 9	Solid-wood panels	ASI	NA 042–02-15 AA
CEN/TC 112/WG 11	Particleboards and fibreboards	DIN	NA 042–02-15 AA

Table 8.12 CEN/TC 124 Timber structures

Committee ID	Name	Secretariat	DIN Mirror committee
CEN/TC 124	Timber structures	AFNOR	NA 005-04-01 AA
CEN/TC 124/WG 1	Test methods	SNV	NA 005-04-01 AA
CEN/TC 124/WG 2	Solid timber	AFNOR	NA 005-04-01 AA
CEN/TC 124/WG 3	Glued-laminated timber	ASI	NA 005-04-01 AA
CEN/TC 124/WG 4	Connectors	DS	NA 005-04-01 AA
CEN/TC 124/WG 5	Prefabricated wall, floor and roof elements	DIN	NA 005-04-01 AA
CEN/TC 124/WG 6	Wood poles	NBN	NA 005-04-07 AA

Table 8.13 CEN/TC 175 Round and sawn timber

Committee ID	Name	Secretariat	DIN Mirror committee
CEN/TC 175	Round and sawn timber	AFNOR	NA 042-01-02 AA
CEN/TC 175/WG 1	General matters, definitions, measurement methods	UNI	NA 042-01-02 AA
CEN/TC 175/WG 2	Sawn timber	AFNOR	NA 042-01-16 AA
CEN/TC 175/WG 4	Round timber	DIN	NA 042-01-15 AA
CEN/TC 175/WG 30	Specific user requirements - Consolidation	SIS	NA 042-01-02 AA
CEN/TC 175/WG 32	Specific user requirements - Timber in joinery	AFNOR	NA 042-01-16 AA
CEN/TC 175/WG 33	Specific user requirements - Timber in flooring	AFNOR	NA 042-01-17 AA
CEN/TC 175/WG 34	Specific user requirements - Timber in packaging and pallets		NA 042-01-16 AA
CEN/TC 175/WG 36	Specific user requirements - Other timber products		NA 042-01-16 AA
CEN/TC 175/WG 37	Specific user requirements - Timber in stairs	AFNOR	NA 042-01-17 AA
CEN/TC 175/WG 38	Specific user requirements - Timber in cladding and panelling	DIN	NA 042-01-16 AA
CEN/TC 175/WG 39	Specific user requirements - Fire retardant treated wood	AFNOR	NA 042-01-02 AA

Table 8.14 CEN/TC 193 Wood adhesives

Committee ID	Name	Secretariat	DIN Mirror committee
CEN/TC 193/SC 1	Adhesives for wood and derived timber products	DIN	NA 042-04-05 AA
CEN/TC 193/SC 1/WG 4	Tests methods, classification and performance requirements for adhesives other than phenolic and aminoplastic, for load- bearing timber structures	DIN	NA 042-04-05 AA
CEN/TC 193/SC 1/WG 6	Test methods and requirements for adhesives for glued-in rods and on-site repair of load-bearing structures	DIN	NA 042-04-05 AA
CEN/TC 193/SC 1/WG 8	Revision of EN 301 and EN 302	SN	NA 042-04-05 AA
CEN/TC 193/SC 1/WG 12	Adhesives for non-loadbearing timber structures	UNI	NA 042-04-05 AA
CEN/TC 193/SC 1/WG 13	Performance of wood adhesives at elevated temperatures - Test methods and evaluation	DIN	NA 042-04-05 AA
CEN/TC 193/SC 1/WG 14	Testing and requirements for hardwood bonding and chemically modified wood substrates	AFNOR	NA 042-04-05 AA

Table 8.15 CEN/TC 351 Construction Products - Assessment of release of dangerous substances

Committee ID	Name	Secretariat	DIN Mirror committee
CEN/TC 351	Construction Products - Assessment of release of dangerous substances	NEN	NA 005-53 FBR
CEN/TC 351/WG 1	Release from construction products into soil, ground water and surface water	DIN	NA 005-53-01 GA
CEN/TC 351/WG 2	Emissions from construction products into indoor air	AFNOR	NA 005-53-02 GA
CEN/TC 351/WG 3	Radiation from construction products	NEN	NA 005-53 FBR
CEN/TC 351/WG 4	Terminology	DIN	NA 005-53 FBR
CEN/TC 351/WG 5	Content and eluate analysis in construction products	NEN	NA 005-53 FBR

8.3.3 Russia

Galina Gorbacheva

Technical Committees of the Federal Technical Regulation and Metrology Agency (Rosstandart) deal with forest, wood and wood products. The standardisation work is divided into 437 Technical Committees (TC), 22 of which are related to forestry and wood sector (Tables 8.16 and 8.17).

Table 8.16 Technical Committees (TC) in Rosstandart

TC number	TC title	TC similar to ISO or CEN	No. of standards
078	Timber	ISO/TC 218	25
121	Wood-based panels	ISO/TC 89	36
135	Furniture	ISO/TC 136	101
144	Building materials and products	ISO/TC 2, TC 59/SC 8, TC 59/SC 16, TC 71, TC 74, TC 77, TC 77/WG 20, TC 86, TC 117, TC 131, TC 142, TC 152, TC 162, TC 163, TC 165, TC 167, TC 179, TC 189, TC 218, TC 219	12
157	Wood pulp. Paper, cardboard and products made of them	ISO/TC 6	27
158	Electrical insulating paper and cardboard		5
177	Pulp, paper, cardboard and industrial and technical materials for various purposes	ISO/TC 6/SC 2, TC 6/SC 5	21
195	Paints and varnishes	ISO/ TC 35	43
202	Chemical methods of plant protection and methods of their analysis	ISO/ TC 47/SC 1	1
214	Protection of products and materials against corrosion, ageing and biological damage	ISO/TC 107, TC 156	19
223	Package	ISO/TC 122, TC 51	39
230	Plastics, polymeric materials, methods of testing	ISO/TC 61/SC 1, TC 61/SC 2, TC 61/SC 4, TC 61/SC 5, TC 61/SC 6, TC 61/SC 9, TC 61/SC 11	159
272	Pulp and paper equipment		
290	Coatings, paint and varnish	ISO/ TC 35/SC 12, TC 35/SC 14	26
326	Biotechnology	ISO/TC 47, TC 134, TC 147, TC 276	71
359	Seeds and planting material		73
361	Forest machines	ISO/ TC 23/SC 2, TC 23/SC 14, TC 23/SC 15,TC 23/SC 18	31
366	Green environmental technologies and green innovative products	ISO/TC 205	11
385	Paint materials on natural binders. Solvents. Desiccants. Auxiliary materials. Container, packaging, labeling and transportation of paints and varnishes	ISO/TC 35	13
441	Nanotechnology	ISO/TC 229	61
449	Forestry and related activities	ISO/TC 23/SC 3, TC 23/SC 14, TC 23/SC 15, TC 23/SC 17	61
497	Composites, structures and products	ISO/TC 61, TC 61/SC 13, TC 79, TC 206, TC 59, TC 163, TC 221, TC 71, TC 77, TC 89, TC 92, TC 98, TC 138/SC 6, TC 162 CEN/TC 33, CEN/TC 88, CEN/TC 99, CEN/TC 104, CEN/TC 125, CEN/TC 127, CEN/TC 134, CEN/TC 165, CEN/TC 166	301

Table 8.17 List of important GOST standards related to the wood sector

No	Name
GOST 11,603–73	Wood. Determination of residual stress
GOST 16,483.0–89 (CMEA ST6470-88)	Wood. General requirements for physical and mechanical tests
GOST 16,483.1–84 (CMEA ST388-76)	Wood. Determination of density
GOST 16,483.2–70	Wood. Determination of conventional ultimate strength in local compression perpendicular to the fibre orientation
GOST 16,483.3–84 (CMEA ST390-76)	Wood. Determination of static bending strength
GOST 16,483.4–73	Wood. Determination of impact bending strength
GOST 16,483.5–73	Wood. Determination of ultimate shearing strength parallel to the fibre orientation
GOST 16,483.6–80	Wood. Selection of sample trees and logs for the determination of physical and mechanical properties of wood plantations
GOST 16,483.7–71	Wood. Determination of moisture content
GOST 16,483.9–73 (CMEA ST1142-78)	Wood. Determination of modulus of elasticity in static bending
GOST 16,483.10–73	Wood. Determination of ultimate strength in compression parallel the fibre orientation
GOST 16,483.11–72	Wood. Determination of conventional ultimate strength in compression perpendicular to the fibre orientation
GOST 16,483.12–72	Wood. Determination of ultimate strength in shearing perpendicular to the fibre orientation
GOST 16,483.13–72	Wood. Determination of ultimate strength in cutting perpendicular to the fibre orientation
GOST 16,483.14–72	Wood. Methods for swelling testing
GOST 16,483.15–72	Wood. Determination of water-tightness
GOST 16,483.16–81 (CMEA ST 2367–80)	Wood. Determination of impact hardness
GOST 16,483.17–81	Wood. Determination of static hardness
GOST 16,483.18–72	Wood. Determination of the number of growth rings per 1 cm and the content of latewood in a growth ring
GOST 16,483.19–72	Wood. Determination of moisture absorption
GOST 16,483.20–72	Wood. Determination of water absorption
GOST 16,483.21–72	Wood. Determination of physico-mechanical characteristics after technological treatment
GOST 16,483.22–81	Wood. Determination of cleavage strength
GOST 16,483.23–73	Wood. Determination of ultimate strength in tension along the fibre orientation
GOST 16,483.24–73	Wood. Determination of modulus of elasticity in compression along fibres
GOST 16,483.25–73	Wood. Determination of modulus of elasticity in compression across fibres
GOST 16,483.26–73	Wood. Determination of modulus of elasticity in tension along fibres
GOST 16,483.27–73	Wood. Determination of modulus of elasticity in tension across the fibre orientation
GOST 16,483.28–73	Wood. Determination of ultimate tensile strength across the fibre orientation
GOST 16,483.29–73	Wood. Determination of factors of cross-sectional deformation
GOST 16,483.30–73	Wood. Determination of modulus of shear
GOST 16,483.31–74 (CMEA ST1143-78)	Wood. Determination of modulus of elasticity and shear and decrement vibrations
GOST 16,483.32–77	Wood. Determination of ultimate hygroscopicity
GOST 16,483.33–77	Wood. Determination of resistivity to nail and woodscrew withdrawal
GOST 16,483.34–77	Wood. Determination of gas permeability
GOST 16,483.35–88	Wood. Determination of swelling
GOST 16,483.37–88	Wood. Determining of shrinkage
GOST 16,483.39–81 (CMEA ST2363-80)	Wood. Determination of wear-proofness index
GOST 18,407–73	Wood. Evaluation of electrical resistances at A.C. voltage
GOST 18,408–73	Wood. Evaluation of electrical resistances at D.C. voltage
GOST 6782.1–75	Coniferous sawn timber. Shrinkage values
GOST 6782.2–75	Broad-leaved sawn timber. Shrinkage values
GOST 16,588–91 (ISO 4470–81)	Sawn products and wooden details. Determination of moisture content
GOST 21,554.1–81 (CMEA ST2790-80)	Sawn timber and component manufactures. Determination of ultimate strength in compression parallel to the fibre

(continued)

Table 8.17 (continued)

No	Name
GOST 21,554.2–81 (CMEA ST2811-80)	Sawn timber and blanks. Determination of ultimate strength in static bending
GOST 21,554.3–82	Sawn timber and blanks. Determination of strength in bending, tension and compression
GOST 21,554.4–78	Sawn timber and wooden component. Determination of the ultimate strength in compression parallel to the the fibre orientation
GOST 21,554.5–78 (CMEA ST2024-79)	Sawn timber and wooden component. Determination of the ultimate strength in tension parallel to the fibre orientation
GOST 21,554.6–78	Sawn timber and wooden component. Determination of the ultimate strength in shearing parallel to the fibre orientation
GOST 21,554.7–78	Sawn timber and wooden component. Determination of ultimate strength in bearing perpendicular the fibre orientation

Table 8.18 List of standards overseen by ASTM Committee D-7: Wood

Fire Performance of Wood	
D2898 - 10(2017)	Standard Practice for Accelerated Weathering of Fire-Retardant-Treated Wood for Fire Testing
D3201 / D3201M-13	Standard Test Method for Hygroscopic Properties of Fire-Retardant Wood and Wood-Based Products
D5516 - 18	Standard Test Method for Evaluating the Flexural Properties of Fire-Retardant Treated Softwood Plywood Exposed to Elevated Temperatures
D5664 - 17	Standard Test Method for Evaluating the Effects of Fire-Retardant Treatments and Elevated Temperatures on Strength Properties of Fire-Retardant Treated Lumber
D6305 - 08(2015)e1	Standard Practice for Calculating Bending Strength Design Adjustment Factors for Fire-Retardant-Treated Plywood Roof Sheathing
D6513-14	Standard Practice for Calculating the Superimposed Load on Wood-frame Walls for Standard Fire-Resistance Tests
D6841 - 16	Standard Practice for Calculating Design Value Treatment Adjustment Factors for Fire-Retardant-Treated Lumber
D7746-11(2016)	Standard Practice for Calculating the Superimposed Load on Wood-frame Floor-Ceiling Assemblies for Standard Fire-Resistance Tests
D7857 - 16	Standard Test Method for Evaluating the Flexural Properties and Internal Bond Strength of Fire-Retarded Mat-Formed Wood Structural Composite Panels Exposed to Elevated Temperatures
	D8223 - 19 Standard Practice for Evaluation of Fire-Retardant Treated Laminated Veneer Lumber
	E69 - 15 Standard Test Method for Combustible Properties of Treated Wood by the Fire-Tube Apparatus
Forests	
D5536 -	17 Standard Practice for Sampling Forest Trees for Determination of Clear Wood Properties
D7480 - 08(2014)	Standard Guide for Evaluating the Attributes of a Forest Management Plan
D7612 - 10(2015)	Standard Practice for Categorizing Wood and Wood-Based Products According to Their Fiber Sources
Fundamental Test Methods and Properties	
D143 - 14	Standard Test Methods for Small Clear Specimens of Timber
D198 - 15	Standard Test Methods of Static Tests of Lumber in Structural Sizes
D1102 - 84(2013)	Standard Test Method for Ash in Wood
D1105 - 96(2013)	Standard Test Method for Preparation of Extractive-Free Wood
D1106 - 96(2013)	Standard Test Method for Acid-Insoluble Lignin in Wood
D1107 - 96(2013)	Standard Test Method for Ethanol-Toluene Solubility of Wood
D1108 - 96(2013)	Standard Test Method for Dichloromethane Solubles in Wood
D1109 - 84(2013)	Standard Test Method for 1% Sodium Hydroxide Solubility of Wood
D1110 - 84(2013)d	Standard Test Methods for Water Solubility of Woo
D1166 - 84(2013)	Standard Test Method for Methoxyl Groups in Wood and Related Materials
D1666 - 17	Standard Test Methods for Conducting Machining Tests of Wood and Wood-Base Panel Materials
D1762 - 84(2013)	Standard Test Method for Chemical Analysis of Wood Charcoal
D2394 - 17	Standard Test Methods for Simulated Service Testing of Wood and Wood-Based Finish Flooring
D2395 - 17	Standard Test Methods for Density and Specific Gravity (Relative Density) of Wood and Wood-Based Materials
D4442 - 16	Standard Test Methods for Direct Moisture Content Measurement of Wood and Wood-Based Materials
D4444 - 13(2018)	Standard Test Method for Laboratory Standardization and Calibration of Hand-Held Moisture Meters
D4445-10(2015)	Standard Test Method for Fungicides for Controlling Sapstain and Mold on Unseasoned Lumber (Laboratory Method)
D4761 - 19	Standard Test Methods for Mechanical Properties of Lumber and Wood-Based Structural Materials
D4933 - 16	Standard Guide for Moisture Conditioning of Wood and Wood-Based Materials
D6782 - 13	Standard Test Methods for Standardization and Calibration of In-Line Dry Lumber Moisture Meters
D6815 - 09(2015)	Standard Specification for Evaluation of Duration of Load and Creep Effects of Wood and Wood-Based Products
D6874 - 12	Standard Test Methods for Nondestructive Evaluation of Wood-Based Flexural Members Using Transverse Vibration
D6958 - 03(2014)	Standard Test Methods for Evaluating Side-Bonding Potential of Wood Coatings
D7438 - 13	Standard Practice for Field Calibration and Application of Hand-Held Moisture Meters
D7469 - 16	Standard Test Methods for End-Joints in Structural Wood Products

(continued)

Table 8.18 (continued)

Fire Performance of Wood	
Lumber and Engineered Wood Products	
D245 - 06(2011)	Standard Practice for Establishing Structural Grades and Related Allowable Properties for Visually Graded Lumber
D1990 - 16	Standard Practice for Establishing Allowable Properties for Visually-Graded Dimension Lumber from In-Grade Tests of Full-Size Specimens
D2555 - 17a	Standard Practice for Establishing Clear Wood Strength Values
D2915 - 17	Standard Practice for Sampling and Data-Analysis for Structural Wood and Wood-Based Products
D3737 - 18e1	Standard Practice for Establishing Allowable Properties for Structural Glued Laminated Timber (Glulam)
D3957 - 09(2015)	Standard Practices for Establishing Stress Grades for Structural Members Used in Log Buildings
D5055 - 19	Standard Specification for Establishing and Monitoring Structural Capacities of Prefabricated Wood I-Joists
D5456 - 19	Standard Specification for Evaluation of Structural Composite Lumber Products
D5457 - 19b	Standard Specification for Computing Reference Resistance of Wood-Based Materials and Structural Connections for Load and Resistance Factor Design
D6570 - 18a	Standard Practice for Assigning Allowable Properties for Mechanically Graded Lumber
D7031 - 11(2019)	Standard Guide for Evaluating Mechanical and Physical Properties of Wood-Plastic Composite Products
D7032 - 17	Standard Specification for Establishing Performance Ratings for Wood-Plastic Composite and Plastic Lumber Deck Boards, Stair Treads, Guards, and Handrails
D7199 - 07(2012)	Standard Practice for Establishing Characteristic Values for Reinforced Glued Laminated Timber (Glulam) Beams Using Mechanics-Based Models
D7341 - 14	Standard Practice for Establishing Characteristic Values for Flexural Properties of Structural Glued Laminated Timber by Full-Scale Testing
D7374 - 08(2015)	Standard Practice for Evaluating Elevated Temperature Performance of Adhesives Used in End-Jointed Lumber
D7470 - 08(2015)	Standard Practice for Evaluating Elevated Temperature Performance of End-Jointed Lumber Studs
D7612-10(2015)	Standard Practice for Categorizing Wood and Wood-Based Products According to Their Fiber Sources
D7770-12	Standard Test Method for Collection of Volatile Organic Compounds Emitted During Simulated Manufacturing of Engineered Wood Products Via a Sealed Caul Plate Method
E1333-14	Standard Test Method for Determining Formaldehyde Concentrations in Air and Emission Rates from Wood Products Using a Large Chamber
Panel Products	
D1037 - 12	Standard Test Methods for Evaluating Properties of Wood-Base Fiber and Particle Panel Materials
D1038 - 19	Standard Terminology Relating to Veneer, Plywood, and Wood Structural Panels
D1554 - 10(2016)	Standard Terminology Relating to Wood-Base Fiber and Particle Panel Materials
D1666-17	Standard Test Methods for Conducting Machining Tests of Wood and Wood-Base Panel Materials
D2718 - 18	Standard Test Methods for Structural Panels in Planar Shear (Rolling Shear)
D2719 - 19	Standard Test Methods for Wood Structural Panels in Shear Through-the-Thickness
D3043 - 17	Standard Test Methods for Structural Panels in Flexure
D3044 - 16	Standard Test Method for Shear Modulus of Wood-Based Structural Panels
D3499 - 19	Standard Test Method for Toughness of Wood Structural Panels
D3500 - 14	Standard Test Methods for Structural Panels in Tension
D3501 - 05a(2018)	Standard Test Methods for Wood-Based Structural Panels in Compression
D5582 - 14	Standard Test Method for Determining Formaldehyde Levels from Wood Products Using a Desiccator
D5651 - 13	Standard Test Method for Surface Bond Strength of Wood-Base Fiber and Particle Panel Materials
D6007 - 14	Standard Test Method for Determining Formaldehyde Concentrations in Air from Wood Products Using a Small-Scale Chamber
D6643 - 01(2016)	Standard Test Method for Testing Wood-Base Panel Corner Impact Resistance
D7033 - 14	Standard Practice for Establishing Design Capacities for Oriented Strand Board (OSB) Wood-Based Structural-Use Panels
D7433 - 19	Standard Test Method for Measuring Surface Water Absorption of Overlaid Wood-Based Panels
D7519 - 11	Standard Test Method for Internal Bond Strength and Thickness Swell of Cellulosic-Based Fiber and Particle Panels After Repeated Wetting
D7770 - 12	Standard Test Method for Collection of Volatile Organic Compounds Emitted During Simulated Manufacturing of Engineered Wood Products Via a Sealed Caul Plate Method

(continued)

Table 8.18 (continued)

Fire Performance of Wood	
D7989-18	Standard Practice for Demonstrating Equivalent In-Plane Lateral Seismic Performance to Wood-Frame Shear Walls Sheathed with Wood Structural Panels
E1333 - 14	Standard Test Method for Determining Formaldehyde Concentrations in Air and Emission Rates from Wood Products Using a Large Chamber
Pole and Pile Products	
D25 - 12(2017)	Standard Specification for Round Timber Piles
D1036 - 99(2017)	Standard Test Methods of Static Tests of Wood Poles
D2899 - 12(2017)	Standard Practice for Establishing Allowable Stresses for Round Timber Piles
D3200 - 74(2017)	Standard Specification and Test Method for Establishing Recommended Design Stresses for Round Timber Construction Poles
D7381 - 07(2013)	Standard Practice for Establishing Allowable Stresses for Round Timbers for Piles from Tests of Full-Size Material
Terminology and Editorial	
D9 - 12	Standard Terminology Relating to Wood and Wood-Based Products
D1165 - 18	Standard Nomenclature of Commercial Hardwoods and Softwoods
Treatments for Wood Products	
D3345 - 17	Standard Test Method for Laboratory Evaluation of Solid Wood for Resistance to Termites
D4445 - 10(2015)	Standard Test Method for Fungicides for Controlling Sapstain and Mold on Unseasoned Lumber (Laboratory Method)
Wood Assemblies	
D1761 - 12	Standard Test Methods for Mechanical Fasteners in Wood
D5124 - 96(2018)	Standard Practice for Testing and Use of a Random Number Generator in Lumber and Wood Products Simulation
D5652 - 15	Standard Test Methods for Single-Bolt Connections in Wood and Wood-Based Products
D5764 - 97a(2018)	Standard Test Method for Evaluating Dowel-Bearing Strength of Wood and Wood-Based Products
D5933 - 19	Standard Specification for 258-in. and 4-in. Diameter Metal Shear Plates for Use in Wood Constructions
D6513 - 14	Standard Practice for Calculating the Superimposed Load on Wood-frame Walls for Standard Fire-Resistance Tests
D6555 - 17	Standard Guide for Evaluating System Effects in Repetitive-Member Wood Assemblies
D7147 - 11(2018)	Standard Specification for Testing and Establishing Allowable Loads of Joist Hangers
D7374-08(2015)	Standard Practice for Evaluating Elevated Temperature Performance of Adhesives Used in End-Jointed Lumber
D7469-16 Standard	Test Methods for End-Joints in Structural Wood Products
D7470-08(2015)	Standard Practice for Evaluating Elevated Temperature Performance of End-Jointed Lumber Studs
D7672 - 14e1	Standard Specification for Evaluating Structural Capacities of Rim Board Products and Assemblies
D7746 - 11(2016)	Standard Practice for Calculating the Superimposed Load on Wood-frame Floor-Ceiling Assemblies for Standard Fire-Resistance Tests
D7989 - 18	Standard Practice for Demonstrating Equivalent In-Plane Lateral Seismic Performance to Wood-Frame Shear Walls Sheathed with Wood Structural Panels
D8023 - 17	Standard Specification for Round Wood Dowels (Pegs) for Use in Wood Construction

Table 8.19 ASTM Committee D–7: Wood subcommittees and sections

Subcommittees	Sections
D07.01 Fundamental test methods and properties	
	D07.01.01 Physical test methods and properties
	D07.01.03 Mechanical test methods and properties
	D07.01.04 Methods and evaluation of duration of load and creep effects
D07.02 Lumber and engineered wood products	
	D07.02.01 Solid sawn timber
	D07.02.02 Glulam
	D07.02.03 Structural composite lumber
	D07.02.04 Prefabricated wood I-joists
	D07.02.06 Reliability-based design
	D07.02.07 Wood plastic composites
D07.03 Panel products	
	D07.03.01 Structural use panels
	D07.03.02 Fiber and particle panel materials
D07.04 Pole and pile products	
D07.05 Wood assemblies	
	D07.05.02 Wood connections
	D07.05.03 Repetitive member assemblies
	D07.05.04 Floor system assemblies
D07.06 Treatments for wood products	
	D07.06.01 Specifications and chemical analysis of wood preservatives
	D07.06.02 Product standards and treatment methods
	D07.06.03 Evaluation of preservatives

8.3.4 United States

Robert J. Ross, Christopher Adam Senalik

The most important wood-related ASTM standards are listed in Table 8.19.

References

Berg WG, Cummin JH, Forman J et al (1907) Strength of bridge and trestle timbers. In: Proceedings of the annual convention of American International Association of Railway Superintendents of Bridges and Buildings, New Orleans LA, October 16, 1895, p 14–63
Fulweiler WH (1955) 50 years of committee D-7 on wood. ASTM Bulletin. No. 206. American Society for Testing and Materials, West Conshohocken
Green DW, Ethington RL. King EG et al (2004) ASTM Committee D-7: Wood: promoting safety and standardization for 100 years. For Prod J 54(9):8–18

Dr. Johannes Welling was the director (acting) of the Thünen-Institute of Wood Research until 2020. Since his retirement, he has worked as a freelance expert in the woodworking sector. He has more than 40 years of experience in process engineering. His scientific career started with optimising the drying quality of sawn timber. Later, his activities widened to include wood composites, life-cycle assessment, and many other topics. As a founding member of the European Drying Group (EDG), he initiated various COST Actions. He coordinated and participated in several EU projects. For over 20 years, he was an active member of various CEN and DIN standardisation committees. He travelled the world as an expert on behalf of GTZ/GIZ and other development aid organisations and acted as principal scientist in international cooperation projects.

Prof.Dr. Galina Gorbacheva is an Associate Professor of the Department of Wood Science and Technology, Mytishchi Branch, Bauman Moscow State Technical University. She received her PhD in wood physics from the Moscow State Forest University (MSFU) in 2004. Dr. Gorbacheva is Scientific Secretary of Regional Coordinating Council on Wood Science, and a leading researcher in projects funded by the Ministry of Education and Science of the Russian Federation, a Member of TC 144 "Building materials and constructions", TC 078 "Timber" at the Federal Agency on Technical Regulating and Metrology of the Russian Federation, and a consultant for Federal Customs Service of Russia, the federal and regional bodies of state of the Russian Federation, the objects of cultural and historical heritage.

Prof.Dr. Robert J. Ross has served as Project Leader for several research work units at the USDA Forest Service, Forest Products Laboratory since 1988. He currently serves as Assistant Director (Acting) at FPL. His current research focuses on developing and using nondestructive evaluation technologies for various wood products and structures. He has written or co-authored more than 300 technical reports and articles about nondestructive evaluation and jointly holds two dozen U.S. and foreign patents. He served as team leader and editor of the Centennial Edition of Wood Handbook – Wood as an Engineering Material, FPL's flagship publication. He holds adjunct faculty appointments at Beijing Forestry University and Mississippi State University and is a research professor at Michigan Technological University.

Prof.Dr. Jiali Jiang has been a researcher for wood physics research work units at the Chinese Academy of Forestry since 2009. She currently serves as an associate professor at RIWI. Her current research focuses are the characterisation and evaluation of wood viscoelastic properties and the wood–water relationship. Dr. Jiang received the IUFRO Outstanding Doctoral Research Award (ODRA) in 2010. She holds a BSc (Wood Science and Engineering) from Beijing Forestry University and an MSc and PhD in Wood Science and Technology from the Chinese Academy of Forestry.

Res.Eng. Christopher Adam Senalik is a Research General Engineer with the USDA Forest Service, Forest Products Laboratory (FPL). In 2005, he left a career in heavy truck collision forensic engineering to pursue his doctorate in nondestructive evaluation of wood products and utility poles. While at FPL, his focus has been on the nondestructive testing of wood, the analysis of timber bridge structures and abutments, and moisture cycling effects on the strength of wood composites. He is currently involved in evaluating the use of ground-penetrating radar for inspecting wood structures. Adam is a technical advisor to the American Lumber Standards Committee (ALSC) Board of Review.

Correction to: Wood Material and Processing Data

Peter Niemz, Alfred Teischinger and Dick Sandberg

Correction to:

P. Niemz et al. (eds.), *Wood Material and Processing Data*, https://doi.org/10.1007/978-3-031-38329-8

The author's name, Prof. Dr. Galina Gorbacheva, was incorrectly typed in Chapters 2 and 8. This error has now been corrected. The correction has no impact on the content of either chapter.

The updated versions of these chapters can be found at
https://doi.org/10.1007/978-3-031-38329-8_2
https://doi.org/10.1007/978-3-031-38329-8_8

P. Niemz et al. (eds.), *Wood Material and Processing Data*, https://doi.org/10.1007/978-3-031 38329-8_9

Appendix

Common and Binomial Names of Tree Species

Sorted by Common Name

Common/vernacular name	Binomial name
abachi	*Triplochiton scleroxylon* K.Schum
abura	*Mitragyna ciliata* Aubrév. & Pellegr
acacia	*Acacia* spp.
acacia hybrid	*Acacia mangium* × *auriculiformis*
afara	*Terminalia superba* Engl. & Diels
African baobab	*Adansonia digitata* L
African coralwood	*Pterocarpus soyauxii* Taub
African limba wood	*Terminalia superba* Engl. & Diels
African locust bean	*Parkia biglobosa* (Jacq.) R. Br. ex G. Don
African oil palm (monocot tree)	*Elaeis guineensis* Jacq
African padauk	*Pterocarpus soyauxii* Taub
African whitewood	*Triplochiton scleroxylon* K. Schum
Alaska cedar	*Cupressus nootkatensis* D. Don
Alaska yellow cedar	*Cupressus nootkatensis* D. Don
albizzia	A genus of more than 160 species
albizzia	*Paraserienthes falcata* Becker
alder	*Alnus* spp.
Aleppo pine	*Pinus halepensis* Mill
alligatorwood	*Liquidambar styraciflua* L
alpine ash	*Eucalyptus delegatensis* R. T. Baker
American ash	*Fraxinus americana* L
American aspen	*Populus tremuloides* Michx
American basswood	*Tilia americana* L
American beech	*Fagus grandifolia* Ehrh
American chestnut	*Castanea dentata* (Marsh.) Borkh
American elm	*Ulmus americana* L
American hackberry	*Celtis occidentalis* L
American hornbeam	*Carpinus caroliniana* Walter

P. Niemz et al. (eds.), *Wood Material and Processing Data*, https://doi.org/10.1007/978-3-031-38329-8

Common/vernacular name	Binomial name
American larch	*Larix laricina* (Du Roi) K. Koch
American linden	*Tilia americana* L
American pepper	*Schinus molle* L
American planetree	*Platanus occidentalis* L
American storax	*Liquidambar styraciflua* L
American sweetgum	*Liquidambar styraciflua* L
American sycamore	*Platanus occidentalis* L
American tulip tree	*Liriodendron tulipifera* L
American white birch	*Betula papyrifera* Marshall
anacahuita	*Schinus molle* L
Andaman padauk	*Pterocarpus dalbergioides* DC
Andaman redwood	*Pterocarpus dalbergioides* DC
Angelique	*Dicorynia guianensis* Amshoff
Angelique batard	*Dicorynia guianensis* Amshoff
Angelique gris	*Dicorynia guianensis* Amshoff
apple	*Malus* spp.
apple	*Malus domestica* Borkh
Arolla pine	*Pinus cembra* L
aromatic cedar	*Juniperus virginiana* L
ash	*Fraxinus* spp.
Asian white birch	*Betula platyphylla*Sukaczev
aspen	*Populus tremula* L
Australian silver oak	*Grevillea robusta* A.Cunn. ex R. Br
Austrian pine	*Pinus nigra* Arnold
Austrian stone pine	*Pinus cembra* L
Ayan	*Distemonanthus benthamianus* Baill
ayous	*Triplochiton scleroxylon* K. Schum
azobé	*Lophira alata* Banks ex Gaertn
bahia	*Mitragyna ciliata* Aubrév. & Pellegr
balau	*Shorea* spp.
bald cypress	*Taxodium distichum* (L.) Rich
baldcypress	*Taxodium distichum* (L.) Rich
Balkan beech	*Fagus moesiaca* C
balsa	*Ochroma pyramidale* (Cav. ex Lam.) Urb
balsam fir	*Abies balsamea* (L.) Mill
balsam poplar	*Populus balsamifera L*
Baltic pine	*Pinus sylvestris* L
bam	Populus balsamifera L
bamboo (monocot tree)	*Dendrocalamus asper* (Schult.) Backer
bamboo (monocot tree)	*Dendrocalamus* spp.
bamtree	Populus balsamifera L

Common/vernacular name	Binomial name
bangkirai	*Shorea* spp.
Barakaroeballi	*Dicorynia guianensis* Amshoff
Basralokus	*Dicorynia guianensis* Amshoff
Baton Rouge	*Juniperus virginiana* L
bead-tree	*Melia azedarach* L
beaverwood	*Celtis occidentalis* L
beechwood	*Gmelina arborea* Roxb
bilsted	*Liquidambar styraciflua* L
birch	*Betula* spp.
bitter pecan	*Carya aquatica* (F.Michx.) Nutt
bitternut hickory	*Carya cordiformis* (Wangenh.) K. Koch
black alder	*Alnus glutinosa* L
black ash	*Fraxinus nigra* Marshall
black birch	*Betula lenta* L
black cherry	*Prunus serotina* Ehrh
black gum	*Nyssa sylvatica* Marshall
black larch	*Larix laricina* (Du Roi) K. Koch
black locust	*Robinia pseudoacacia* L
black poplar	*Populus nigra var. pyramidalis* L. (*Populus nigra* var. *Betulifolia*)
Black Sea fir	*Abies bornmuelleriana* Mattf. (*Abies nordmanniana* var. *equi-trojani* (Asch. & Sint. ex Boiss.) Coode & Cullen)
black spruce	*Picea mariana* (Mill.) Britton, Sterns & Poggenburg
black tupelo	*Nyssa sylvatica* Marshall
black walnut	*Juglans nigra* L
black wattle	*Acacia mangium* Willd
blackjack pine	*Pinus ponderosa* Douglas ex C.Lawson
blue gum	*Eucalyptus globulus* Labill
boabab	*Adansonia digitata* L
Bornmüller's fir	*Abies bornmuelleriana* Mattf. (*Abies nordmanniana* var. *equi-trojani* (Asch. & Sint. ex Boiss.) Coode & Cullen)
broom hickory	*Carya glabra* (Mill) Sweet
brown top	*Eucalyptus obliqua* L'Hér
brown top stringbark	*Eucalyptus obliqua* L'Hér
brutia pine	*Pinus brutia* Ten
bull pine	*Pinus ponderosa* Douglas ex C. Lawson
bullnut	*Carya tomentosa* Sarg
buna	*Fagus crenata* Blume
burflower-tree	*Anthocephalus chinensis* (Lamk) A. Rich. ex Walp. and *Anthocephalus macrophyllus* (Roxb.) Havil *Anthocephalus cadamba* Miq.
Burmese padauk	*Pterocarpus macrocarpus* Kurz
Burmese teak	*Tectona grandis* L. f
buttonwood	*Platanus occidentalis* L

Common/vernacular name	Binomial name
cadamba	*Anthocephalus chinensis* (Lamk) A. Rich. ex Walp. and *Anthocephalus macrophyllus* (Roxb.) Havil
Calabrian pine	*Pinus brutia* Ten
California black oak	*Quercus kelloggii* Newberry
California incense-cedar	*Calocedrus decurrens* (Torr.) Florin
California mountain pine	*Pinus monticola* Douglas ex D. Don
California pepper tree	*Schinus molle* L
California redwood	*Sequoia sempervirens* (D. Don) Endl
Canadian hemlock	*Tsuga canadensis* (L.) Carrière
canoe birch	*Betula papyrifera* Marshall
Cape lilac	*Melia azedarach* L
Caribbean pine	*Pinus caribaea* Morelet
Carolina cedar	*Juniperus virginiana* L
Carpathian walnut	*Juglans regia* L
Cathay poplar	*Populus cathayana* Rehder
cativo	*Prioria copaifera* Griseb
Caucasian alder	*Alnus subcordata* C.A. Mey
Caucasian fir	*Abies bornmuelleriana* Mattf. (*Abies nordmanniana* var. *equi-trojani* (Asch. & Sint. ex Boiss.) Coode & Cullen)
cedar	*Pinaceae* spp.
cedar elm	*Ulmus crassifolia* Nutt
Central Province teak (CP teak)	*Tectona grandis* L. f
cheesewood	*Endospermum malaccense* L. S. Sm
cherry alder	*Syzygium luehmannii* (F. Muell.) L. A. S. Johnson
cherry birch	*Betula lenta* L
cherry satinash	*Syzygium luehmannii* (F. Muell.) L. A. S. Johnson
chinaberry tree	*Melia azedarach* L
China-fir	*Cunninghamia konishi*i Hayata
Chinese fir	*Cunninghamia konishi*i Hayata
Chinese white poplar	*Populus tomentosa* Carr
Christmas tree	*Abies procera* Rehder
clove Lilli Pilli	*Syzygium luehmannii* (F. Muell.) L. A. S. Johnson
cluster pine	*Pinus pinaster* Aiton
coast Douglas-fir	*Pseudotsuga menziesii* var. menziesii
coast pignut hickory	*Carya glabra* Miller
coast redwood	*Sequoia sempervirens* (D. Don) Endl
coast spruce	*Picea sitchensis* (Bong.) Carr
coastal redwood	*Sequoia sempervirens* (D. Don) Endl
colonial pine	*Araucaria cunninghamii* Mudie
Columbian pine	*Pseudotsuga menziesii* (Mirbel) Franco
common alder	*Alnus glutinosa* L
common ash	*Fraxinus excelsior* L

Common/vernacular name	Binomial name
common aspen	*Populus tremula* L
common beech	*Fagus sylvatica* L
common bur-flower tree	*Anthocephalus chinensis* (Lamk) A. Rich. ex Walp. and *Anthocephalus macrophyllus* (Roxb.) Havil
common hackberry	*Celtis occidentalis* L
common hornbeam	*Carpinus betuls* L
common juniper	*Juniperus communis* L
common larch	*Larix decidua* Mill
common oak	*Quercus robur* L
common walnut	*Juglans regia* L
contorta pine	*Pinus contorta* Douglas
cork elm	*Ulmus thomasii* Sarg
corkwood	*Ochroma pyramidale* (Cav. ex Lam.) Urb
Cornish oak	*Quercus petraea* (Matt.) Lieb
Corsican pine	*Pinus nigra* var. maritima
cottongum	*Nyssa aquatica* L
cottonwood	*Populus nigra* L. (section *Aigeiros*)
cream of tartar tree	*Adansonia digitata* L
creek cherry	*Syzygium luehmannii* (F. Muell.) L. A. S. Johnson
creek maple	*Acer saccharinum* L
Cunninghamia	*Cunninghamia konishii* Hayata
cypress	*Cupressus* spp.
dark-bark spruce	*Picea jezoensis* (Siebold & Zucc.) Carr
dead finish	*Eucalyptus cloeziana* F. Muell
djarraly	*Eucalyptus marginata* Donn ex Sm
dodongba	*Parkia biglobosa* (Jacq.) R. Br. ex G. Don
Dorrigo pine	*Araucaria cunninghamii* Mudie
Douglas fir	*Pseudotsuga menziesii* (Mirbel) Franco
down tree	*Ochroma pyramidale* (Cav. ex Lam.) Urb
downy birch	*Betula pubescens* Ehrh
durmast oak	*Quercus petraea* (Matt.) Lieb
East Asian white birch	*Betula pendula* Roth
East Indian mahogany	*Pterocarpus dalbergioides* DC
East Mediterranean pine	*Pinus brutia* Ten
eastern American black walnut	*Juglans nigra* L
eastern arborvitae	*Thuja occidentalis* L
eastern balsam-poplar	*Populus balsamifera* L
eastern hemlock	*Tsuga canadensis* (L.) Carrière
eastern hemlock-spruce	*Tsuga canadensis* (L.) Carrière
eastern juniper	*Juniperus virginiana* L
eastern larch	*Larix laricina* (Du Roi) K. Koch
eastern red cedar	*Juniperus virginiana* L

Common/vernacular name	Binomial name
eastern redcedar	*Juniperus virginiana* L
eastern white cedar	*Thuja occidentalis* L
eastern white pine	*Pinus strobus* L
ekki	*Lophira alata* Banks ex Gaertn
elm	*Ulmus* sp.
Engelmann spruce	*Picea engelmannii* Parry ex Engelm
English oak	*Quercus robur* L
English walnut	*Juglans regia* L
escobilla	*Schinus molle* L
eucalyptus	*Eucalyptus* spp.
Eurasian aspen	*Populus tremula* L
European alder	*Alnus glutinosa* L
European ash	*Fraxinus excelsior* L
European aspen	*Populus tremula* L
European beech	*Fagus sylvatica* L
European hornbeam	*Carpinus betuls* L
European larch	*Larix decidua* Mill
European oak	*Quercus robur* L
European silver fir	*Abies alba* Mill
European spruce	*Picea abies* (L.) Karst
European white birch	*Betula pubescens* Ehrh
ezo spruce	*Picea jezoensis* (Siebold & Zucc.) Carr
ezomatsu	*Picea jezoensis* (Siebold & Zucc.) Carr
false acacia	*Robinia pseudoacacia* L
false pepper	*Schinus molle* L
fiddletree	*Liriodendron tulipifera* L
fiddletree	*Liriodendron tulipifera* L
filipinus pine	*Pinus ponderosa* Douglas ex C.Lawson
fir	*Abies alba* Mill
flooded gum	*Eucalyptus grandis* W. Hill ex Maiden
forest mangrove	*Acacia mangium* Willd
frake	*Terminalia superba* Engl. & Diels
Gabon	*Aucoumea klaineana* Pierre
Gaboon	*Aucoumea klaineana* Pierre
giant arborvitae	*Thuja plicata* Donn ex D. Don
giant bamboo (monocot tree)	*Dendrocalamus barbatus*
giant cedar	*Thuja plicata* Donn ex D.Don
giant fir	*Abies grandis* (Douglas ex D. Don) Lindley
gmelina	*Gmelina arborea* Roxb
golden aspen	*Populus tremuloides* Michx
golden birch	*Betula alleghaniensis* Britt

Common/vernacular name	Binomial name
goomar teak	*Gmelina arborea* Roxb
grand fir	*Abies grandis* (Douglas ex D. Don) Lindley
great silver fir	*Abies grandis* (Douglas ex D. Don) Lindley
green ash	*Fraxinus pennsylvanica* Marshall
grey alder	*Alnus incana* (L.) Moench
gulf cypress	*Taxodium distichum* (L.) Rich
Gympie messmate	*Eucalyptus cloeziana* F. Muell
hackberry	*Celtis occidentalis* L
hackmatack	*Larix laricina* (Du Roi) K. Koch
hackmatack	*Populus balsamifera L*
hairy birch	*Betula pubescens* Ehrh
hazel pine	*Liquidambar styraciflua* L
hazelnut	*Corylus* spp.
hickory wattle	*Acacia mangium* Willd
hinoki	*Chamaecyparis obtuse* (Siebold & Zucc.) Endl
hinoki cypress	*Chamaecyparis obtuse* (Siebold & Zucc.) Endl
hognut	*Carya tomentosa* Sarg
hoop pine	*Araucaria cunninghamii* Mudie
hybrid poplar	*Populus*ssp.
incense cedar	*Calocedrus decurrens* (Torr.) Florin
Indian lilac	*Melia azedarach* L
insignis pine	*Pinus radiata* D. Don
ipil	*Intsia* spp. (*I. bijuga, I. palembanica*)
iru	*Parkia biglobosa* (Jacq.) R. Br. ex G. Don
Italian cypress	*Cupressus sempervirens* L
Japanese beech	*Fagus crenata* Blume
Japanese cedar	*Cryptomeria japonica* (L.f.) D.Don
Japanese cypress	*Chamaecyparis obtuse* (Siebold & Zucc.) Endl
Japanese pine	*Pinus densiflora* Siebold & Zucc
Japanese red pine	*Pinus densiflora* Siebold & Zucc
Japanese redwood	*Cryptomeria japonica* (L.f.) D.Don
Japanese sugi	*Cryptomeria japonica* (L.f.) D.Don
Japanese white birch	*Betula platyphylla*Sukaczev
jarrah	*Eucalyptus marginata* Donn ex Sm
jezo spruce	*Picea jezoensis* (Siebold & Zucc.) Carr
Johnstone river teak	*Intsia bijuga* (Colebr.) Kuntze
kadam	*Anthocephalus chinensis* (Lamk) A. Rich. ex Walp. and *Anthocephalus macrophyllus* (Roxb.) Havil
Kashmir tree	*Gmelina arborea* Roxb
Kazdagi fir	*Abies bornmuelleriana* Mattf. (*Abies nordmanniana* var. *equi-trojani* (Asch. & Sint. ex Boiss.) Coode & Cullen)

Common/vernacular name	Binomial name
kelempayan	*Anthocephalus chinensis* (Lamk) A. Rich. ex Walp. and *Anthocephalus macrophyllus* (Roxb.) Havil
Kellogg oak	*Quercus kelloggii* Newberry
Korean red pine	*Pinus densiflora* Siebold & Zucc
korina	*Terminalia superba* Engl. & Diels
Kwila	*Intsia* spp. (*I. bijuga, I. palembanica*)
laran	*Anthocephalus chinensis* (Lamk) A. Rich. ex Walp. and *Anthocephalus macrophyllus* (Roxb.) Havil
large-leafed lime	*Tilia platyphyllos* Scop. (*Tilia grandifolia* Ehrh.)
large-leafed linden	*Tilia platyphyllos* Scop. (*Tilia grandifolia* Ehrh.)
large maple	*Acer saccharinum* L
large tupelo	*Nyssa aquatica* L
lauan	*Shorea* spp.
lawaan	*Shorea* spp.
Leichhardt pine	*Anthocephalus chinensis* (Lamk) A. Rich. ex Walp. and *Anthocephalus macrophyllus* (Roxb.) Havil
lemon tree	*Citrus limon* (L.) Osbeck
limba	*Terminalia superba* Engl. & Diels
linden	*Tilia* spp.
littleleaf linden	*Tilia cordata* Mill
loblolly pine	*Pinus taeda* L
lodgepole pine	*Pinus contorta* Douglas
longleaf pine	*Pinus palustris* Mill
lowland white fir	*Abies grandis* (Douglas ex D. Don) Lindley
luan	*Shorea* spp.
macaw-fat (monocot tree)	*Elaeis guineensis* Jacq
Madeira walnut	*Juglans regia* L
magnolia	*Magnolia* spp.
mahogany birch	*Betula lenta* L
Malay beechwood	*Gmelina arborea* Roxb
mangium	*Acacia mangium* Willd
manna gum	*Eucalyptus viminalis* Labill
Mantsurian poplar	*Populus cathayana* Rehder
maple	*Acer* spp.
maritime pine	*Pinus pinaster* Aiton
Masson's pine	*Pinus massoniana* Lamb
M'Boy	*Mitragyna ciliata* Aubrév. & Pellegr
Mediterranean cypress	*Cupressus sempervirens* L
meranti	*Shorea* spp.
merbau	*Intsia* spp. (*I. bijuga, I. palembanica*)
messmate	*Eucalyptus obliqua* L'Hér
messmate stringybark	*Eucalyptus obliqua* L'Hér

Common/vernacular name	Binomial name
Mexican Douglas fir	*Pseudotsuga menziesii* var. Lindleyana (Roezl) Carrière
mindi	*Melia azedarach* L
mockernut	*Carya tomentosa* Sarg
mockernut hickory	*Carya tomentosa* Sarg
molle del Peru	*Schinus molle* L
monkey-bread tree	*Adansonia digitata* L
Monterey pine	*Pinus radiata* D. Don
moor birch	*Betula pubescens* Ehrh
Moreton bay pine	*Araucaria cunninghamii* Mudie
mountain ash	*Sorbus aucuparia* L
mountain aspen	*Populus tremuloides* Michx
mountain spruce	*Picea engelmannii* Parry ex Engelm
movingui	*Distemonanthus benthamianus* Baill
Nagpur teak	*Tectona grandis* L. f
narrow-leaved ash	*Fraxinus angustifolia* Vahl
néré	*Parkia biglobosa* (Jacq.) R. Br. ex G. Don
netetou	*Parkia biglobosa* (Jacq.) R. Br. ex G. Don
nettletree	*Celtis occidentalis* L
Nigerian Satinwood	*Distemonanthus benthamianus* Baill
noble fir	*Abies procera* Rehder
Nootka cedar	*Cupressus nootkatensis* D. Don
North American beech	*Fagus grandifolia* Ehrh
northern hackberry	*Celtis occidentalis* L
northern red oak	*Quercus rubra* L
northern white pine	*Pinus strobus* L
northern white-cedar	*Thuja occidentalis* L
Norway maple	*Acer platanoides* L
Norway pine	*Pinus resinosa* Sol. ex Aiton
Norway spruce	*Picea abies* (L.) Karst
oak	*Quercus* spp.
obeche	*Triplochiton scleroxylon* K. Schum
occidental plane	*Platanus occidentalis* L
oil palm	*Elaeis guineensis* Jacq
oleaster-leafed pear	*Pyrus elaeagnifolia* Pall
Olga bay larch	*Larix olgensis* A.Henry
Olgan larch	*Larix olgensis* A.Henry
Oregon fir	*Abies grandis* (Douglas ex D. Don) Lindley
Oregon pine	*Pseudotsuga menziesii* (Mirbel) Franco
Oriental beech	*Fagus orientalis* Lipsky
osiers	*Salix* spp.
Pacific red cedar	*Thuja plicata* Donn ex D. Don

Common/vernacular name	Binomial name
Pacific silver fir	*Abies amabilis* Douglas ex J. Forbes
Pacific teak	*Intsia bijuga* (Colebr.) Kuntze
padauk	*Pterocarpus* spp.
paper birch	*Betula papyrifera* Marshall
Pará rubber tree	*Hevea brasiliensis* Müll. Arg
paulownia	*Paulownia elongata* S. Y. Hu
pedunculate oak	*Quercus robur* L
pencil cedar	*Juniperus virginiana* L
pencil pine	*Cupressus sempervirens* L
pepper tree	*Schinus molle* L
peppercorn tree	*Schinus molle* L
pepperina	*Schinus molle* L
Persian cypress	*Cupressus sempervirens* L
Persian lilac	*Melia azedarach* L
Persian walnut	*Juglans regia* L
Peruvian pepper	*Schinus molle* L
Peruvian peppertree	*Schinus molle* L
Philippine mahogany	*Shorea* spp.
pignut	*Carya glabra* Miller
pignut hickory	*Carya glabra* Miller
pine	*Pinus* spp.
pirul Peruvian mastic	*Schinus molle* L
pitchpine	*Pinus palustris* Mill
Polish larch	*Larix decidua var. polonica* (Racib. ex Wóycicki)
Pondcypress	*Taxodium distichum* (L.) Rich
ponderosa pine	*Pinus ponderosa* Douglas ex C. Lawson
poplar	*Populus* spp.
Port Orford-cedar	*Chamaecyparis lawsoniana* (A. Murray) Parl
Pride of India	*Melia azedarach* L
pruche du Canada	*Tsuga canadensis* (L.) Carrière
quaking aspen	*Populus tremula* L
quaking aspen	*Populus tremuloides* Michx
Queensland pine	*Araucaria cunninghamii* Mudie
radiata pine	*Pinus radiata* D. Don
red alder	*Alnus rubra* Bong
red ash	*Fraxinus pennsylvanica* Marshall
red cedar	*Juniperus virginiana* L
red cypress	*Taxodium distichum* (L.) Rich
red fir	*Abies procera* Rehder
red grandis	*Eucalyptus grandis* W. Hill ex Maiden
red hickory	*Carya ovalis* (Wangenh.) Sarg

Common/vernacular name	Binomial name
red ironwood tree	*Lophira alata* Banks ex Gaertn
red juniper	*Juniperus virginiana* L
red larch	*Larix laricina* (Du Roi) K. Koch
red oak	*Quercus rubra* L
red pine	*Pinus resinosa* Sol. ex Aiton
red savin	*Juniperus virginiana* L
red stringybark	*Eucalyptus macrorhyncha* F. Muell. ex Benth
red-bud maple	*Acer trautvetteri* Medw
redgum	*Liquidambar styraciflua* L
redwood	*Sequoia sempervirens* (D. Don) Endl
ribbon gum	*Eucalyptus viminalis* Labill
riberry	*Syzygium luehmannii* (F. Muell.) L. A. S. Johnson
Richmond river pine	*Araucaria cunninghamii* Mudie
Riga pine	*Pinus sylvestris* L
river red gum	*Eucalyptus camaldulensis* Dehnh
rock elm	*Ulmus thomasii* Sarg
rock maple	*Acer saccharum* Marshall
Rocky Mountain Douglas-fir	*Pseudotsuga menziesii* var. Glauca (Mayr) Franco
rose gum	*Eucalyptus grandis* W. Hill ex Maiden
rowan	*Sorbus aucuparia* L
rubberwood	*Hevea brasiliensis* Müll. Arg
Russian larch	*Larix sibirica* Ledeb
sallows	*Salix* spp.
sambawawa	*Triplochiton scleroxylon* K. Schum
sand hickory	*Carya pallida* (Ashe) Engelm. & Graebn
sapele	*Entandrophragma cylindricum* Harms
sapelli	*Entandrophragma cylindricum* Harms
satin-walnut	*Liquidambar styraciflua* L
Scots pine	*Pinus sylvestris* L
scrub cherry	*Syzygium luehmannii* (F. Muell.) L. A. S. Johnson
scrub mahogany	*Intsia bijuga* (Colebr.) Kuntze
seraya	*Shorea* spp.
seringueira	*Hevea brasiliensis* Müll. Arg
sesendok	*Endospermum malaccense* L. S.Sm
sessile oak	*Quercus petraea* (Matt.) Lieb
sharinga tree	*Hevea brasiliensis* Müll. Arg
shinglewood	*Thuja plicata* Donn ex D. Don
shining gum	*Eucalyptus nitens* (H. Deane & Maiden) Maiden
shore pine	*Pinus contorta* Douglas
shortleaf pine (Southern yellow pine)	*Pinus echinata* Mill
Siberian larch	*Larix sibirica* Ledeb

Common/vernacular name	Binomial name
Siberian silver birch	*Betula platyphylla* Sukaczev
Siebold's beech	*Fagus crenata* Blume
silk oak	*Grevillea robusta* A. Cunn. ex R. Br
silky oak	*Grevillea robusta* A. Cunn. ex R. Br
silver birch	*Betula pendula* Roth
silver fir	*Abies alba* Mill
silver maple	*Acer saccharinum* L
silver oak	*Grevillea robusta* A. Cunn. ex R. Br
silver pine	*Pinus monticola* Douglas ex D. Don
silver spruce	*Picea engelmannii* Parry ex Engelm
silvertop	*Eucalyptus nitens* (H. Deane & Maiden) Maiden
silverleaf maple	*Acer saccharinum* L
Sitka spruce	*Picea sitchensis* (Bong.) Carr
slash pine	*Pinus elliottii* Engelm
small-leaved Lilly Pilly	*Syzygium luehmannii* (F. Muell.) L. A. S. Johnson
small-leaved lime	*Tilia cordata* Mill
small-leaved linden	*Tilia cordata* Mill
small-leaved watergum	*Syzygium luehmannii* (F. Muell.) L. A. S. Johnson
smoothbark hickory	*Carya glabra* Miller
soft maple	*Acer saccharinum* L
soft pine	*Pinus strobus* L
sour gum	*Nyssa sylvatica* Marshall
southern blue gum	*Eucalyptus globulus* Labill
Southern cypress	*Taxodium distichum* (L.) Rich
southern red oak	*Quercus rubra* L
Southern silky oak	*Grevillea robusta* A.Cunn. ex R. Br
Southern yellow pine	A number of conifer species which tend to grow in similar plant communities in the USA and yield similar high-strength timber
Spanish oak	*Quercus falcata* Michx
speckled alder	*Alnus incana* (L.) Moench
spice birch	*Betula lenta* L
spotted gum	*Corymbia maculata* (Hook.) K. D. Hill & L. A. S.Johnson
spruce	*Picea* spp.
star-leaved gum	*Liquidambar styraciflua* L
stone pine	*Pinus cembra* L
stringybark	*Eucalyptus obliqua* L'Hér
sugar cone pine	*Pinus lambertiana* Douglas
sugar maple	*Acer saccharum* Marshall
sugar pine	*Pinus lambertiana* Douglas
sugarberry	*Celtis occidentalis* L
sugi	*Cryptomeria japonica* (L.f.) D. Don
sumbala	*Parkia biglobosa* (Jacq.) R. Br. ex G. Don

Common/vernacular name	Binomial name
swamp black-gum	*Nyssa biflora* Walter
swamp cypress	*Taxodium distichum* (L.) Rich
swamp hickory	*Carya glabra* Miller
swamp mahogany	*Eucalyptus robusta* Sm
swamp maple	*Acer saccharinum* L
swamp messmate	*Eucalyptus robusta* Sm
swamp tupelo	*Nyssa biflora* Walter
swan river mahogany	*Eucalyptus marginata* Donn ex Sm
sweet birch	*Betula lenta* L
sweet chestnut	*Castanea sativa* Mill
sweet pignut	*Carya glabra* Miller
sweet pignut hickory	*Carya ovalis* (Wangenh.) Sarg
sweetgum	*Liquidambar styraciflua* L
Swiss pine	*Pinus cembra* L
Swiss stone pine	*Pinus cembra* L
sycamore	*Acer pseudoplatanus* L
sycamore maple	*Acer pseudoplatanus* L
syringa berrytree	*Melia azedarach* L
tacamahac poplar	*Populus balsamifera L*
tacamahaca	*Populus balsamifera L*
tamarack	*Larix laricina* (Du Roi) K. Koch
tanbark-oak	*Notholithocarpus densiflorus* (Hook. & Arn.) Manos, Cannon & S. H. Oh
tanoak	*Notholithocarpus densiflorus* (Hook. & Arn.) Manos, Cannon & S. H. Oh
Tasmanian oak	*Eucalyptus obliqua* L'Hér
teak	*Tectona grandis* L. f
Texas cedar elm	*Ulmus crassifolia* Nutt
tidewater red cypress	*Taxodium distichum* (L.) Rich
tidewater spruce	*Picea sitchensis* (Bong.) Carr
trembling aspen	*Populus tremuloides* Michx
trembling poplar	*Populus tremuloides* Michx
Trojan fir	*Abies bornmuelleriana* Mattf. (*Abies nordmanniana* var. *equi-trojani* (Asch. & Sint. ex Boiss.) Coode & Cullen)
tuart	*Eucalyptus gomphocephala* DC
tulip poplar	*Liriodendron tulipifera* L
tulip tree	*Liriodendron tulipifera* L
tulipwood	*Liriodendron tulipifera* L
tupelo	*Nyssa* spp.
tupelo-gum	*Nyssa aquatica* L
Turkey fir	*Abies bornmuelleriana* Mattf. (*Abies nordmanniana* var. *equi-trojani* (Asch. & Sint. ex Boiss.) Coode & Cullen)

Common/vernacular name	Binomial name
Turkish filbert	*Corylus colurna* L
Turkish hazel	*Corylus colurna* L
Turkish pine	*Pinus brutia* Ten
Tuscan cypress	*Cupressus sempervirens* L
twisted pine	*Pinus contorta* Douglas
upside-down tree	*Adansonia digitata* L
Ussuri poplar	*Populus ussuriensis* Komarov
walnut	*Juglans* spp.
Vancouver fir	*Abies grandis* (Douglas ex D. Don) Lindley
warty birch	*Betula pendula* Roth
water beech	*Platanus occidentalis* L
water elm	*Ulmus americana* L
water hickory	*Carya aquatica* (F.Michx.) Nutt
water maple	*Acer saccharinum* L
water oak	*Quercus nigra* L
water tupelo	*Nyssa aquatica* L
water-gum	*Nyssa aquatica* L
watergum	*Syzygium luehmannii* (F. Muell.) L. A. S. Johnson
wawa	*Triplochiton scleroxylon* K. Schum
western arborvitae	*Thuja plicata* Donn ex D. Don
western hemlock	*Tsuga heterophylla* (Raf.) Sarg
western hemlock-spruce	*Tsuga heterophylla* (Raf.) Sarg
western larch	*Larix occidentalis* Nutt
western plane	*Platanus occidentalis* L
western red cedar	*Thuja plicata* Donn ex D. Don
western white fir	*Abies grandis* (Douglas ex D. Don) Lindley
Western white pine	*Pinus monticola* Douglas ex D. Don
western yellow-pine	*Pinus ponderosa* Douglas ex C. Lawson
Weymouth pine	*Pinus strobus* L
white ash	*Fraxinus americana* L
white birch	*Betula pubescens* Ehrh
white cedar	*Melia azedarach* L
white cypress	*Taxodium distichum* (L.) Rich
white elm	*Ulmus americana* L
white fir	*Abies concolor* (Gordon) Lindley ex Hildebrand
white gum	*Eucalyptus viminalis* Labill
white hickory	*Carya tomentosa* Sarg
white maple	*Acer saccharinum* L
white milkwood	*Endospermum malaccense* L. S. Sm
white oak	*Quercus alba* L
white pine	*Pinus strobus* L

Common/vernacular name	Binomial name
white poplar	*Populus tremuloides* Michx
white spruce	*Picea engelmannii* Parry ex Engelm
white teak	*Gmelina arborea* Roxb
white willow	*Salix alba* L
whiteheart hickory	*Carya tomentosa* Sarg
whitewood	*Endospermum malaccense* L. S. Sm
whitewood	*Liriodendron tulipifera* L
wild olive	*Nyssa aquatica* L
wild pear	*Pyrus* spp.
willow oak	*Quercus phellos* L
willows	*Salix* spp.
Virginian juniper	*Juniperus virginiana* L
wormwood	*Artemisia monosperma* L
yellow birch	*Betula alleghaniensis* Britt
yellow cedar	*Cupressus nootkatensis* D. Don
yellow cypress	*Taxodium distichum* (L.) Rich
yellow pine	A number of conifer species which tend to grow in similar plant communities in the USA and yield similar high-strength timber
yellow-poplar	*Liriodendron tulipifera* L
yemane	*Gmelina arborea* Roxb
yezo spruce	*Picea jezoensis* (Siebold & Zucc.) Carr

Sorted by Binomial Name

Binomial name	Common/vernacular name
Abies alba Mill	European silver fir, fir, silver fir
Abies amabilis Douglas ex J. Forbes	Pacific silver fir
Abies balsamea (L.) Mill	balsam fir
Abies bornmuelleriana Mattf	Black Sea fir, Bornmüller's fir, Caucasian fir, Kazdagi fir, Trojan fir, Turkey fir
Abies concolor (Gordon) Lindley ex Hildebrand	white fir
Abies grandis (Douglas ex D. Don) Lindley	giant fir, grand fir, great silver fir, lowland white fir, Oregon fir,Vancouver fir, western white fir
Abies nordmanniana var. *equi-trojani* (Asch. & Sint. ex Boiss.) Coode & Cullen)	Black Sea fir, Bornmüller's fir, Caucasian fir, Kazdagi fir, Trojan fir, Turkey fir
Abies procera Rehder	Christmas tree, noble fir, red fir
Acacia mangium Willd	black wattle, forest mangrove, hickory wattle, mangium
Acacia mangium × *auriculiformis*	acacia hybrid
Acer platanoides L	Norway maple
Acer pseudoplatanus L	sycamore, sycamore maple
Acer saccharinum L	creek maple, large maple, silver maple, silverleaf maple, soft maple, swamp maple, water maple, white maple

Binomial name	Common/vernacular name
Acer saccharum Marshall	rock maple, sugar maple
Acer trautvetteri Medw	red-bud maple
Adansonia digitata L	African baobab, boabab, cream of tartar tree, monkey-bread tree, upside-down tree
Alnus glutinosa L	European alder, common alder, black alder
Alnus incana (L.) Moench	grey alder, speckled alder
Alnus rubra Bong	red alder
Alnus subcordata C. A. Mey	Caucasian alder
Anthocephalus chinensis (Lamk) A. Rich. ex Walp. and *Anthocephalus macrophyllus* (Roxb.) Havil *Anthocephalus cadamba Miq.*	burflower-tree, cadamba, common bur-flower tree, kadam, kelempayan, laran, Leichhardt pine
Araucaria cunninghamii Mudie	colonial pine, Dorrigo pine, hoop pine, Moreton bay pine, Queensland pine, Richmond river pine
Artemisia monosperma L	wormwood
Aucoumea klaineana Pierre	Gabon, Gaboon
Betula alleghaniensis Britt	golden birch, yellow birch
Betula lenta L	black birch, cherry birch, mahogany birch, spice birch, sweet birch
Betula papyrifera Marshall	American white birch, canoe birch, paper birch
Betula pendula Roth	East Asian white birch, silver birch, warty birch
Betula platyphylla Sukaczev	Asian white birch, Japanese white birch, Siberian silver birch
Betula pubescens Ehrh	downy birch, European white birch, hairy birch, moor birch, white birch
Calocedrus decurrens (Torr.) Florin	California incense-cedar, incense cedar
Carpinus betuls L	common hornbeam, European hornbeam
Carpinus caroliniana Walter	American hornbeam
Carya aquatica (F.Michx.) Nutt	bitter pecan, water hickory
Carya cordiformis (Wangenh.) K.Koch	bitternut hickory
Carya glabra Miller	broom hickory, coast pignut hickory, pignut, pignut hickory, smoothbark hickory, swamp hickory, sweet pignut
Carya ovalis (Wangenh.) Sarg	red hickory, sweet pignut hickory
Carya pallida (Ashe) Engelm. & Graebn	sand hickory
Carya tomentosa Sarg	bullnut, hognut, mockernut, mockernut hickory, white hickory, whiteheart hickory
Castanea dentata (Marsh.) Borkh	American chestnut
Castanea sativa Mill	sweet chestnut
Celtis occidentalis L	American hackberry,beaverwood, common hackberry, hackberry, nettletree, northern hackberry, sugarberry
Chamaecyparis lawsoniana (A. Murray) Parl	Port Orford-cedar
Chamaecyparis obtuse (Siebold & Zucc.) Endl	hinoki, hinoki cypress, Japanese cypress
Citrus limon (L.) Osbeck	lemon tree
Corylus colurna L	Turkish filbert, Turkish hazel

Binomial name	Common/vernacular name
Corymbia maculata (Hook.) K. D. Hill & L. A. S.Johnson	spotted gum
Cryptomeria japonica (L. f.) D. Don	Japanese cedar, Japanese redwood, Japanese sugi, sugi
Cunninghamia konishii Hayata	China-fir, Chinese fir, Cunninghamia
Cupressus nootkatensis D. Don	Alaska cedar, Alaska yellow cedar, Nootka cedar, yellow cedar
Cupressus sempervirens L	Italian cypress, Mediterranean cypress, pencil pine, Persian cypress, Tuscan cypress
Dendrocalamus asper (Schult.) Backer	bamboo
Dendrocalamus barbatus	giant bamboo
Dendrocalamus spp.	bamboo
Dicorynia guianensis Amshoff	Angelique, Angelique batard, Angelique gris, Barakaroeballi, Basralokus
Distemonanthus benthamianus Baill	Ayan, movingui, Nigerian Satinwood
Elaeis guineensis Jacq. (monocot tree)	African oil palm, macaw-fat, oil palm
Endospermum malaccense L. S. Sm	cheesewood, sesendok, white milkwood, whitewood
Entandrophragma cylindricum Harms	sapele, sapelli
Eucalyptus camaldulensis Dehnh	river red gum
Eucalyptus cloeziana F. Muell	dead finish, Gympie messmate
Eucalyptus delegatensis R. T. Baker	alpine ash
Eucalyptus globulus Labill	blue gum, southern blue gum
Eucalyptus gomphocephala DC	tuart
Eucalyptus grandis W. Hill ex Maiden	flooded gum, red grandis, rose gum
Eucalyptus macrorhyncha F. Muell. ex Benth	red stringybark
Eucalyptus marginata Donn ex Sm	djarraly, jarrah, swan river mahogany
Eucalyptus nitens (H. Deane & Maiden) Maiden	shining gum, silvertop
Eucalyptus obliqua L'Hér	brown top, brown top stringbark, messmate, messmate stringybark, Tasmanian oak
Eucalyptus robusta Sm	swamp mahogany, swamp messmate
Eucalyptus viminalis Labill	manna gum, ribbon gum, white gum
Fagus crenata Blume	buna, Japanese beech, Siebold's beech
Fagus grandifolia Ehrh	American beech, North American beech
Fagus moesiaca C	Balkan beech
Fagus orientalis Lipsky	Oriental beech
Fagus sylvatica L	common beech, European beech
Fraxinus americana L	American ash, white ash
Fraxinus angustifolia Vahl	narrow-leaved ash
Fraxinus excelsior L	common ash, European ash
Fraxinus nigra Marshall	black ash
Fraxinus pennsylvanica Marshall	green ash, red ash
Gmelina arborea Roxb	beechwood, gmelina, goomar teak, Kashmir tree, Malay beechwood, white teak, yemane
Grevillea robusta A. Cunn. ex R. Br	Australian silver oak, silk oak, silky oak, silver oak, Southern silky oak

Binomial name	Common/vernacular name
Hevea brasiliensis Müll. Arg	Pará rubber tree, rubberwood, seringueira, sharinga tree
Intsia bijuga (Colebr.) Kuntze	Johnstone river teak, Pacific teak, scrub mahogany
Intsia spp. (*I. bijuga, I. palembanica*)	ipil, Kwila, merbau
Juglans nigra L	black walnut, eastern American black walnut
Juglans regia L	Carpathian walnut, common walnut, English walnut, Madeira walnut, Persian walnut
Juniperus communis L	common juniper
Juniperus virginiana L	aromatic cedar, Baton Rouge, Carolina cedar, eastern juniper, eastern red cedar, eastern redcedar, pencil cedar, red cedar, red juniper, red savin, Virginian juniper
Larix decidua Mill	European larch, common larch
Larix decidua var. polonica (Racib. ex Wóycicki)	Polish larch
Larix laricina (Du Roi) K. Koch	American larch, black larch, eastern larch, hackmatack, red larch, tamarack
Larix occidentalis Nutt	western larch
Larix olgensis A.Henry	Olga bay larch, Olgan larch
Larix sibirica Ledeb	Russian larch, Siberian larch
Liquidambar styraciflua L	alligatorwood, American storax, American sweetgum, bilsted, hazel pine, redgum, satin-walnut, star-leaved gum, sweetgum
Liriodendron tulipifera L	American tulip tree, fiddletree, fiddletree, tulip poplar, tulip tree, tulipwood, whitewood, yellow-poplar
Lophira alata Banks ex Gaertn	azobé, ekki, red ironwood tree,
Magnolia spp.	magnolia
Malus domestica Borkh	apple
Melia azedarach L	bead-tree, Cape lilac, chinaberry tree, Indian lilac, mindi, Persian lilac, Pride of India, syringa berrytree, white cedar
Mitragyna ciliata Aubrév. & Pellegr	abura, bahia, M'Boy
Notholithocarpus densiflorus (Hook. & Arn.) Manos, Cannon & S. H. Oh	tanbark-oak, tanoak
Nyssa aquatica L	cottongum, large tupelo, tupelo-gum, water tupelo, water-gum, wild olive
Nyssa biflora Walter	swamp black-gum, swamp tupelo
Nyssa sylvatica Marshall	black gum, black tupelo, sour gum
Ochroma pyramidale (Cav. ex Lam.) Urb	balsa, corkwood, down tree
Paraserienthes falcata Becker	albizzia
Parkia biglobosa (Jacq.) R. Br. ex G. Don	African locust bean, dodongba, iru, néré, netetou, sumbala
Paulownia elongata S. Y. Hu	paulownia
Picea abies (L.) Karst	European spruce, Norway spruce
Picea engelmannii Parry ex Engelm	Engelmann spruce, mountain spruce, silver spruce, white spruce
Picea jezoensis (Siebold & Zucc.) Carr	dark-bark spruce, ezo spruce, ezomatsu, jezo spruce, yezo spruce
Picea mariana (Mill.) Britton, Sterns & Poggenburg	black spruce

Binomial name	Common/vernacular name
Picea sitchensis (Bong.) Carr	coast spruce, Sitka spruce, tidewater spruce
Pinaceae spp.	cedar
Pinus brutia Ten	brutia pine, Calabrian pine, East Mediterranean pine, Turkish pine
Pinus caribaea Morelet	Caribbean pine
Pinus cembra L	Arolla pine, Austrian stone pine, stone pine, Swiss pine, Swiss stone pine
Pinus contorta Douglas	contorta pine, lodgepole pine, shore pine, twisted pine
Pinus densiflora Siebold & Zucc	Japanese pine, Japanese red pine, Korean red pine
Pinus echinata Mill	shortleaf pine (Southern yellow pine)
Pinus elliottii Engelm	slash pine
Pinus halepensis Mill	Aleppo pine
Pinus lambertiana Douglas	sugar cone pine, sugar pine
Pinus massoniana Lamb	Masson's pine
Pinus monticola Douglas ex D. Don	California mountain pine, silver pine, Western white pine
Pinus nigra Arnold	Austrian pine
Pinus nigra var. maritima	Corsican pine
Pinus palustris Mill	longleaf pine, pitchpine
Pinus pinaster Aiton	cluster pine, maritime pine
Pinus ponderosa Douglas ex C. Lawson	blackjack pine, bull pine, filipinus pine, ponderosa pine, western yellow-pine
Pinus radiata D. Don	insignis pine, Monterey pine, radiata pine
Pinus resinosa Sol. ex Aiton	Norway pine, red pine
Pinus strobus L	eastern white pine, northern white pine, soft pine, Weymouth pine, white pine
Pinus sylvestris L	Baltic pine, Riga pine, Scots pine
Pinus taeda L	loblolly pine
Platanus occidentalis L	American planetree, American sycamore, buttonwood, occidental plane, water beech, western plane
Populus balsamifera L	balsam poplar, bam, bamtree, eastern balsam-poplar, hack-matack, tacamahac poplar, tacamahaca
Populus cathayana Rehder	Cathay poplar, Mantsurian poplar
Populus nigra L. (section *Aigeiros*)	cottonwood
Populus nigra var. *Betulifolia*	black poplar
Populus nigra var. pyramidalis L	black poplar
Populus spp.	hybrid poplar
Populus tomentosa Carr	Chinese white poplar
Populus tremula L	aspen, common aspen, Eurasian aspen, European aspen, quaking aspen,
Populus tremuloides Michx	American aspen, golden aspen, mountain aspen, quaking aspen, trembling aspen, trembling poplar, white poplar
Populus ussuriensis Komarov	Ussuri poplar
Prioria copaifera Griseb	cativo

Binomial name	Common/vernacular name
Prunus serotina Ehrh	black cherry
Pseudotsuga menziesii (Mirbel) Franco	Columbian pine, Douglas fir, Douglas-fir, Oregon pine
Pseudotsuga menziesii var. Glauca (Mayr) Franco	Rocky Mountain Douglas-fir
Pseudotsuga menziesii var. Lindleyana (Roezl) Carrière	Mexican Douglas fir
Pseudotsuga menziesii var. menziesii	coast Douglas-fir
Pterocarpus dalbergioides DC	Andaman padauk, Andaman redwood, East Indian mahogany
Pterocarpus macrocarpus Kurz	Burmese padauk
Pterocarpus soyauxii Taub	African coralwood, African padauk
Pterocarpus spp.	padauk
Pyrus spp.	wild pear
Pyrus elaeagnifolia Pall	oleaster-leafed pear
Quercus alba L	white oak
Quercus falcata Michx	Spanish oak
Quercus kelloggii Newberry	California black oak, Kellogg oak
Quercus nigra L	water oak
Quercus petraea (Matt.) Lieb	Cornish oak, durmast oak, sessile oak
Quercus phellos L	willow oak
Quercus robur L	common oak, English oak, European oak, pedunculate oak
Quercus rubra L	northern red oak, red oak, southern red oak
Robinia pseudoacacia L	black locust, false acacia
Salix alba L	white willow
Salix spp.	osiers, sallows, willows
Schinus molle L	American pepper, anacahuita, California pepper tree, escobilla, false pepper, molle del Peru, pepper tree, peppercorn tree, pepperina, Peruvian pepper, Peruvian peppertree, pirul Peruvian mastic
Sequoia sempervirens (D. Don) Endl	California redwood, coast redwood, coastal redwood, redwood
Shorea spp.	balau, bangkirai, lauan, lawaan, luan, meranti, Philippine mahogany, seraya
Sorbus aucuparia L	mountain ash, rowan
Syzygium luehmannii (F. Muell.) L. A. S. Johnson	cherry alder, cherry satinash, clove Lilli Pilli, creek cherry, riberry, scrub cherry, small-leaved Lilly Pilly, small-leaved watergum, watergum
Taxodium distichum (L.) Rich	bald cypress, baldcypress, gulf cypress, Pondcypress, red cypress, Southern cypress, swamp cypress, tidewater red cypress, white cypress, yellow cypress
Tectona grandis L. f	Burmese teak, Central Province teak (CP teak), Nagpur teak, teak
Terminalia superba Engl. & Diels	afara, African limba wood, frake, korina, limba
Thuja occidentalis L	eastern arborvitae, eastern white cedar, northern white-cedar

Binomial name	Common/vernacular name
Thuja plicata Donn ex D. Don	giant arborvitae, giant cedar, Pacific red cedar, shingle-wood, western arborvitae, western red cedar
Tilia americana L	American basswood, American linden
Tilia cordata Mill	littleleaf linden, small-leaved lime, small-leaved linden
Tilia grandifolia Ehrh	large-leafed lime, large-leafed linden
Tilia platyphyllos Scop	large-leafed lime, large-leafed linden
Triplochiton scleroxylon K. Schum	abachi, African whitewood, ayous, obeche, sambawawa, wawa
Tsuga canadensis (L.) Carrière	Canadian hemlock, eastern hemlock, eastern hem-lock-spruce, pruche du Canada
Tsuga heterophylla (Raf.) Sarg	western hemlock, western hemlock-spruce
Ulmus americana L	American elm, water elm, white elm
Ulmus crassifolia Nutt	cedar elm, Texas cedar elm
Ulmus thomasii Sarg	cork elm, rock elm

Grouped by Genus

Group	Binomial name	Common/vernacular name
Acasia	***Acacia* spp.**	
	Acacia mangium Willd	black wattle, forest mangrove, hickory wattle, mangium
	Acacia mangium × auriculiformis	acacia hybrid
Alder	***Alnus* spp.**	
	Alnus glutinosa L	European alder, common alder, black alder
	Alnus incana (L.) Moench	grey alder, speckled alder
	Alnus rubra Bong	red alder
	Alnus subcordata C. A. Mey	Caucasian alder
Apple	***Malus* spp.**	
	Malus domestica Borkh	apple
Ash	***Fraxinus* spp.**	
	Fraxinus americana L	American ash, white ash
	Fraxinus angustifolia Vahl	narrow-leaved ash
	Fraxinus excelsior L	common ash, European ash
	Fraxinus nigra Marshall	black ash
	Fraxinus pennsylvanica Marshall	green ash, red ash
Beech	***Fagus* spp.**	
	Fagus crenata Blume	buna, Japanese beech, Siebold's beech
	Fagus grandifolia Ehrh	American beech, North American beech
	Fagus moesiaca C	Balkan beech
	Fagus orientalis Lipsky	Oriental beech
	Fagus sylvatica L	common beech, European beech
Birch	***Betula* spp.**	

Group	Binomial name	Common/vernacular name
	Betula alleghaniensis Britt	golden birch, yellow birch
	Betula lenta L	black birch, cherry birch, mahogany birch, spice birch, sweet birch
	Betula papyrifera Marshall	American white birch, canoe birch, paper birch
	Betula pendula Roth	East Asian white birch, silver birch, warty birch
	Betula platyphylla Sukaczev	Asian white birch, Japanese white birch, Siberian silver birch
	Betula pubescens Ehrh	downy birch, European white birch, hairy birch, moor birch, white birch
Chestnut	***Castanea* spp.**	
	Castanea dentata (Marsh.) Borkh	American chestnut
	Castanea sativa Mill	sweet chestnut
Cypress	***Cupressus* spp.**	
	Cupressus nootkatensis D. Don	Alaska cedar, Alaska yellow cedar, Nootka cedar, yellow cedar
	Cupressus sempervirens L	Italian cypress, Mediterranean cypress, pencil pine, Persian cypress, Tuscan cypress
Elm	***Ulmus* spp.**	
	Ulmus americana L	American elm, water elm, white elm
	Ulmus crassifolia Nutt	cedar elm, Texas cedar elm
	Ulmus thomasii Sarg	cork elm, rock elm
Eucalyptus	***Eucalyptus* spp.**	
	Eucalyptus camaldulensis Dehnh	river red gum
	Eucalyptus cloeziana F. Muell	dead finish, Gympie messmate
	Eucalyptus delegatensis R. T. Baker	alpine ash
	Eucalyptus globulus Labill	blue gum, southern blue gum
	Eucalyptus gomphocephala DC	tuart
	Eucalyptus grandis W. Hill ex Maiden	flooded gum, red grandis, rose gum
	Eucalyptus macrorhyncha F. Muell. ex Benth	red stringybark
	Eucalyptus marginata Donn ex Sm	djarraly, jarrah, swan river mahogany
	Eucalyptus nitens (H. Deane & Maiden) Maiden	shining gum, silvertop
	Eucalyptus obliqua L'Hér	brown top, brown top stringbark, messmate, messmate stringybark, Tasmanian oak
	Eucalyptus robusta Sm	swamp mahogany, swamp messmate
	Eucalyptus viminalis Labill	manna gum, ribbon gum, white gum
Fir	***Abies* spp.**	
	Abies alba Mill	European silver fir, fir, silver fir
	Abies amabilis Douglas ex J. Forbes	Pacific silver fir
	Abies balsamea (L.) Mill	balsam fir
	Abies bornmuelleriana Mattf	Black Sea fir, Bornmüller's fir, Caucasian fir, Kazdagi fir, Trojan fir, Turkey fir
	Abies concolor (Gordon) Lindley ex Hildebrand	white fir

Group	Binomial name	Common/vernacular name
	Abies grandis (Douglas ex D. Don) Lindley	giant fir, grand fir, great silver fir, lowland white fir, Oregon fir,Vancouver fir, western white fir
	Abies nordmanniana var. *equi-trojani* (Asch. & Sint. ex Boiss.) Coode & Cullen)	Black Sea fir, Bornmüller's fir, Caucasian fir, Kazdagi fir, Trojan fir, Turkey fir
	Abies procera Rehder	Christmas tree, noble fir, red fir
Hazelnut	***Corylus* spp.**	
	Corylus colurna L	Turkish filbert, Turkish hazel
Hemlock	***Tsuga* spp.**	
	Tsuga canadensis (L.) Carrière	Canadian hemlock, eastern hemlock, eastern hem-lock-spruce, pruche du Canada
	Tsuga heterophylla (Raf.) Sarg	western hemlock, western hemlock-spruce
	Tsuga heterophylla (Raf.) Sarg	
Hickory	***Carya* spp.**	
	Carya aquatica (F. Michx.) Nutt	bitter pecan, water hickory
	Carya cordiformis (Wangenh.) K.Koch	bitternut hickory
	Carya glabra (Mill.) Sweet	broom hickory, coast pignut hickory, pignut, pignut hickory, smoothbark hickory, swamp hickory, sweet pignut
	Carya ovalis (Wangenh.) Sarg	red hickory,sweet pignut hickory
	Carya pallida (Ashe) Engelm. & Graebn	sand hickory
	Carya tomentosa Sarg	bullnut, hognut, mockernut, mockernut hickory, white hickory, whiteheart hickory
Hornbeam	***Carpinus* spp.**	
	Carpinus betuls L	common hornbeam, European hornbeam
	Carpinus caroliniana Walter	American hornbeam
Juniper	***Juniperus* spp.**	
	Juniperus communis L	common juniper
	Juniperus virginiana L	aromatic cedar, Baton Rouge, Carolina cedar, eastern juniper, eastern red cedar, eastern redce-dar, pencil cedar, red cedar, red juniper, red savin, Virginian juniper
Larch	***Larix* spp.**	
	Larix decidua Mill	European larch, common larch
	Larix decidua var. polonica *(Racib. ex Wóycicki)*	Polish larch
	Larix laricina (Du Roi) K. Koch	American larch, black larch, eastern larch, hack-matack, red larch, tamarack
	Larix occidentalis Nutt	western larch
	Larix olgensis A. Henry	Olga bay larch, Olgan larch
	Larix sibirica Ledeb	Russian larch, Siberian larch
Linden	***Tilia* spp.**	
	Tilia americana L	American basswood, American linden
	Tilia cordata Mill	littleleaf linden, small-leaved lime, small-leaved linden
	Tilia grandifolia Ehrh	large-leafed lime, large-leafed linden

Group	Binomial name	Common/vernacular name
	Tilia platyphyllos Scop	large-leafed lime, large-leafed linden
Maple	***Acer* spp.**	
	Acer platanoides L	Norway maple
	Acer pseudoplatanus L	sycamore, sycamore maple
	Acer saccharinum L	creek maple, large maple, silver maple, silverleaf maple, soft maple, swamp maple, water maple, white maple
	Acer saccharum Marshall	rock maple, sugar maple
	Acer trautvetteri Medw	red-bud maple
Merbau	***Intsia* spp.**	
	Intsia spp. (*I. bijuga, I. palembanica*)	Ipil, Kwila, merbau
Oak	***Quercus* spp.**	
	Quercus alba L	white oak
	Quercus falcata Michx	Spanish oak
	Quercus kelloggii Newberry	California black oak, Kellogg oak
	Quercus nigra L	water oak
	Quercus petraea (Matt.) Lieb	Cornish oak, durmast oak, sessile oak
	Quercus phellos L	willow oak
	Quercus robur L	common oak, English oak, European oak, pedunculate oak
	Quercus rubra L	northern red oak, red oak, southern red oak
Pine	***Pinus* spp.**	
	A genus of more than 160 species	albizzia
	A number of conifer species which tend to grow in similar plant communities in the USA and yield similar high-strength timber	Southern yellow pine, yellow pine
	Pinus brutia Ten	brutia pine, Calabrian pine, East Mediterranean pine, Turkish pine
	Pinus caribaea Morelet	Caribbean pine
	Pinus cembra L	Arolla pine, Austrian stone pine, stone pine, Swiss pine, Swiss stone pine
	Pinus contorta Douglas	contorta pine, lodgepole pine, shore pine, twisted pine
	Pinus densiflora Siebold & Zucc	Japanese pine, Japanese red pine, Korean red pine
	Pinus echinata Mill	shortleaf pine (Southern yellow pine)
	Pinus elliottii Engelm	slash pine
	Pinus halepensis Mill	Aleppo pine
	Pinus lambertiana Douglas	sugar cone pine, sugar pine
	Pinus massoniana Lamb	Masson's pine
	Pinus monticola Douglas ex D. Don	California mountain pine, silver pine, Western white pine
	Pinus nigra Arnold	Austrian pine
	Pinus nigra var. maritima	Corsican pine

Group	Binomial name	Common/vernacular name
	Pinus palustris Mill	longleaf pine, pitchpine
	Pinus pinaster Aiton	cluster pine, maritime pine
	Pinus ponderosa Douglas ex C. Lawson	blackjack pine, bull pine, filipinus pine, ponderosa pine, western yellow-pine
	Pinus radiata D. Don	insignis pine, Monterey pine, radiata pine
	Pinus resinosa Sol. ex Aiton	Norway pine, red pine
	Pinus strobus L	eastern white pine, northern white pine, soft pine, Weymouth pine, white pine
	Pinus sylvestris L	Baltic pine, Riga pine, Scots pine
	Pinus taeda L	loblolly pine
Poplar	***Populus* spp.**	
	Populus balsamifera L	balsam poplar, bam, bamtree, eastern balsam-poplar, hackmatack, tacamahac poplar, tacamahaca
	Populus cathayana Rehder	Cathay poplar, Mantsurian poplar
	Populus nigra L. (section *Aigeiros*)	cottonwood
	Populus nigra var. *Betulifolia*	black poplar
	Populus nigra var. pyramidalis L	black poplar
	Populus tomentosa Carr	Chinese white poplar
	Populus tremula L	aspen, common aspen, Eurasian aspen, European aspen, quaking aspen,
	Populus tremuloides Michx	American aspen, golden aspen, mountain aspen, quaking aspen, trembling aspen, trembling poplar, white poplar
	Populus ussuriensis Komarov	Ussuri poplar
	Populus spp.	hybrid poplar
Rubberwood	***Hevea* spp.**	
	Hevea brasiliensis Müll. Arg	Pará rubber tree, rubberwood, seringueira, sharinga tree
Spruce	***Picea* spp.**	
	Picea abies (L.) Karst	European spruce, Norway spruce
	Picea engelmannii Parry ex Engelm	Engelmann spruce, mountain spruce, silver spruce, white spruce
	Picea jezoensis (Siebold & Zucc.) Carr	dark-bark spruce, ezo spruce, ezomatsu, jezo spruce, yezo spruce
	Picea mariana (Mill.) Britton, Sterns & Poggenburg	black spruce
	Picea sitchensis (Bong.) Carr	coast spruce, Sitka spruce, tidewater spruce
Thuja	***Thuja* spp.**	
	Thuja occidentalis L	eastern arborvitae, eastern white cedar, northern white-cedar
	Thuja plicata Donn ex D. Don	giant arborvitae, giant cedar, Pacific red cedar, shinglewood, western arborvitae, western red cedar
Tupelo	***Nyssa* spp.**	
	Nyssa aquatica L	cottongum, large tupelo, tupelo-gum, water tupelo, water-gum, wild olive

Group	Binomial name	Common/vernacular name
	Nyssa biflora Walter	swamp black-gum, swamp tupelo
Walnut	***Juglans* spp.**	
	Juglans nigra L	black walnut, eastern American black walnut
	Juglans regia L	Carpathian walnut, common walnut, English walnut, Madeira walnut, Persian walnut
Willow	***Salix* spp.**	
	Salix alba L	white willow
	Salix spp.	osiers, sallows, willows
Others		
	Adansonia digitata L	African baobab, boabab, cream of tartar tree, monkey-bread tree, upside-down tree
	Anthocephalus chinensis (Lamk) A. Rich. ex Walp. and *Anthocephalus macrophyllus* (Roxb.) Havil *Anthocephalus cadamba* Miq.	burflower-tree, cadamba, common bur-flower tree, kadam, kelempayan, laran, Leichhardt pine
	Araucaria cunninghamii Mudie	colonial pine, Dorrigo pine, hoop pine, Moreton bay pine, Queensland pine, Richmond river pine
	Artemisia monosperma L	wormwood
	Aucoumea klaineana Pierre	Gabon, Gaboon
	Calocedrus decurrens (Torr.) Florin	California incense-cedar, incense cedar
	Celtis occidentalis L	American hackberry,beaverwood, common hackberry, hackberry, nettletree, northern hackberry, sugarberry
	Chamaecyparis lawsoniana (A. Murray) Parl	Port Orford-cedar
	Chamaecyparis obtuse (Siebold & Zucc.) Endl	hinoki, hinoki cypress, Japanese cypress
	Citrus limon (L.) Osbeck	lemon tree
	Corymbia maculata (Hook.) K. D. Hill & L. A. S. Johnson	spotted gum
	Cryptomeria japonica (L. f.) D. Don	Japanese cedar, Japanese redwood, Japanese sugi, sugi
	Cunninghamia konishii Hayata	China-fir, Chinese fir, Cunninghamia
	Dendrocalamus asper (Schult.) Backer	bamboo
	Dendrocalamus barbatus	giant bamboo
	Dendrocalamus spp.	bamboo
	Dicorynia guianensis Amshoff	Angelique, Angelique batard, Angelique gris, Barakaroeballi, Basralokus
	Distemonanthus benthamianus Baill	Ayan, movingui, Nigerian Satinwood
	Elaeis guineensis Jacq. (monocot tree)	African oil palm, macaw-fat, oil palm
	Endospermum malaccense L. S. Sm	cheesewood, sesendok, white milkwood, whitewood
	Entandrophragma cylindricum Harms	sapele, sapelli
	Gmelina arborea Roxb	beechwood, gmelina, goomar teak, Kashmir tree, Malay beechwood, white teak, yemane

Group	Binomial name	Common/vernacular name
	Grevillea robusta A. Cunn. ex R. Br	Australian silver oak, silk oak, silky oak, silver oak, Southern silky oak
	Intsia bijuga (Colebr.) Kuntze	Johnstone river teak, Pacific teak, scrub mahogany
	Liquidambar styraciflua L	alligatorwood, American storax, American sweet-gum, bilsted, hazel pine, redgum, satin-walnut, star-leaved gum, sweetgum
	Liriodendron tulipifera L	American tulip tree, fiddletree, fiddletree, tulip poplar, tulip tree, tulipwood, whitewood, yellow-poplar
	Lophira alata Banks ex Gaertn	azobé, ekki, red ironwood tree,
	Magnolia spp.	magnolia
	Melia azedarach L	bead-tree, Cape lilac, chinaberry tree, Indian lilac, mindi, Persian lilac, Pride of India, syringa berrytree, white cedar
	Mitragyna ciliata Aubrév. & Pellegr	abura, bahia, M'Boy
	Notholithocarpus densiflorus (Hook. & Arn.) Manos, Cannon & S. H. Oh	tanbark-oak, tanoak
	Nyssa sylvatica Marshall	black gum, black tupelo, sour gum
	Ochroma pyramidale (Cav. ex Lam.) Urb	balsa, corkwood, down tree
	Paraserienthes falcata Becker	albizzia
	Parkia biglobosa (Jacq.) R. Br. ex G. Don	African locust bean, dodongba, iru, néré, netetou, sumbala
	Paulownia elongata S. Y. Hu	paulownia
	Pinaceae spp.	cedar
	Platanus occidentalis L	American planetree, American sycamore, buttonwood, occidental plane, water beech, western plane
	Prioria copaifera Griseb	cativo
	Prunus serotina Ehrh	black cherry
	Pseudotsuga menziesii (Mirbel) Franco	Columbian pine, Douglas fir, Douglas-fir, Oregon pine
	Pseudotsuga menziesii var. Glauca (Mayr) Franco	Rocky Mountain Douglas-fir
	Pseudotsuga menziesii var. Lindleyana (Roezl) Carrière	Mexican Douglas fir
	Pseudotsuga menziesii var. menziesii	coast Douglas-fir
	Pterocarpus dalbergioides DC	Andaman padauk, Andaman redwood, East Indian mahogany
	Pterocarpus macrocarpus Kurz	Burmese padauk
	Pterocarpus soyauxii Taub	African coralwood, African padauk
	Pterocarpus spp.	padauk
	Pyrus spp.	wild pear
	Pyrus elaeagnifolia Pall	oleaster-leafed pear
	Robinia pseudoacacia L	black locust, false acacia

Group	Binomial name	Common/vernacular name
	Schinus molle L	American pepper, anacahuita, California pepper tree, escobilla, false pepper, molle del Peru, pepper tree, peppercorn tree, pepperina, Peruvian pepper, Peruvian peppertree, pirul Peruvian mastic
	Sequoia sempervirens (D. Don) Endl	California redwood, coast redwood, coastal redwood, redwood
	Shorea spp.	balau, bangkirai, lauan, lawaan, luan, meranti, Philippine mahogany, seraya
	Sorbus aucuparia L	mountain ash, rowan
	Syzygium luehmannii (F. Muell.) L. A. S. Johnson	cherry alder, cherry satinash, clove Lilli Pilli, creek cherry, riberry, scrub cherry, small-leaved Lilly Pilly, small-leaved watergum, watergum
	Taxodium distichum (L.) Rich	bald cypress, baldcypress, gulf cypress, Pondcypress, red cypress, Southern cypress, swamp cypress, tidewater red cypress, white cypress, yellow cypress
	Tectona grandis L. f	Burmese teak, Central Province teak (CP teak), Nagpur teak, teak
	Terminalia superba Engl. & Diels	afara, African limba wood, frake, korina, limba
	Triplochiton scleroxylon K. Schum	abachi, African whitewood, ayous, obeche, sambawawa, wawa

The manufacturer's authorised representative in the EU is Springer Nature Customer Service Centre GmbH, Europaplatz 3, 69115 Heidelberg, Germany. If you have any concerns regarding our products, please contact ProductSafety@springernature.com

d and bound by CPI Group (UK) Ltd, Croydon, CR0 4YY
01/10/2026
02225990-0006